Mathematical Modeling for Design of Machine Components (TK-Integrated)

Suryaji R. Bhonsle
Michigan Technological University

Klaus J. Weinmann
Michigan Technological University

PRENTICE HALL, Upper Saddle River, New Jersey 07458

Library of Congress Cataloging-in-Publication Data

Bhonsle, S. R. (Suryaji R.)
   Mathematical modeling for design of machine components (TK
-integrated) / S.R. Bhonsle, K.J. Weinmann.
      p.   cm.
   Includes bibliological references and index.
   ISBN: 0-13-727231-6
   1. Machine design--Mathematical models.   I. Weinmann, Klaus J.
   II. Title.
   TJ233.B55   1999
   621.8'15'015118--dc21   99-13195
                                                 CIP

Acquisition Editor: *Bill Stenquist*
Editorial/Production Supervision: *Barbara A. Till*
Editor-in-Chief: *Marcia Horton*
Assistant Vice President of Production and Manufacturing: *David W. Riccardi*
Managing Editor: *Eileen Clark*
Manufacturing Buyer: *Donna Sullivan*
Manufacturing Manager: *Trudy Pisciotti*
Creative Director: *Paula Maylahn*
Art Director: *Jayne Conte*
Cover Designer: *Bruce Kenselaar*
Editorial Assistant: *Meg Weist*
Copy Editor: *Brian Baker*
Compositor: *Craig Little*

©1999 by Prentice-Hall, Inc.
Simon & Schuster / A Viacom Company
Upper Saddle River, New Jersey 07458

The author and publisher of this book have used their best efforts in preparing this book. These efforts include the
development, research, and testing of the theories and programs to determine their effectiveness. The author and
publisher make no warranty of any kind, expressed or implied, with regard to these programs or the documentation
contained in this book. The author and publisher shall not be liable in any event for incidental or consequential damages
in connection with, or arising out of, the furnishing, performance, or use of these programs.

Printed in the United States of America

10 9 8 7 6 5 4 3 2 1

ISBN 0-13-727231-6

Prentice-Hall International (UK) Limited, *London*
Prentice-Hall of Australia Pty. Limited, *Sydney*
Prentice-Hall Canada Inc., *Toronto*
Prentice-Hall Hispanoamericana, S.A., *Mexico*
Prentice-Hall of India Private Limited, *New Delhi*
Prentice-Hall of Japan, Inc., *Tokyo*
Simon & Schuster Asia Pte. Ltd., *Singapore*
Editora Prentice-Hall do Brasil, Ltda., *Rio de Janeiro*

*I*ntegration of computers in Machine Component Design has become possible because of the availability of computers in universities and colleges. Design engineers working in industry have to work on their design projects using computers. Hence the integration of computers in design courses is of prime importance.

Mathematical Modeling for Design of Machine Components is written to be a supplementary text for a junior machine component courses in a traditional mechanical engineering curriculum. The intent is to introduce and integrate computers in machine component design courses. Emphasis is placed on the practical aspects of creating lead models to design various machine components. Lead models are rule-based mathematical models used in solving specific engineering problems. Most of the lead models presented can be modified, rewritten, or newly created once the design concepts are clear. To clarify the theoretical aspects, each chapter has a short but concise presentation on analysis and design of machine component theory. Sample problems based on the theory are solved by using conventional techniques and with the aid of lead models.

Computer-aided machine component design is a need for present and future engineers. The computer is a powerful tool that can be used to analyze, design, or optimize a machine component with the aid of proper software. However, use of such a powerful tool does not guarantee correct results. Most software uses numerical procedures involving approximations of theoretical behavior. In order to produce the correct results, the lead model must be designed correctly and be able to reach numerical convergencies. Correct results depend upon the software and the user's ability to utilize the tool. This text in no way supplants the engineer's responsibility to do hand calculations of engineering problems, to use good design practice, or to apply engineering judgment to the problem. Instead, it is expected that this text will supplement these skills to ensure that the best design is obtained.

Many lead models in the accompanying disk are extensive in nature and effort should not be expended to create them anew. Users should modify them wherever and whenever necessary.

All lead models require a clear understanding of the fundamentals involved in designing a particular machine component. Lead models cannot be used as a

"Black Box." For this reason, the theory involved in designing of a machine component is explained first. Then an example problem based on the developed theory is presented and then the lead model developed is used to solve such a problem with the aid of the TK-Solver software. The questions at the end of each chapter can be solved by carrying out hand calculations or with the aid of TK-Solver and a lead model.

The text became possible because of the sincere editing and extensive typing and retyping of the manuscript by Ms. Carol Janisse, which is appreciated by both the authors. We thank all who helped us make this text possible. In our classroom teaching, we used texts by authors like Shigley & Junivall. Their influence on our understanding design of machine components has been extensive and very useful.

Introduction

1.1 Design of Machine Components

A machine is a device built to augment human muscle power. Today we can find devices in common use that supplement human brain power and control systems. Thus, machinery may include a wide range of devices, from a simple lever to a complex robot or computer.

In this text, we will limit our discussion to mechanical machines. The textbook will provide sufficient theoretical approaches to the design of machines. Every machine is made of one or more components. The design of components is the primary task performed when one builds a machine. Design can be carried out by well-developed mathematical models. In developing mathematical models, certain assumptions are made for simplification. Finally, the governing mathematical model describes the behavior of a machine component. The model usually consists of a set of linear or nonlinear simple equations or differential equations along with boundary conditions.

Such equations for the design of a machine component are often quite involved, as are the calculations required to complete the design. The natural approach to improving the process of design is to use computers and mathematical models. For this reason, integrating computers and computer software into undergraduate engineering design courses has been a major goal of most colleges and universities. Computer-assisted design analysis is no longer exclusive to a few privileged engineers; engineering students and engineers alike now have access to better computers and software. This provides educators an opportunity to focus on teaching basic concepts and principles, while placing less emphasis on the tools and techniques of numerical computation. Trivial activities such as performing iterative processes, back solving equations, solving lists, plotting results, and even making long mathematical calculations should not be a time consuming process. Before computers became commonplace, students were

1

assigned long, arduous mathematical procedures that provided them with an overall understanding of basic concepts and principles, but did not allow them to delve deeper into the *"What if...?"* and *"How can I...?"* questions that are especially relevant in mechanical engineering design courses.

Computers are shaping new modes of teaching and learning in engineering courses. Specialized software allows teachers to teach and students to learn detailed concepts and principles underlying an engineering theory. Such theory, when investigated further with the aid of computers, makes education more like an exploration. Knowledge of programming is very important, and the understanding of a computer language is essential in helping students learn software packages. However, it is exceedingly time consuming for students to write, for example, a FORTRAN program and to go through the stages of debugging the program for logic and syntax errors. To minimize the time needed to learn a software and use computers efficiently, the proper blending of software and computers has to take place. One of the ways to achieve this goal is to write *Lead Models* (LMs), which can then be used to solve design problems.

Lead models are rule-based mathematical models used in solving specific engineering problems. The purpose of this text is to show efficient ways to integrate computers into design engineering by creating lead models using an appropriate software.

1.2 Use of Lead Models

Lead Models can be used to:

a. Solve design engineering problems in a short amount of time.

b. Solve more than one type of design problem using the same model or with only minor changes, if any.

c. Solve problems without writing a formal program or solve problems using very few programming procedures.

d. Obtain alternative solutions with ease.

e. Solve the same problem for incremental values of the same input variable.

f. Back solve a problem, even if it requires back solving for multiple variables. As a simple example, if the stress is given then find the area, but if the area is given, then find the stress without changing the program or modifying the program.

g. Plot results efficiently.

h. Solve problems irrespective of the order of the stipulated equations, with the unknowns on the left side or the right side of the equation. (This is possible as long as the number of equations equals the number of unknowns. Solutions are still possible if the number of unknowns exceeds that of the equations, but some values just need to be set.) If the values of the necessary input variables are given, the model should solve for the output variables.

i. Solve linear or nonlinear simultaneous equations or solve transcendental equations without writing an elaborate program.

j. Obtain solutions and other results that can be presented in table format without an inordinate amount of time spent formatting or plotting elaborate graphs.

k. Have the results in the form of graphs, tables, equations, solutions, etc., which can then be presented to the reading audience in report form.

l. Ask *"What if...?"* and *"How can I...?"* questions and get numerical answers for such questions within a few seconds.

m. Optimize design in a very short time.

n. Make the user proficient in the use of Unix-, Macintosh-, or Windows-based workstations.

Many software packages are on the market that try to satisfy these requirements. However, satisfying all of the requirements is possible using an easy-to-learn package called TK Solver, which is an equation solver and knowledge management software. TK Solver not only allows students to synthesize solutions represented by equations, but also shows the students what solutions might look like when numerical values change. In short, TK Solver is a rule-based declarative programming language that solves equations and performs mathematical modeling. Systems of equations and programs can be written in TK Solver with extraordinary ease and speed.

To obtain maximum benefit from a given piece of software, it is necessary to understand the software well and learn to use it effectively. This can be achieved within a short time. Lead models that exemplify the versatility and power of the software have been devised for various machine components for student study and use. A way to accelerate the learning process is to solve sample problems using TK Solver software and lead models created by an expert. This text provides the fundamentals of machine design and the design of machine elements using mathematical models presented in the form of lead models.

1.3 Presentation of Lead Models

Lead Models are presented for the following reasons:

a. To solve specific design problems using TK Solver software as an equation-solving engine.

b. To study a possible approach to understanding a theory, its application(s), and its results.

c. To know how the model can be used to back solve if necessary.

d. To study the effect of list solving.

e. To obtain alternative solutions without going through a repetitive process.

f. To ask *"What if...?"* and *"How can I...?"* questions and get answers to these questions quickly. In so doing, we learn how to design a machine component that is reliable, robust, and economical.

g. To realize that the lead model created using TK Solver software does not allow students to avoid learning the subject matter or supplant the need to read the text. Nor does it relieve them from the responsibility of understanding what is involved in a problem solving assignment.

h. To change or rewrite a lead model if it is inadequate.

i. To realize that the laws and limitations which apply in getting correct solutions to a design problem prevail in TK Solver software, just as they prevail in any other engineering software. In other words, the models should not be used as black boxes; rather, one must peek inside and understand them.

j. To use a lead model or create a new model from it according to students' requirements and specifications. This will enable the students to solve their problems without spending too much time and effort on them and without becoming frustrated.

In sum, computer-aided design using TK Solver will be fun for students because it will reduce the amount of time they spend in arriving at a solution to a design engineering problem. This is even more so in the design of machine components, which requires an iterative process to arrive at an appropriate solution.

If students gain some knowledge by using the TK Solver software, and if the software can be used without learning new rules, then we have succeeded in making learning by computer rewarding and successful. This is the motive for creating this text.

The text has 19 chapters, each chapter is divided as shown in Table 1–1.

Learning the theory presented in each chapter is sufficient to know the basic assumptions of, and to solve many problems related to, the topic presented. Students who understand the theory can go directly to the TK lead model and solve a given problem either as a project or as homework. The model can be studied, improved, modified, or rewritten to solve specific problems. There are some models (e.g., lead models for the design of columns, curved beams, as well as theories of failures and the design of journal bearings) that students should try to use in their present form or with minor modifications if necessary. Recreating the sophisticated models that have been provided is not recommended, as it would be both time consuming and pointless to try to duplicate these efforts. The

Table 1–1 Arrangement of each chapter

A CHAPTER						
Short Exposition of Theory Related to the Topic		TK Lead Model for a Sample Problem Based on Theory			Homework/Project	
Theory used to develop equations for the topic	Summary of derived equations for use in lead model	Rules sheet	Variabes sheet	Tables/ graphs sheet	Hand-calculated solution	Solution with use of lead model or modification thereof

use and modification of the lead models presented will make students proficient in using TK Solver as an equation-solving engine. This experience will allow anyone to use TK Solver to solve extensive engineering problems in courses in the design of machine components or in other science and engineering courses.

The arrangement of the chapters is such that simple lead models are initially presented to give the student experience with easily understandable examples, the use of computers, and the TK Solver software in solving design problems. Progressively more sophisticated lead models are provided on a disk. The use of these lead models will save time and allow the reader to ask *"What if...?"* and *"How can I...?"* questions. Answers to such questions will accelerate the analysis or design optimization of a machine component, which is the goal of the text.

The accompanying disk contains lead models with the filenames shown in Table 1–2. (by knowing the file name and its description, the user can decide which model to use).

The models listed in Table 1–2 are for a Unix-based platform; however, you will find the same models for Macintosh (with extension .mtk) and PC (with extension .ptk) platforms.

In using TK Solver and lead models, it is very important that the display unit notation in the Variables Sheet match the units in the Units Sheet. For example, if the Units Sheet has "lb" for pounds then you must use "lb" in the display unit column of the Variables Sheet and not "lbs." For this reason, a sample Units Sheet is presented on pages 7 and 8. Always have this sheet available while developing new or modified models.

Table 1–2 List of filenames and description of lead models on accompanying disk

Model No.	Filename	Description
1.	stress.tk or ex3-1.tk	Three- and two-dimensional stress strain relations
2.	ststrain.tk or ex3-2.tk	Stress-strain relations
3.	stgage.tk or ex4-1.tk	Experimental stress analysis
4.	bstadiag.tk or ex4-2.tk	Analytical stress analysis
5.	cast.tk or ex4-3.tk	Castigliano's theorem deflections
6.	castind.tk or ex5-5.tk	Castigliano's indeterminate structure
7.	coldes.tk or ex6-1.tk	Design of column
8.	thfail.tk or ex7-1.tk	Theories of failure
9.	statnd.tk or ex8-1.tk	Normal distribution
10.	weibul.tk or ex8-2.tk	Weibull distribution
11.	fatig.tk or ex9-2.tk	Metal fatigue, S–N curve
12.	curvedb.tk or ex10-1.tk	Curved-beam design
13.	twc.tk or ex11-1.tk	Design of thick-walled cylinders
14.	pscrew.tk or ex13-1.tk	Design of power screw
15.	bolt.tk or ex13-2.tk	Design of bolted connections
16.	boltf.tk or ex13-3.tk	Design of bolts based on fatigue
17.	spring.tk or ex14-1.tk	Design of helical compression springs
18.	brake.tk or ex16-1.tk	Design of internally expanding drum brakes
19.	shaft1.tk or ex17-1.tk	Shaft design
20.	shaft.tk or ex18-1.tk	Shaft design
21.	jbearing.tk or ex19-1.tk	Journal-bearing design
22.	jbearing.tk or ex19-2.tk	Journal-bearing design
23.	jbearing.tk or ex19-3.tk	Journal-bearing design iterative solution
24.	jbearing.tk or ex19-4.tk	Journal-bearing design/optimization

UNITS SHEET

Units

From	To	Multiply By	Add Offset	Comment
in/in	m/m	▓▓▓▓▓		
mm/mm	m/m			
cm/cm	m/m			
m	cm	100		Meters to centimeters
m	mm	1000		Meters to millimeters
m	in	39.3700787402		Meters to inches
ft	in	12		Feet to inches
mile	ft	5280		Mile to feet
mile	m	1609.344		Mile to meter
in^2	mm^2	645.16		Area
in^2	m^2	.00064516		" "
ft^3	m^3	.028316846592		Volume
ft^3	in^3	1728		" "
m^3	mm^3	1E9		" "
m^3	cm^3	1000000		" "
yard^3	m^3	.764554857984		" "
gal	m^3	.003785411748		Liquid gallon
m^4	mm^4	1E12		Second moment of area
m^4	cm^4	1E8		" " " " " "
m^4	in^4	22402509.61		" " " " " "
ft^4	in^4	20736		" " " " " "
N	lbf	.2248089431		Newtons to pound force
kN	N	1000		Kilo newtons to newtons
Nm	lb-ft	.737562149277		Newton meters to foot pound
Nm	Nmm	1000		Newton meters to newton millim

Units

From	To	Multiply By	Add Offset	Comment
lb-ft	lb-in	12		Pound feet to pound inches
kPa	Pa	1000		Kilo pascals to pascals
MPa	Pa	1000000		Mega pascals to pascals
GPa	Pa	1E9		Giga pascals to pascals
Pa	psi	.000145037738		Pascals to pounds per sq.inch
ksi	psi	1000		Kilo pounds per sq.in to pound
ft/s^2	m/s^2	.3048		Acceleration
in/s	m/s	.0254		Velocity
ft/s	in/s	12		" "
m/s	mm/s	1000		" "
km/hr	m/s	.2777777778		" "
mph	ft/s	1.466666667		" "
km/hr	mph	.621371192146		" "
g/cm^3	kg/m^3	1000		Density
lbm/in^3	kg/m^3	27679.905		" "
lbm/ft^3	kg/m^3	16.018463		" "
slug/ft^3	kg/m^3	515.379		" "
Btu	J	1055.87		Energy
ft-lb	J	1.3558179		" "
Kwh	J	3600000		" "
klb	lb	1000		Force
klb	N	9.80665		" "
kip	N	4448.22161526		" "
lbm	kg	.45359237		Mass
slug	kg	14.5939029		" "

UNITS SHEET

From	To	Multiply By	Add Offset	Comment
slug	lbm	32.1740484744		" "
ton	kg	907.18474		" "
ton	lbm	2000		" "
Btu/s	W	1054.35026449		Power
ft-lb/min	W	.022596966		" "
ft-lb/s	W	1.3558179		" "
Hp	ft-lb/s	550		" "
Hp	W	745.699845		" "
Hp	kW	.745699845		" "
W	kW	.001		" "
atm	N/m^2	101325		Pressure
bar	N/m^2	100000		" "
R	K	1.8		Temperature
C	F	1.8	32	" "
day	s	86400		Time
hr	s	3600		" "
min	s	60		" "
ft^2/s	m^2/s	.09290304		Viscosity
Hz	rad/s	.159154943092		Frequency to radians per secon
rad/s	rpm	9.54929658551		Radians per second to revoluti
deg	rad	.0174		Degrees to radians
lbm-ft^2	kg-m^2	.0421		Moment of inertia
lbm-in^2	kg-mm^2	293		" " " "
deg	rad	.01745329252		Degree's to radians

Getting to Know TK Solver

Lead models are implemented in the software package TK Solver, developed and marketed by Universal Technical System, Inc. TK Solver, or simply TK, provides a user-friendly environment for the iterative solution of systems of (linear or nonlinear) simultaneous equations and has back-solving capabilities.

With back-solving capabilities, one need not isolate an unknown on the left side of an equation, as in most other programs; TK Solver software will find the value of an unknown even if it is on the right side of an equation. Similarly, equations can be listed in any order and need not be in a format that can be represented by a set of matrices, such as $[A]\{X\} = \{B\}$, where X is a vector of the unknowns. Also, TK Solver software has many built-in functions—which makes solving problems much easier.

TK Solver software is organized into eight sheets or windows (see Fig. 2–1), each of which is used to enter or read a specific type of information. These sheets are called the Rules, Variables, Lists, Tables, Plots, Functions, Units, and Formats sheets. In addition to the sheets, a menu bar (see Fig. 2–3) is located at the top of the screen to access various commands, most of which have keyboard equivalents.

Together, the Rules and Variables Sheets make up the core of the program. Equations to be solved, which may include calls to built-in or user-defined functions, are entered into the Rules Sheet. (See Fig. 2–3.) Unlike structured programming in traditional high-level languages such as FORTRAN, PASCAL, or C, neither the format of the equations nor the order in which they are entered is important.

As equations are entered into the Rules Sheet, all variables used in the equations will appear in the Variables Sheet. (See Fig. 2–4.) The Variables Sheet contains fields in which input values can be entered and output values can be read. There are also fields for the status of variables and for display units and comments. (See Fig. 2–4.) All other sheets are shown in Figs. 2–5 through 2–12.

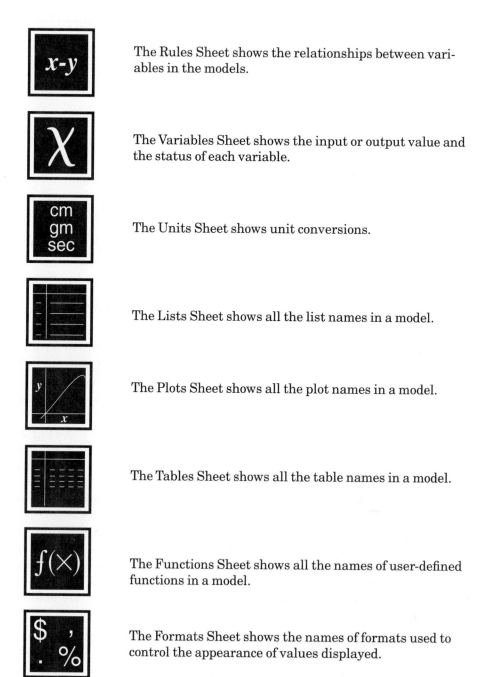

The Rules Sheet shows the relationships between variables in the models.

The Variables Sheet shows the input or output value and the status of each variable.

The Units Sheet shows unit conversions.

The Lists Sheet shows all the list names in a model.

The Plots Sheet shows all the plot names in a model.

The Tables Sheet shows all the table names in a model.

The Functions Sheet shows all the names of user-defined functions in a model.

The Formats Sheet shows the names of formats used to control the appearance of values displayed.

Figure 2–1   Summary of Sheet Structure

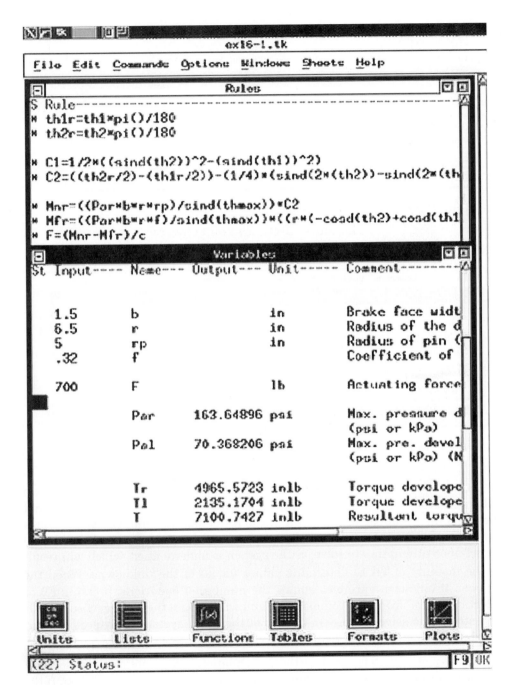

Figure 2–2 The TK window (see menu bar at top)

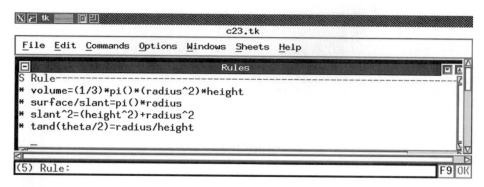

Figure 2–3   Sample Rules Sheet

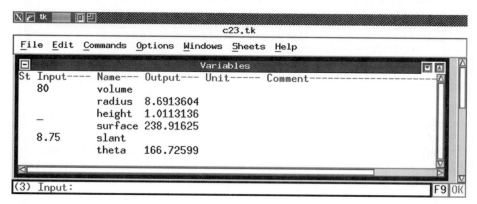

Figure 2–4   Sample Variables Sheet

Once the equations have been entered into the Rules Sheet and the known values entered into the Variables Sheet, it is advisable to verify manually that the remaining number of unknown variables equals the number of equations used. If this is not the case, then the system is either overdefined, which TK will report upon attempting to solve, or the system is underdefined, which will result in the inability of TK to determine values for all of the unknowns. When the number of unknown variables equals the number of equations, pressing F9 or selecting "Solve" from the Commands Menu will initiate the TK direct solver. TK will attempt to solve for the unknowns of the given system by internally rearranging the equations provided.

For some well-structured systems of equations, this kind of direct solution will be successful. However, it is often necessary to use the iterative solver. If the direct solver fails, the iterative solver can be initiated by inserting a guessed value for one or more of the unknown variables and pressing F9 to solve again. To do this, type "G" into the Variables Sheet status field corresponding to the

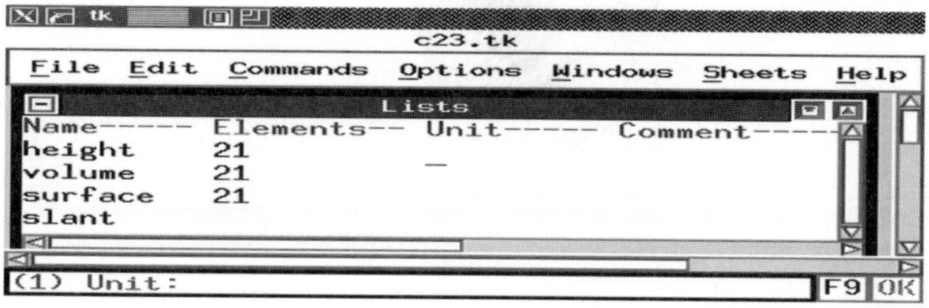

Figure 2–5 Sample Lists

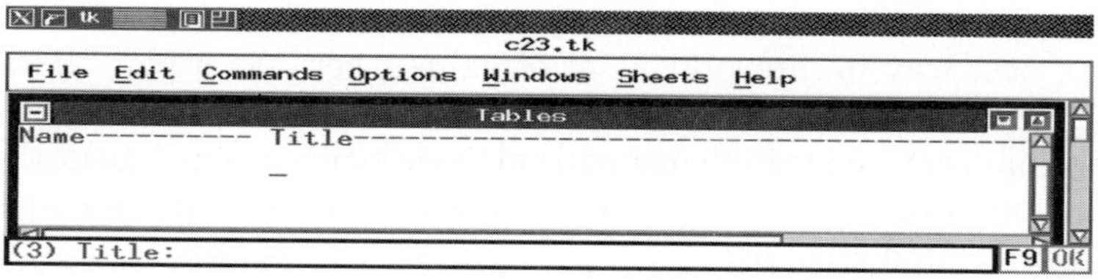

Figure 2–6 Sample Tables Sheet

variable for which the guessed value is to be used. The input field will display a default number, which may be retained or changed to a more appropriate value.

Apart from indicating guessed values, the status field in the Variables Sheet can associate a list with a given variable. This is done by entering an "L" and typing an "I" over it (the "I" will not appear on the screen) into the status field corresponding to that variable. The name of the variable will then appear in the Lists Sheets. (See Fig. 2–5.) This creates a list in which a set of input values can be specified or a set of resulting values can be read after a List Solve has been completed. By this means, in addition to solving for a single value of each unknown, as discussed earlier, a range of values for each unknown can be determined corresponding to a range of known values by typing "L" and "O" in the status field corresponding to the desired unknowns. Once again, the "O" will not appear on the screen. After the input list has been filled, either by manually inputting values or by selecting Fill List from the Commands Menu, a List Solve can be indicated by pressing SHIFT-F9 or by selecting List Solve from the Commands Menu. These solving options are often useful to determine trends in data and relationships between different variables in the system.

While the results of a List Solve can be seen directly by looking at the numbers in the output lists, it is often helpful to arrange these lists into tables or plots for easier interpretation. TK allows for both of these options. The Tables Sheet (see Fig. 2–6) permits one or more lists to be printed side by side to show large amounts of data in a more compact and organized form. These tables can also be used to adjust the values in the input lists. After reexecuting a List Solve, the effects of such changes can be seen.

Figure 2–7   Sample Plots Sheet

The Plot Sheet (see Fig. 2–7) allows one list of values to be plotted against one or more different lists as a line chart or bar chart. Creating some type of plot is often the best way to interpret a set of values, because a plot allows for the easy recognition of trends, which are more difficult to see by looking at numbers alone. Line chart options include logarithmic axis scaling, axis labels, the display of grids, and other options. It is also possible to create pie charts for a given data list and an associated label list.

As mentioned earlier, users may custom define their functions, which can be called from a Rules Sheet or from other functions. Custom-defined functions, which are defined in the Function Sheet (see Fig. 2–8), can have one of the following three basic forms:

1. The *Rule Function* (see Fig. 2–10) uses the same free structure as in the Rules Sheet and is treated as an extension of it. All variables specified may be passed to or from the function, regardless of their designation.

2. The *Procedure Function* is written in a special language similar to traditional high-level languages. The advantage of this type of function is that it allows for the use of conditional and loop structures resembling those found in other languages. However, because of its structured format, the function will be executed only when all of the specified input values are known. This requirement can severely cripple the back-solving capabilities that make TK so flexible. Therefore, procedure functions should be used sparingly if back-solving is desirable.

3. The *List Function* (see Fig. 2–11) is defined by domain values and a corresponding list of range values. When either a domain or a range value is passed to the function, a value for the other will be returned. If the input value lies between two values in the corresponding list, one of the four mapping options described in TK Solver can be specified, and that option will determine how intermediate values are calculated.

The Units Sheet (see Fig. 2–12) is used to add flexibility to a model by allowing units other than those for which the equations were developed to be

```
X tk ⬛ □ꔹ                                                    ▓
              /home/mefac/srbhonsl/me-desbo/book1tk/jbrnews.tk
 File  Edit  Commands  Options  Windows  Sheets  Help
┌────────────────────────────────────────────────────────────────────┐
│ ◻                          Functions                          ▽ ▢   │
│Name--------- Type----  Arguments-- Comment------------------------  │
│                                    **************************************│
│                                    DO NOT CHANGE ANY OF THE FOLLOWING│
│                                    FUNCTIONS!  THEY CONTAIN THE DATA FOR│
│                                    THE RAIMONDI AND BOYD GRAPHS!     │
│                                    **************************************│
│q13          List      1:1          DO NOT TOUCH Figure 13.13 L/D = 1/4│
│h13          List      1:1          DO NOT TOUCH Figure 13.13 L/D = 1/2│
│o13          List      1:1          DO NOT TOUCH Figure 13.13 L/D = 1.0│
│i13          List      1:1          DO NOT TOUCH Figure 13.13 L/D = Infinite│
│q14          List      1:1          DO NOT TOUCH Figure 13.14 L/D = 1/4│
│h14          List      1:1          DO NOT TOUCH Figure 13.14 L/D = 1/2│
│o14          List      1:1          DO NOT TOUCH Figure 13.14 L/D = 1.0│
│i14          List      1:1          DO NOT TOUCH Figure 13.14 L/D = Infinite│
│q15          List      1:1          DO NOT TOUCH Figure 13.15 L/D = 1/4│
│h15          List      1:1          DO NOT TOUCH Figure 13.15 L/D = 1/2│
│o15          List      1:1          DO NOT TOUCH Figure 13.15 L/D = 1.0│
│i15          List      1:1          DO NOT TOUCH Figure 13.15 L/D = Infinite│
│q16          List      1:1          DO NOT TOUCH Figure 13.16 L/D = 1/4│
│h16          List      1:1          DO NOT TOUCH Figure 13.16 L/D = 1/2│
│o16          List      1:1          DO NOT TOUCH Figure 13.16 L/D = 1.0│
│i16          List      1:1          DO NOT TOUCH Figure 13.16 L/D = Infinite│
│q17a         List      1:1          DO NOT TOUCH Figure 13.17 L/D = 1/4│
│h17a         List      1:1          DO NOT TOUCH Figure 13.17 L/D = 1/2│
│o17a         List      1:1          DO NOT TOUCH Figure 13.17 L/D = 1.0│
│i17a         List      1:1          DO NOT TOUCH Figure 13.17 L/D = Infinite│
│q17b         List      1:1          DO NOT TOUCH Figure 13.17 L/D = 1/4│
│h17b         List      1:1          DO NOT TOUCH Figure 13.17 L/D = 1/2│
│o17b         List      1:1          DO NOT TOUCH Figure 13.17 L/D = 1.0│
│i17b         List      1:1          DO NOT TOUCH Figure 13.17 L/D = Infinite│
│q18          List      1:1          DO NOT TOUCH Figure 13.17 L/D = 1/4│
│h18          List      1:1          DO NOT TOUCH Figure 13.17 L/D = 1/2│
│o18          List      1:1          DO NOT TOUCH Figure 13.17 L/D = 1.0│
│i18          List      1:1          DO NOT TOUCH Figure 13.17 L/D = Infinite│
│q19          List      1:1          DO NOT TOUCH Figure 13.17 L/D = 1/4│
│h19          List      1:1          DO NOT TOUCH Figure 13.17 L/D = 1/2│
│o19          List      1:1          DO NOT TOUCH Figure 13.17 L/D = 1.0│
│◁                                                                  ▷ │
└────────────────────────────────────────────────────────────────────┘
 (5) Name:                                                    F9 OK
```

```
X tk ⬛ □ꔹ □ꔹ                                                  ▓
              /home/mefac/srbhonsl/me-desbo/book1tk/jbrnews.tk
 File  Edit  Commands  Options  Windows  Sheets  Help
┌────────────────────────────────────────────────────────────────────┐
│ ◻                          Functions                          ▽ ▢   │
│Name----------  Type-----  Arguments--  Comment-------------------   │
│o19            List       1:1          DO NOT TOUCH Figure 13.17 L/D = 1.0│
│i19            List       1:1          DO NOT TOUCH Figure 13.17 L/D = Infinite│
│mu10           List       1:1          DO NOT TOUCH Figure 13.6 Grade=10│
│mu20           List       1:1          DO NOT TOUCH Figure 13.6 Grade=20│
│mu30           List       1:1          DO NOT TOUCH Figure 13.6 Grade=30│
│mu40           List       1:1          DO NOT TOUCH Figure 13.6 Grade=40│
│mu50           List       1:1          DO NOT TOUCH Figure 13.6 Grade=50│
│mu60           List       1:1          DO NOT TOUCH Figure 13.6 Grade=60│
│mu70           List       1:1          DO NOT TOUCH Figure 13.6 Grade=70│
│mu             Rule       2:1          DO NOT TOUCH Figure 13.6 (all)  │
│Smin           List       1:1          DO NOT TOUCH Figure 13.13 Minimum f│
│Smax           List       1:1          DO NOT TOUCH Figure 13.13 Maximum LOAD│
│◁                                                                  ▷ │
└────────────────────────────────────────────────────────────────────┘
 (1) Name:                                                    F9 OK
```

Figure 2–8   Sample Function Sheet

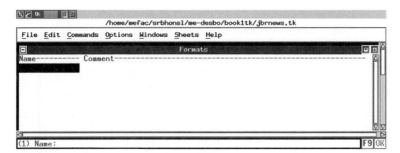

Figure 2–9   Sample Formats Subsheet

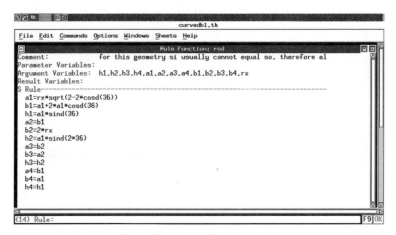

Figure 2–10   Sample Rule Function Subsheet

used for specifying input values and reading the results. This is especially useful for models that have applications in both English and SI units. Switching between units is accomplished by specifying a calculation unit and a display unit for each of the variables in the Variables Sheet. These units are assumed to be identical for any variable for which one of them is not specified. Upon solving, TK converts the values in the input fields from the display units to the calculation units according to the conversion factors entered into the Units Sheet. Once a solution is found, all results are converted from calculation units, which can also be specified for lists and are treated similarly.

The Formats Sheet allows for global formatting of the numerical display of numbers. This sheet includes such options as right or left justification, scientific or decimal notation, margin widths, etc. Many of these options can also be specified individually for specific sheets.

```
┌─────────────────────────────────────────────────────────────────┐
│ ▣ ⌐ tk    ▣ ⍗                                                    │
│                    /home/mefac/srbhonsl/me-desbo/book1tk/jbrnews.tk│
│  File  Edit  Commands  Options  Windows  Sheets  Help            │
│ ┌──────────────────── List Function: q13 ──────────────────┐▣▣│
│ Comment:         DO NOT TOUCH Figure 13.13 L/D = 1/4    ho/C vs. S│
│ Domain List:     Sfit                                            │
│ Mapping:         Cubic                                           │
│ Range List:      q13                                             │
│ Element-- Domain------------- Range---------------              │
│ 1         0                   0                                 │
│ 2         .002                .005                             │
│ 3         .004                .01                              │
│ 4         .006                .015                             │
│ 5         .008                .02                              │
│ 6         .01                 .028                             │
│ 7         .02                 .045                             │
│ 8         .03                 .06                              │
│ 9         .04                 .075                             │
│ 10        .05                 .085                             │
│ 11        .06                 .092                             │
│ 12        .07                 .1                               │
│ 13        .08                 .105                             │
│ 14        .09                 .11                              │
│ 15        .1                  .115                             │
│ 16        .2                  .17                              │
│ (1) Domain: 0                                          F9 OK    │
└─────────────────────────────────────────────────────────────────┘
```

```
┌─────────────────────────────────────────────────────────────────┐
│ ▣ ⌐ tk    ▣ ⍗                                                    │
│                    /home/mefac/srbhonsl/me-desbo/book1tk/jbrnews.tk│
│  File  Edit  Commands  Options  Windows  Sheets  Help            │
│ ┌──────────────────── List Function: q13 ──────────────────┐▣▣│
│ Element-- Domain------------- Range---------------              │
│ 16        .2                  .17                              │
│ 17        .3                  .21                              │
│ 18        .4                  .25                              │
│ 19        .5                  .285                             │
│ 20        .6                  .31                              │
│ 21        .7                  .335                             │
│ 22        .8                  .36                              │
│ 23        .9                  .38                              │
│ 24        1                   .395                             │
│ 25        2                   .525                             │
│ 26        3                   .605                             │
│ 27        4                   .66                              │
│ 28        5                   .705                             │
│ 29        6                   .745                             │
│ 30        7                   .775                             │
│ 31        8                   .805                             │
│ 32        9                   .83                              │
│ 33        10                  .85                              │
│ (1) Domain: 0                                          F9 OK    │
└─────────────────────────────────────────────────────────────────┘
```

Figure 2–11   Sample List Function Sheet

A detailed reference manual, along with a user's guide, is provided with the TK Solver software. Consult your laboratory assistant for details. The next section contains easy hands-on instructions for working with UNIX and DOS operating systems for beginners. For further information, refer to a laboratory manual on TK Solver.

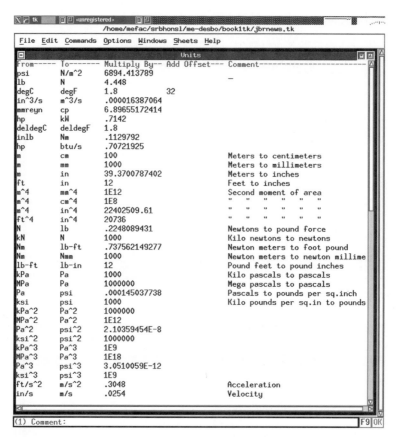

Figure 2–12   Sample Units Sheet

Getting Started with TK Solver Plus (UNIX Environment)

Login into a Sun workstation and type x11. While in X-Windows, open a medium window (by left-clicking in the rectangle written "medium 80 x 38"), and type

```
tk<return>
```

(If you do not succeed, it is advisable to ask your laboratory instructor how to log onto a specific terminal.)

The TK Solver software will now be loaded. To close the copyright window, left-click the pointer in that window. You will see several options on the top of the TK screen and also two windows on the screen. These windows contain the *Variables Sheet* and *Rules Sheet*. You can make these two sheets occupy the whole screen by selecting "Windows" from the top menu and left-clicking the "Tile" sub-option.

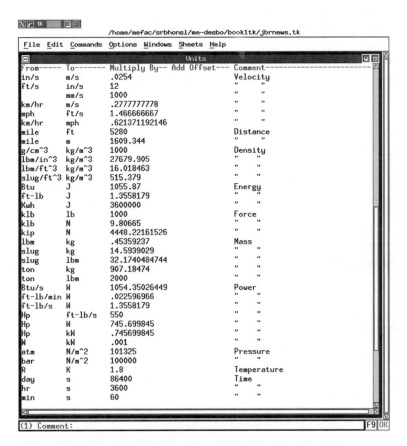

Figure 2-12 (continued) Sample Units Sheet

TK Solver has eight sheets. You can access any sheet by selecting the "Select Sheet" option found under the "Window" menu.

To move within any sheet, scroll bars are provided both at the right side to scroll vertically and at the bottom to scroll horizontally. Other sheets can be opened by left-clicking icons at the bottom of the screen. Right now, you will work with the Rules Sheet and Variables Sheet.

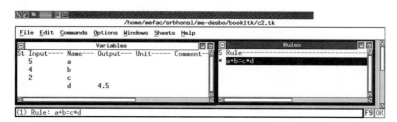

Figure 2–13

Example 2-1

Given: $a + b = c * d$ if $a = 5$, $b = 4$, and $c = 2$

Find: d

* Move the cursor to the Rules Sheet, left-click, then enter the equation, and then press [Enter]. You will see that TK has automatically entered the names of the variables on the Variables Sheet. (**Note**: If the Variables Sheet is not already open, users will not see variable names until step #2.)

* Switch to the Variables Sheet by moving the cursor into that sheet and left-clicking.

* Left-click in the appropriate "input" field, and enter the given inputs shown in Fig. 2–13.

* Select the "Command" option and left-click on the "Solve" suboption or press the F9 key.

* You will see the value of d (= 4.5) appear in the output column. (See Fig. 2–13.)

Given: $d = 4$, $a = 5$, $b = 4$

Find: c

* Enter 4 for d in the Input column.

* Delete the value for c. (Press B or the space bar while in the Status field of the variable c).

* Solve as you did for d. You will see an answer of 2.25 for c in the output column.

Given: $a = 23.2*5.6$, $d = 4$, $b = 4$

Find: c

* Type 23.2*5.6 in the Input field of the variable a.

* Solve as before.

* The answer ($d = 33.48$) will appear in the output column.

Save the model as follows:

* Select the "File" option and left-click on the "Save" suboption.

* Type a file name in the "File Name:" box and left-click the "OK" button.

To access on-line help:

* Select the "Help" option and left-click any option you require.

* Left click at "Cancel" to quit Help.

* To reset TK, select the "File" option and left-click the "New" suboption.

Example 2-2

This example illustrates some built-in functions of TK, such as "pi()" for "π" and "^" for "raise to the power of."
Let the following be geometric characteristics of a cone:

volume = 1/3 * pi()* radius^2 * height

surface/slant = pi()* radius

slant^2 = height^2 + radius^2

tand(theta/2) = radius/height ("tand" denotes the tangent of an angle in degrees, and "tan" refers to the tangent of an angle in radians)

Enter the preceding equations into TK Solver:

1. Given *slant* = 8.75 and *radius* = 3, what are the remaining characteristics of the cone?

* Solve this problem as was done in Example 2-1.

Note: An asterisk (*) in the Status column of the Rules Sheet indicates that the equation was not solved. (Hint: What other value must be entered into the Variable Sheet to completely solve for all unknowns?)

Results: Volume = 77.4660, height = 8.21964, surface = 82.4643

Figure 2–14

Solve using iteration:

2.   Now, given *slant* = 8.75 and *volume* = 80, enter these values into the Variables Sheet and find the remaining characteristics.

TK will not be able to solve this problem immediately, as there are two unknowns in an equation. But the problem can be solved by using TK's built-in iterative solver:

*   Enter the inputs for *slant* and *volume*: *slant* = 8.75 and *volume* = 80.

*   Take an initial guess at the value of either *radius* or *height* and solve. In this case, we can use 3 as an initial guess for *radius*. Make this a guess by typing G into the Status field for *radius*.

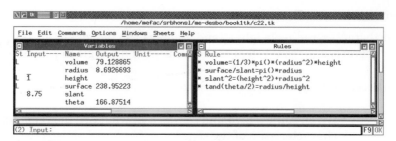

Figure 2–15

* Solve as before. You will get the required values of the characteristics of the cone.

* Try other numbers for a guess for *radius* and see what happens.

* Also try guessing values for *height* instead of *radius*.

List Solving and Plotting

3. Suppose the slant is constrained to have a value of 6.2 in the design of the cone. We next investigate what effects changes in the height have on the volume and surface area of the cone.

We can do this by entering the suggested value for *slant* and creating a list of values for *height*. TK can use those values as inputs and generate a list of solutions for both *volume* and *surface area*. Let us specify a range of values for *height* from 1 to 5 in steps of 0.2.

* Input the given value for *slant*: *slant* = 6.2.

* Type L and O in the status field for *surface* and *volume*, respectively.

* Type L and I in the status field of the *height* variable.

Remember, you will not see I or O on the screen, but it is important to type them.

* Test the maximum and minimum values by entering maximum and minimum values of *height* one after the other into the Input Field of the Variable Sheet and solving for each value. This is done to check whether solving over the specified range is possible.

Next, to view a range of results corresponding to heights between 1 and 5 units, create a table and enter the range of input values as follows:

* Select the "Sheet" option and left-click the "Tables" suboption.

* To display a single sheet, left-click the extreme right top square at the upper right corner of the tables window.

* Name the table 'table1' by left-clicking the "Display Subsheet" sub-option in the "Window" menu.

* Enter the names of the lists on the Table Subsheet:

  height

  volume

  surface

* Enter into the Interactive Table by selecting the "Window" option and left-clicking the "Display Subsheet" suboption.

* Place the cursor in the cell under the list name *height*.

* Select the "Command" option and left-click the "List Fill" suboption.

* Enter the first value as 1, last value as 5, and step size as 0.2 in appropriate boxes. Then left-click [Fill List].

* Select the "Command" option and left-click the "List Solve" suboption. TK fills the table with values for the volume and surface area corresponding to the range of heights given.

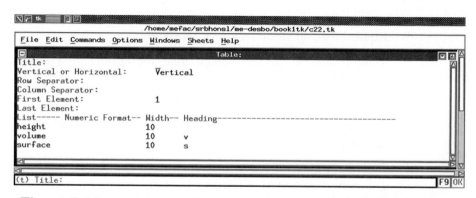

Figure 2–16

* To change the format of the table, go back to "Table Subsheet" and make the necessary modifications by changing the column size. This table can be sent to a printer or file.

To create the plot(s):

* Select the "Sheet" option and then select the "Plot" suboption.

* Enter the plot name "plot1."

* Define the plot type: type L into the plot type field for Line charts.

* Enter into the plot subsheet by selecting the "Window" option and then left-clicking the "Open Subsheet" suboption.

* Enter the *X*-Axis list name *height* and *Y*-Axis list names *surface* and *volume*. Any number of different functions can be included in a single plot.

* Enter the letters "v" for *volume* and "s" for *surface* in the appropriate "character" column to make the plot more readable.

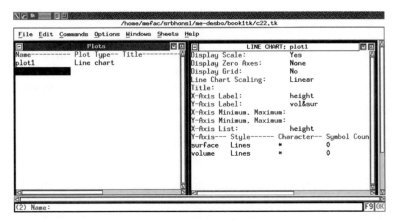

Figure 2–17

* Select the "Command" option and then the "Display Plot" suboption to display the plot.

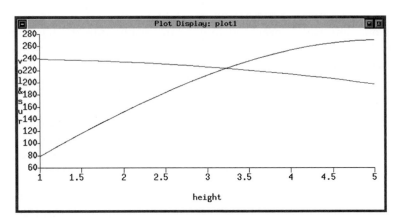

Figure 2–18

* You try the procedure now. Define the *X*-Axis minimum and maximum as 3.5 and 4, respectively. Also, define the *Y*-Axis minimum and maximum as 90 and 100, respectively.

*   Select the "Command" option and the "Display Plot" suboption.
This will display the plot with the given maximum and minimum
values on the axes.

*   You can print the plot by selecting the "Print" suboption in the
"File" option.

The "Save As" command of the "File" menu opens the Save file command
window and prompts the user for a file name. This saves the work done so far.
Any work done later on this same file will not be saved unless it is saved again.

The "Print" command of the "File" menu prints the selected text or graphics
from the active window. If a selection is highlighted by left-clicking, only the
selection will be printed; otherwise, the entire active window will be printed. The
"Print Table" command will print a table to a printer or file.

The "New" command of the "File" menu resets TK by removing the contents
of the current model. When you reset TK with the "New" command, the information in the current sheets is lost unless it is saved before resetting.

Using the "Open" command of the "File" menu, you can open any of your
files to work on. The "Open" command prompts for a file name. You can type in a
name or double-click on the file you want to select.

Final Note

The previous material is prepared to give an insight into the basic commands
necessary for using TK Solver Plus in this course. You can learn more about the
package while using it. This chapter only gives an overview. Different options are
available to get the output in a format of the user's liking, such as making proper
entries in the Table Subsheet, Plot Subsheet, etc.

It is also appropriate to caution users of any software to recognize that
changes are being made constantly by the software developers, and therefore it
should not surprise one to see some deviations from the commands explained
above for newly developed TK Solver packages.

It is also appropriate to realize that if one is inclined to have Units Sheet as
part of model then one should not have any constants in Rules Sheet having
dimensions, unless such constants are presented as having dimensions in Variables Sheet.

For further study on options available, the reader is requested to study the
manual provided by the TK Solver software company.

CHAPTER 3

Stresses, Strains, and Stress-Strain Relations

3.1 Stresses in Three Dimensions

In a course in basic mechanics, one learns various means of determining stresses due to axial load, bending moment, shear load, torques, and pressures. Also, one examines means of analyzing such stresses to determine principal stresses and maximum shear stresses in a plane of a stressed element for plane-stress or plane-strain conditions. Mohr's circle or derived equations are used to determine such principal stresses and shear stresses.

In this chapter, however, problems will be considered in which the assumption of plane stress or plane strain cannot be made. The procedures presented will quantify the stresses to which a machine element is subject in a general, three-dimensional state of stress.

Let us assume that an element, as shown in Fig. 3–1, is subjected to a general, three-dimensional (3-D) state of stress. Such a state consists of three normal stresses, σ_x, σ_y, and σ_z, and three shear stresses, τ_{xy}, τ_{yz}, and τ_{zx}. (Note that the other three shear stresses are $\tau_{yx} = \tau_{xy}$, $\tau_{zy} = \tau_{yz}$, and $\tau_{xz} = \tau_{zx}$.)

The sign convention for 3-D stresses is different than that for plane-stress problems. However, once it is understood, it can be used freely. Normal stresses are considered positive if they produce tension in a material. Shear stresses are positive if the axis perpendicular to the plane on which the shear stress acts is positive and the direction of the shear stress is in the positive axis direction, or if the axis is negative and the shear stress acts in the negative axis direction.

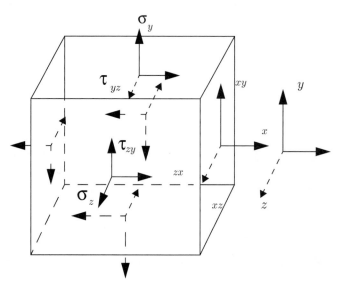

Figure 3-1   Element showing 3-D stresses

Now, a 3-D state of stress can be represented in a 3×3 matrix form as

$$\left[\sigma_{i,\,j}\right] = \begin{bmatrix} \sigma_x & \tau_{xy} & \tau_{xz} \\ \tau_{yx} & \sigma_y & \tau_{yz} \\ \tau_{zx} & \tau_{zy} & \sigma_z \end{bmatrix} \tag{3-1}$$

The designer may be interested in knowing the plane on which the normal stresses are maximum or minimum, the magnitudes of the stresses, and the orientation of such a plane as a result of a general state of stress. The orientation of this principal plane is determined by knowing the direction of the unit vector normal to such a plane. For example, let the plane IJK shown in Fig. 3–2 have a unit (normal) vector $\overline{N} = \cos\alpha\, i + \cos\beta\, \bar{j} + \cos\gamma\, \bar{k}$ where α, β, and γ are the angles between $\overline{N}$ and the x, y, and z axes, respectively. (The plane IJK and the unit vector $\overline{N}$ are perpendicular to each other; hence, vector $\overline{N}$ is called a normal unit vector.)

The angles α, β, and γ are defined by $\cos\alpha = A_x/A$, $\cos\beta = A_y/A$ and $\cos\gamma = A_z/A$, where A_x, A_y, A_z, and A are also shown in Fig. 3–2. It can now be seen that the magnitude of $\overline{N}$ is unity, that is, $\cos^2\alpha + \cos^2\beta + \cos^2\gamma = 1$ and thus $A^2 = A_x{}^2 + A_y{}^2 + A_z{}^2$. Let $l = \cos\alpha$, $m = \cos\beta$, and $n = \cos\gamma$. Then $\overline{N} = l\,i + m\,\bar{j} + n\,\bar{k}$, where l, m, and n are called *direction cosines* for the plane IJK.

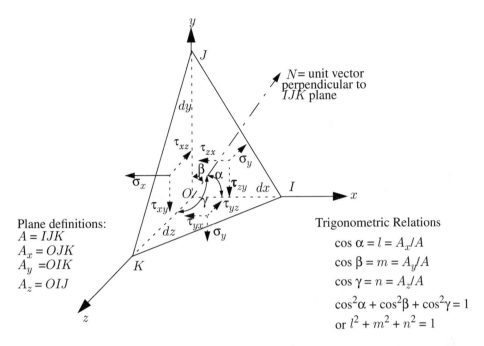

Plane definitions:
$A = IJK$
$A_x = OJK$
$A_y = OIK$
$A_z = OIJ$

Trigonometric Relations

$\cos \alpha = l = A_x/A$

$\cos \beta = m = A_y/A$

$\cos \gamma = n = A_z/A$

$\cos^2\alpha + \cos^2\beta + \cos^2\gamma = 1$

or $l^2 + m^2 + n^2 = 1$

Figure 3–2 Cut element showing the unit normal, vector and the direction of stresses acting on the element

With this background material, and knowing that an element subjected to a general state of stress is in equilibrium when $\Sigma \overline{F} = 0$, it is possible to determine stresses on a plane whose unit normal is $\overline{N} = \cos\alpha \overline{i} + \cos\beta \overline{j} + \cos\gamma \overline{k}$. Such an element is illustrated in Fig. 3–3 with the resultant forces dR_x, dR_y, dR_z, and dR_1, dR_2, dR_3 shown, rather than the stresses.

Since the cut element must be in static equilibrium, the necessary equations can be developed using the laws of statics. However, before proceeding to use static equilibrium equations, let us determine the traction forces dR_x, dR_y, and dR_z on the planes whose unit normals are $\overline{i}$, $\overline{j}$, and $\overline{k}$ (that is, on the planes OJK, OIK, and OIJ; note that the traction forces are not necessarily normal or tangent to a given plane).

For plane OJK, the force $\sigma_x A_x$ (see Fig. 3–2) is acting in the negative x direction, the force $\tau_{xy} A_x$ is acting in the negative y direction and the force $\tau_{xz} A_x$ is acting in the negative z direction; therefore,

$$dR_x = -\sigma_x A_x \overline{i} - \tau_{xy} A_x \overline{j} - \tau_{xz} A_x \overline{k} \tag{3–2}$$

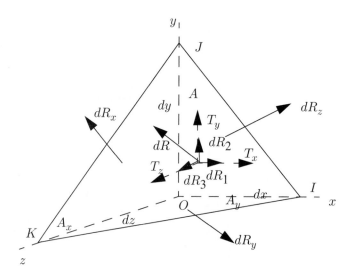

Figure 3–3　Cut element showing tractions acting on the surface IJK

Note that $d\overline{R}_x$ is a traction force acting not in the x-direction, but on plane <u>OJK</u>, whose unit normal is in the x direction.

Similarly, one can show that

$$d\overline{R}_y = -\tau_{yx}A_y\overline{i} - \sigma_y A_y\overline{j} - \tau_{yz}A_y\overline{k} \qquad (3\text{--}3)$$

and

$$d\overline{R}_z = -\tau_{zx}A_z\overline{i} - \tau_{zy}A_z\overline{j} - \sigma_z A_z\overline{k} \qquad (3\text{--}4)$$

Equations (3–2), (3–3), and (3–4) can be represented in matrix form as

$$\left[d\overline{R}_E\right] = \left[d\overline{R}_x \; d\overline{R}_y \; d\overline{R}_z\right] = \left[-\overline{i} - \overline{j} - \overline{k}\right]\begin{bmatrix} \sigma_x A_x & \tau_{xy}A_y & \tau_{xz}A_z \\ \tau_{yx}A_x & \sigma_y A_y & \tau_{yz}A_z \\ \tau_{zx}A_x & \tau_{zy}A_y & \sigma_z A_z \end{bmatrix} \qquad (3\text{--}5)$$

Now, on plane IJK (which defines the area A), the force $d\overline{R}$ must be acting to balance the foregoing forces; thus, taking $d\overline{R} = dR_1\overline{i} + dR_2\overline{j} + dR_3\overline{k}$, where dR_1, dR_2 and dR_3 are the components in the x, y and z directions, respectively, and since $\Sigma\overline{F} = 0$, we obtain

$$dR + d\bar{R}_E = 0 \tag{3-6}$$

For a vector to be equal to another vector, their respective components must be equal, that is, we must have

$$dR_1 = -dR_x$$

$$dR_2 = -dR_y$$

$$dR_3 = -dR_z$$

Substituting for dR_x, dR_y, and dR_z from Equation (3–5), one gets

$$dR_1 = \sigma_x A_x + \tau_{xy} A_y + \tau_{xz} A_z \tag{3-7}$$

$$dR_2 = \tau_{yx} A_x + \sigma_y A_y + \tau_{yz} A_z \tag{3-8}$$

$$dR_3 = \tau_{zx} A_x + \tau_{zy} A_y + \sigma_z A_z \tag{3-9}$$

or, in matrix form,

$$\begin{bmatrix} dR_1 \\ dR_2 \\ dR_3 \end{bmatrix} = \begin{bmatrix} \sigma_x & \tau_{xy} & \tau_{xz} \\ \tau_{yx} & \sigma_y & \tau_{yz} \\ \tau_{zx} & \tau_{zy} & \sigma_z \end{bmatrix} \begin{Bmatrix} A_x \\ A_y \\ A_z \end{Bmatrix} \tag{3-10}$$

Next, we define a new term called the *traction*, T, in the x, y, and z directions as

$$T_x = \frac{dR_1}{A}, \ T_y = \frac{dR_2}{A}, \ T_z = \frac{dR_3}{A} \tag{3-11}$$

Based on Equation (3–11), one can understand traction as stress because it has dimensions of force per unit area. However, this force per unit area is not necessarily normal or tangent to the surface on which it acts.

Substituting for dR_1, dR_2, and dR_3 from Equation (3–11), into Equation (3–10) and recalling that the ratios of A_x/A, A_y/A, and A_z/A are the direction cosines l, m, and n, respectively, we can write the traction in matrix form as

$$\begin{bmatrix} T_x \\ T_y \\ T_z \end{bmatrix} = \begin{bmatrix} \sigma_x & \tau_{xy} & \tau_{xz} \\ \tau_{yx} & \sigma_y & \tau_{yz} \\ \tau_{zx} & \tau_{zy} & \sigma_z \end{bmatrix} \begin{Bmatrix} l \\ m \\ n \end{Bmatrix}$$

or, in shorthand notation,

$$[T] = [\sigma_{ij}] \begin{Bmatrix} l \\ m \\ n \end{Bmatrix} \tag{3–12}$$

Equation (3–12) relates the tractions in the x, y, and z directions (on the plane IJK whose unit normal is $\overline{N}$) to the stress matrix $\begin{bmatrix} \sigma_{ij} \end{bmatrix}$ and the direction cosines l, m, n.

Principal Stresses

Now we are ready to find (1) the direction of a plane on which the normal stress is maximum or minimum and (2) the magnitude of that stress.

Let us take the IJK plane as a principal plane—that is, a plane on which only a normal stress σ exists and no shear stresses exist. Then such a principal stress produces a normal force σA.

For equilibrium, the components of force due to the principal stress σ in the x, y, and z directions must be equal to the traction forces in those same directions, that is,

$$\sigma A \overline{N} \cdot i = T_x A$$

but $\overline{N} \cdot i = l$ (where $l = \cos \alpha$)

$$\therefore \sigma A l = T_x A \tag{3–13}$$

Similarly, summing components in the y and z directions, one gets

$$\sigma A m = T_y A \tag{3–14}$$

and

$$\sigma A n = T_z A \tag{3–15}$$

Substituting for T_x, T_y and T_z from Equation (3–12) results in

$$\sigma_x l + \tau_{xy} m + \tau_{xz} n = \sigma l \tag{3–16}$$

$$\tau_{yx} l + \sigma_y m + \tau_{yz} n = \sigma m \tag{3–17}$$

$$\tau_{zx} l + \tau_{zy} m + \sigma_z n = \sigma n \tag{3–18}$$

or, in matrix form,

$$\begin{bmatrix} \sigma_x - \sigma & \tau_{xy} & \tau_{xz} \\ \tau_{yx} & \sigma_y - \sigma & \tau_{yz} \\ \tau_{zx} & \tau_{zy} & \sigma_z - \sigma \end{bmatrix} \begin{Bmatrix} l \\ m \\ n \end{Bmatrix} = 0 \qquad (3\text{--}19)$$

Equation (3–19) is a set of simultaneous linear *homogeneous* equations having a non-trivial solution if and only if the determinant of the coefficient matrix is equal to zero—that is,

$$\begin{vmatrix} \sigma_x - \sigma & \tau_{xy} & \tau_{xz} \\ \tau_{yx} & \sigma_y - \sigma & \tau_{yz} \\ \tau_{zx} & \tau_{zy} & \sigma_z - \sigma \end{vmatrix} \equiv 0 \qquad (3\text{--}20)$$

Expanding the determinant of Equation (3–20) one gets what is generally called the *Characteristic Equation*:

$$\sigma^3 - I_1 \sigma^2 + I_2 \sigma - I_3 = 0 \qquad (3\text{--}21)$$

where

$$I_1 = \sigma_x + \sigma_y + \sigma_z \qquad (3\text{--}22)$$

$$I_2 = \sigma_x \sigma_y + \sigma_y \sigma_z + \sigma_z \sigma_x - \tau_{xy}^2 - \tau_{yz}^2 - \tau_{zx}^2 \qquad (3\text{--}23)$$

$$I_3 = \begin{vmatrix} \sigma_x & \tau_{xy} & \tau_{xz} \\ \tau_{yx} & \sigma_y & \tau_{yz} \\ \tau_{zx} & \tau_{zy} & \sigma_z \end{vmatrix} = \sigma_x \sigma_y \sigma_z + 2\tau_{xy} \tau_{yz} \tau_{zx} - \sigma_x \tau_{yz}^2 - \sigma_y \tau_{zx}^2 - \sigma_z \tau_{xy}^2 \qquad (3\text{--}24)$$

I_1, I_2, and I_3 are referred to as *stress invariants*, where one can see that, for example, $I_1 = \sigma_x + \sigma_y + \sigma_z$ and I_1 is also equal to $\sigma_1 + \sigma_2 + \sigma_3$ (i.e., the sum of the three orthogonal normal stresses is a constant). Therefore, I_1 is called an invariant; similarly we can show that I_2, and I_3 are invariants.

Note that Equation (3–21) is a cubic equation, the solution of which will yield three real roots of σ, called σ_a, σ_b, and σ_c, which are principal stresses.

If these three principal stress values are called σ_1, σ_2, and σ_3 and are arranged such that $\sigma_1 > \sigma_2 > \sigma_3$, then the maximum shear stress is given by

$$\tau_{max} = \frac{\sigma_1 - \sigma_3}{2} \tag{3-25}$$

The directions of σ_1, σ_2, or σ_3 are required to locate the planes on which such principal stresses act. They are obtained using the method shown in the next section.

3.2 Method of Determining Principal Stresses and Their Directions

Once the three roots are obtained from Equation (3–21), then using two independent equations from a set of equations such as that shown in Equation (3–19) and knowing that $l^2 + m^2 + n^2 = 1$, one could solve for the direction cosines (l_i, m_i, n_i) of the principal stress σ_1, σ_2, or σ_3; however, this method is cumbersome. Instead, the following method will simplify finding the three principal stresses (roots) and their directions. The method is easy to use and will directly yield the values of σ_1, σ_2, σ_3 and their directions, say l_i, m_i, n_i (i = 1, 2, 3), without going through a root-finding method. Root-finding methods require iterative trial-and-error calculations, followed by the solution of three simultaneous nonhomogeneous nonlinear equations.

Procedure to Solve for σ_1, σ_2, σ_3 and l_i, m_i, n_i

Consider the stress matrix

$$[\sigma_{i, j}] = \begin{bmatrix} \sigma_x & \tau_{xy} & \tau_{xz} \\ \tau_{yx} & \sigma_y & \tau_{yz} \\ \tau_{zx} & \tau_{zy} & \sigma_z \end{bmatrix}$$

where all stress components are considered known. Knowing I_1, I_2, and I_3, determine

$$R = (1/3) \, I_1^2 - I_2$$

$$Q = (1/3) \, I_1 \, I_2 - I_3 - (2/27)I_1^3$$

$$S = [(1/3) \, R]^{1/2}$$

$$T = [(1/27) \, R^3]^{.5}$$

$$\alpha = \cos^{-1}[-Q/(2T)]$$

The angle α, obtained in radians from the preceding expression, is converted into degrees for use in the equations that follow. (For I_1, I_2, I_3 see Equations 3–22 to 3–24.) One can now find the principal stresses (called *eigen values*) σ_a, σ_b, and σ_c using the equations

$$\sigma_a = 2S\,[\cos(\alpha/3)] + (1/3)\,I_1$$

$$\sigma_b = 2S\,[\cos(\alpha/3 + 120°)] + (1/3)\,I_1$$

$$\sigma_c = 2S\,[\cos(\alpha/3 + 240°)] + (1/3)\,I_1$$

The resulting principal stresses are renamed σ_1, σ_2, σ_3 and then reordered such that $\sigma_1 > \sigma_2 > \sigma_3$.

With all of the principal stresses determined, we can now find the maximum shear stress, which is given by

$$\tau_{max} = \frac{\sigma_1 - \sigma_3}{2}$$

To find the direction cosines, i.e. to find the direction of each principal stress, the following procedure can be used.

Let us say that it is first required to calculate direction cosines for σ_1. We begin by calculating a, b, c, and k, where

$$a = (\sigma_y - \sigma_1)(\sigma_z - \sigma_1) - \tau_{yz}\,\tau_{yz} \tag{3–26}$$

$$b = -(\sigma_z - \sigma_1)\,\tau_{xy} + \tau_{xz}\,\tau_{yz} \tag{3–27}$$

$$c = \tau_{xy}\,\tau_{yz} - \tau_{xz}\,(\sigma_y - \sigma_1) \tag{3–28}$$

$$k = 1/(a^2 + b^2 + c^2)^{1/2} \tag{3–29}$$

Then

$$l_1 = ak,\ m_1 = bk,\ \text{and}\ n_1 = ck \tag{3–30}$$

This process can be repeated to find new values for a, b, c, and k using Equations 3–26 to 3–30 but with σ_2 in place of σ_1, and then finding

$$l_2 = ak,\ m_2 = bk,\ \text{and}\ n_2 = ck,$$

the directional cosines for the principal stress σ_2.

Again, the process can be repeated with σ_3 in place of σ_2 to find

$$l_3 = ak,\ m_3 = bk,\ \text{and}\ n_3 = ck$$

The method can also be used to determine principal stresses for plane-stress conditions ($\sigma_z = \tau_{xz} = \tau_{yz} = 0$) or plane-strain conditions ($\gamma_{zx} = \gamma_{zy} = \varepsilon_z = 0$). Let us study an example problem to further understand the above methods.

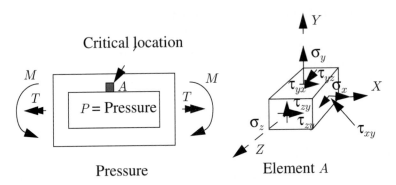

Figure 3–4 Pressure cylinder and stressed element

TK Model to Determine Principal Stresses and Their Directions

Example 3-1

The cylinder shown in Fig. 3–4 is subjected to bending, torsional, and vertical shear in addition to an internal pressure. At a critical location A, the stresses are determined using stress equations for thick-walled cylinders. The 3-D stress matrix is

$$[\sigma_{i, j}] = \begin{bmatrix} 10 & 5 & 3 \\ 5 & 10 & 4 \\ 3 & 4 & 15 \end{bmatrix} \text{ ksi}$$

Remember, in general

$$[\sigma_{i, j}] = \begin{bmatrix} \sigma_x & \tau_{xy} & \tau_{xz} \\ \tau_{yx} & \sigma_y & \tau_{yz} \\ \tau_{zx} & \tau_{zy} & \sigma_z \end{bmatrix}$$

Determine the principal stresses and their directions, using a TK lead model called *ex3-1.tk* or *stress3.tk* as shown on the Rules Sheet on page 37.

Solution

Type all required equations into the Rules Sheet. Enter the stress matrix values as input into the Variables Sheet and press F9. You will see that TK has solved for σ_1, σ_2, and σ_3, the principal stresses.

($\sigma_1 = 19.97$ ksi, $\sigma_2 = 10.10$ ksi, $\sigma_3 = 4.93$ ksi) (see Variables Sheet output column)

RULES SHEET

To determine the direction cosines corresponding to σ_1, enter $Sp = \sigma_1$ in an input column of the Variables Sheet and press F9. You will obtain l_1, m_1, and n_1

$l_1 = .4743$, $m_1 = .5226$, $n_1 = .7083$ (Note: $l_i^2 + m_i^2 + n_i^2 = 1$ see Variables Sheet output column).

Similarly, to determine the direction cosines corresponding to σ_2, enter $Sp = \sigma_2$ into the Variables Sheet and press F9. You will get $l_2 = -.5688$, $m_2 = -.4314$, $n_2 = .7003$ (the Variables Sheet for this set of results is not shown)

Finally, enter $Sp = \sigma_3$ into the Variables Sheet and press F9 (or use the Solve command from the command window) to obtain the direction cosines l_3, m_3, n_3, corresponding to σ_3. You will obtain

$$l_3 = .6713, \; m_3 = -.7354, \; n_3 = .0922$$

Try to find l_2, m_2, and n_2 yourself using the lead model.

VARIABLES SHEET

```
┌─────────────────────────────────────────────────────────────────┐
│ X ⌐ tk ▓▓▓ ⌐ ⊡                                                    │
├─────────────────────────────────────────────────────────────────┤
│                              ex3-1.tk                             │
├─────────────────────────────────────────────────────────────────┤
│  File  Edit  Commands  Options  Windows  Sheets  Help            │
├─────────────────────────────────────────────────────────────────┤
│ ▣                           Variables                        ⊡ ▲ │
│St Input---- Name--- Output--- Unit----- Comment----------------▲  │
│L   19.96    Sp                 psi       Principal stress (psi,or ksi or Pa, or│
│             I1      35         psi          invariants          │
│             I2      350        psi^2        invariants          │
│             I3      995        psi^3        invariants          │
│                                                                 │
│    10       Sx                 psi       NORMAL stress (psi,or ksi or Pa, orMPa│
│    10       Sy                 psi       NORMAL stress (psi,or ksi or Pa, orMPa│
│    15       Sz                 psi       NORMAL stress (psi,or ksi or Pa, orMPa│
│                                                                 │
│    5        Txy                psi       SHEAR stress (psi,or ksi or Pa, orMPa)│
│    4        Tyz                psi       SHEAR stress (psi,or ksi or Pa, orMPa)│
│    3        Txz                psi       SHEAR stress (psi,or ksi or Pa, orMPa)│
│                                                                 │
│             Sa      19.96648   psi       PRINCIPAL stress (psi,or ksi or Pa, or│
│             Sb      4.9344818  psi       PRINCIPAL stress (psi,or ksi or Pa, or│
│             Sc      10.099039  psi       PRINCIPAL stress (psi,or ksi or Pa, or│
│                                                                 │
│             S       30403.022            Constant for given stress matrix│
│             l       59.283402            Constant for given stress matrix│
│             R       2.77303E9            Constant for given stress matrix│
│             Q       -2.871E13            Constant for given stress matrix│
│             T       2.8103E13            Constant for given stress matrix│
│                                                                 │
│L            ERR     -.960064   psi^3     Error term             │
│                                                                 │
│             a       1.58783E9            Constant for given stress matrix and p│
│             b       1.74939E9            Constant for given stress matrix and p│
│             c       2.37118E9            Constant for given stress matrix and p│
│             k       2.988E-10            Constant for given stress matrix and p│
│                                                                 │
│             li      .47437026  rad       Direction cosine li for Si principal s│
│             mi      .52263442  rad       Direction cosine mi for Si principal s│
│             ni      .70839687  rad       Direction cosine ni for Si principal s│
│             S1      19.96648   psi       PRINCIPAL stress (psi,or ksi or Pa, or│
│             S2      10.099039  psi       PRINCIPAL stress (psi,or ksi or Pa, or│
│             S3      4.9344818  psi       PRINCIPAL stress (psi,or ksi or Pa, or│
│                                                                ▼ │
│◁                                                               ▷ │
├─────────────────────────────────────────────────────────────────┤
│ (38)  Input:                                              F9 OK  │
└─────────────────────────────────────────────────────────────────┘
```

The computer provides an error-free solution in a straightforward manner. This method can be extended to simpler 2-D plane-stress problems (i.e., $\sigma_z = 0$, $\tau_{zx} = \tau_{zy} = 0$), or plane–strain problems (i.e., $\varepsilon_z = 0$).

The use of a computer and software such as TK Solver makes solving 3-D stress problems simple and quick. Engineering problems always involve 3-D stresses. In exceptional cases, one can assume a plane-stress or plane-strain condition. In machine design, when one has to deal with 3-D stresses an understanding of the preceding material will allow you to design a machine component even if it is subjected to a 3-D state of stress. Also, if one is knowledgeable about 3-D stresses, it is then easy to understand the theories of failure presented later for a 3-D stress state.

In the past, Mohr's circle solutions gave us results for a 2-D state of stress only. Therefore, it is advisable to use the equations in this chapter to obtain results for 3-D stresses, to which almost all machine elements are subjected. It is

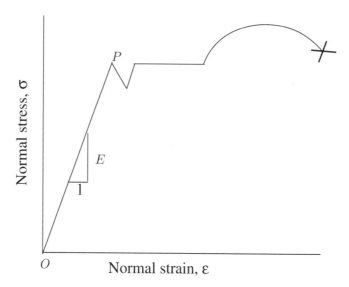

Figure 3–5   Stress–strain diagram for simple tension test

also advisable to solve plane-stress (2-D) problems or plane-strain problems ($\varepsilon_z =$ 0) using the same lead model.

3.3 Stress–Strain Relations

Uniaxial Stress-Strain Relation

A simple tension test yields the stress–strain diagram shown in Fig. 3–5. In the range from O to P, the stress is proportional to the strain; that is, $\sigma_x \propto \varepsilon_x$.

One can therefore define a term called the *Modulus of Elasticity* in the range from O to P as

$$E = \sigma_x / \varepsilon_x \qquad (3\text{--}31)$$

This Equation (3–31) was derived by Hooke and therefore is called Hooke's law.

The units of E are pounds per square inch in the FPS system or Newtons per square meter in SI units. We can rearrange Equation (3–31) to the form

$$\varepsilon_x = \sigma_x / E \qquad (3\text{--}32)$$

However, the stress in the x direction produces strain in the y and z directions, an effect called *Poisson's effect*. Strain in the y and z directions is proportional to strain in the x direction, according to the equations

$$\varepsilon_y = -\nu\varepsilon_x \qquad (3\text{--}33)$$

$$\varepsilon_z = -\nu\varepsilon_x \qquad (3\text{--}34)$$

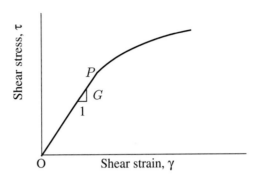

Figure 3-6   Shear stress–shear strain curve

where ε_y and ε_z are strains in the y and z directions, respectively, and ν is called *Poisson's ratio*, which is a material property. (Note that ε_y and ε_z will be opposite in sign to ε_x.)

This fundamental concept based on Hooke's Law and Poisson's ratio can be extended to 3-D stresses and corresponding strains by applying the method of superposition, which states that the triaxial strain is related to triaxial stresses by the equations

$$\varepsilon_x = \frac{\sigma_x}{E} - \frac{\nu}{E}(\sigma_y + \sigma_z) \tag{3–35}$$

$$\varepsilon_y = \frac{\sigma_y}{E} - \frac{\nu}{E}(\sigma_x + \sigma_z) \tag{3–36}$$

$$\varepsilon_z = \frac{\sigma_z}{E} - \frac{\nu}{E}(\sigma_x + \sigma_y) \tag{3–37}$$

A similar study on a test specimen subjected to shear stress shows that

$$\tau \propto \gamma$$

where τ is the shear stress and γ is the shear strain. The shear stress is proportional to the shear strain in the range O to P as shown in Fig. 3–6.

The shear modulus, G, is defined as the ratio of shear stress to shear strain within the region of proportionality as shown in Fig. 3–6. That is,

$$G = \tau/\gamma \tag{3–38}$$

where G has the same units as the modulus of elasticity—namely, lb/in^2 in the FPS system and N/m^2 in the SI system.

Therefore, there are three additional equations for a homogeneous, isotropic material:

$$G = \tau_{xy}/\gamma_{xy} \tag{3–39}$$

$$G = \tau_{yz}/\gamma_{yz} \tag{3–40}$$

$$G = \tau_{zx}/\gamma_{zx} \tag{3–41}$$

Let us study an example to understand the application of the above equation.

Example 3-2

Strain measurements on the surface of a machine element have the values

$\varepsilon_x = 0.001$, $\varepsilon_y = 0.002$, $\gamma_{xy} = 0.003$, $\gamma_{yz} = \gamma_{zx} = \sigma_z = 0$ (note that units for strains can be cancelled: in/in or m/m)

Determine principal stresses σ_1, σ_2, and σ_3 ($\sigma_1 > \sigma_2 > \sigma_3$) if $E = 10 \times 10^6$ psi and $v = 0.29$.

Solution

$$\varepsilon_x = \frac{\sigma_x}{E} - \frac{v}{E}(\sigma_y + \sigma_z)$$

therefore

$$.001 = \frac{\sigma_x}{E} - \frac{v}{E}(\sigma_y + \sigma_z)$$

Similarly using Equation (3–37) and (3–38), we find that

$$.002 = \frac{\sigma_y}{E} - \frac{v}{E}(\sigma_x + \sigma_z)$$

and

$$\varepsilon_z = \frac{\sigma_z}{E} - \frac{v}{E}(\sigma_x + \sigma_y) , \text{(Note: } \sigma_z = 0)$$

Next, using Equation (3–39), we have, after rearranging,

$$\gamma_{xy} = \tau_{xy}/G$$

Therefore,

$$0.003 = \tau_{xy}/G$$

The relation between the G-shear modulus and the E modulus of elasticity is given by $G = \dfrac{E}{2(l + v)}$ therefore

$$G = E/2(1 + v) = \frac{10 \times 10^6}{2(1 + 0.29)} = 3.8759 \times 10^6 \text{ psi}$$

RULES SHEET

Solving the preceding equations in the Rules Sheet using the TK Solver lead model *ex3-2.tk* or *ststrain.tk*, as shown on the next page, yields

$$\sigma_x = 17{,}250.80 \text{ psi}, \ \sigma_y = 25{,}002.73 \text{ psi}, \ \tau_{xy} = 11{,}627.90 \text{ psi}$$

and $\sigma_1 = 33{,}383.65$ psi, $\sigma_2 = 8{,}869.87$ psi, $\sigma_3 = 0$

$$\varepsilon_z = -.0012254 \text{ in/in}$$

(Hint: Guess the values of σ_x and σ_y to solve the equations; see the Variables Sheet for the output.)

Some comments are in order regarding this lead model:

1.   You will see that the model uses not only the equations developed in this section, but also those from the previous lead model called *ex3-1.tk*, to convert normal and shear stresses to principal stresses and to obtain the maximum shear stress. See the Rules Sheet for details.

2.   This particular model can have either FPS or SI units. This means that you can have mixed units. To have mixed units, follow the instructions given next; this will train you to develop

VARIABLES SHEET

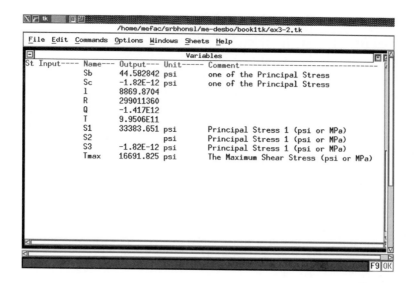

models in the future if you need to have either one set of units or mixed units.

Copy the Unit Sheet from the lead model *ex3-1.tk* by opening the model and going into its Unit Sheet and highlighting it entirely. Then go into the "edit" command and click on "copy." Now open *ex3-2.tk* and open the Unit Sheet by selecting the "sheets" option and clicking with the left mouse button on the Unit Sheet. Then click the cursor in the Units Sheet and go to the "edit" command and select "paste." This will place all the unit values you had copied earlier into the current unit sheet. Next, save *ex3-2.tk* by using the "save" option from the "file" command.

3. Now place the cursor at ε_x for example, and go to the window command and open the subsheet. In that sheet, type **in/in** at the "display units" line and type **m/m** at the "calculation units" line. Repeat this process for every variable in the Variables Sheet, until all variables are accounted for. Remember that the units, in FPS or SI, must be consistent with the Units Sheet. For example, if stress display units are psi, the calculation units for SI must be MPa, Pa, or kPa. Make sure that there is a relation between psi and MPa, Pa, or kPa in the Units Sheet. Completing this exercise will prepare you for building more complex models in the future.

PROBLEMS

3–1)   Given $[\sigma_{i,\,j}] = \begin{bmatrix} 5 & 3 & 2 \\ 3 & 10 & 12 \\ 2 & 12 & 15 \end{bmatrix}$ ksi

find σ_1, σ_2, σ_3 and l_i, m_i, n_i for $i = 1, 2, 3$, once by hand calculations and then using the lead model *stress3.tk*. (Use the sample problem as a test problem when you start using the *stress3.tk* model.) Also, show how easy it is to find σ_1, σ_2, and σ_3 in MPa (i.e., in SI units). (You are now ready to solve any 3-D or 2-D plane stress or plane strain problem using the same lead model.)

3-2)   Using Equation (3–20), prove that for a 2-D plane-stress problem,

$$\sigma_1 = \frac{\sigma_x + \sigma_y}{2} + \left[\left(\frac{\sigma_x - \sigma_y}{2} \right)^2 + \tau_{xy}^2 \right]^{\frac{1}{2}}$$

$$\sigma_2 = \frac{\sigma_x + \sigma_y}{2} - \left[\left(\frac{\sigma_x - \sigma_y}{2} \right)^2 + \tau_{xy}^2 \right]^{\frac{1}{2}}$$

$$\sigma_3 = \sigma_z = 0$$

(This derivation requires only hand calculations; note that $\sigma_z = \tau_{xz} = \tau_{yz} = 0$ for a plane stress condition.)

Now rearrange σ_1, σ_2, σ_3 such that $\sigma_1 > \sigma_2 > \sigma_3$, and show that

$$\tau_{max} = \frac{\sigma_1 - \sigma_3}{2}$$

3-3)   If $\sigma_x = 50$ ksi, $\sigma_y = 20$ ksi and $\tau_{xy} = 20$ ksi (all other stresses $= 0$), determine σ_1, σ_2, and σ_3 such that $\sigma_1 > \sigma_2 > \sigma_3$ and also determine τ_{max}. Use the *stress3.tk* lead model and also the equations derived in Problem 3-2. Compare your results.

3-4)   Given $\sigma_x = 50$ ksi, $\sigma_y = -20$ ksi and $\tau_{xy} = 20$ ksi, determine σ_1, σ_2, and σ_3 if $\varepsilon_z = \gamma_{zx} = \gamma_{zy} = 0$ (plane strain), $V = 0.30$, and $E = 30 \times 10^6$ psi. Use the *stress3.tk* model. (Note: You will need V and E when using the lead model.)

3-5)   Draw and label the Mohr's circle diagram for the following state of stress: $\sigma_x = 100$ ksi, $\sigma_y = 40$ ksi and $\tau_{xy} = 25$ ksi ccw; all other stresses are zero.

Find the three principal stresses and the maximum shear stress for this state of stress. Sketch a stress element showing the nonzero principal stresses and the orientation of the element to the x-axis.

3-6)   The state of stress on a critical element obtained from a crankshaft is given by

$$[\sigma_{ij}] = \begin{bmatrix} 100 & 25 & 0 \\ 25 & 40 & 0 \\ 0 & 0 & 10 \end{bmatrix} \text{ ksi}$$

   a.   Determine the principal stress state (σ_1, σ_2, and σ_3).

   b.   Determine the maximum shear stress.

   c.   Show the orientation of the principal element (subjected to the σ_1, σ_2, and σ_3 stresses) with respect to the x axis.

3-7)   A pressure cylinder having an internal radius of 6" and thickness of 1" is subjected to an internal pressure of 5000 psi, a twisting torque of 50,000 in-lb, and a bending moment of 50,000 in-lb as shown in the following diagram:

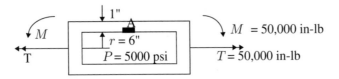

The transverse stress σ_t due to internal pressure is = 32,692.31 psi, the longitudinal stress σ_1 due to internal pressure is = 13,846.15 and the radial compressive stress σ_r due to internal pressure is 5000 psi and all stresses are at $r = r_i = 6$". (These stresses do not include stresses due to bending and torque loads.)

   a.   Determine the bending and torsional stress at $r = 6$" on the element at point A.

   b.   Superimpose the preceding stresses on pressure stresses, and from such stresses determine the principal stresses σ_1, σ_2, σ_3, the maximum shear stress, and the principal stresses.

   c.   Determine the safety factor if the yield-point normal stress S_{yp} = 40,000 psi. (Safety factor = S_{yp}/σ_1.)

3-8)   Draw and label the Mohr's circle diagram for the following plane state of
        stress:

$\sigma_x = 100$ ksi, $\sigma_y = 40$ ksi, $\tau_{xy} = 25$ ksi ccw. All other stresses are zero.

Find the three principal normal stress components and the maximum
three-dimensional shear stress components for this state of stress.
Sketch a stress element, showing the orientation to the x-axis of the non-
zero principal normal stresses on the element.

3-9)   The state of stress on a critical element obtained from a machine is

$$\left[\sigma_{ij}\right] = \begin{bmatrix} 100 & 250 & 0 \\ 25 & 400 & 0 \\ 0 & 0 & 110 \end{bmatrix}$$

   a.   Determine the principal stresses σ_1, σ_2, and σ_3.

   b.   Determine the maximum shear stress.

   c.   Show the orientation of the principal element (subjected to the
         σ_1, σ_2, and σ_3 stresses) with respect to the x-axis.

   d.   Determine the principal strains and the maximum shear strain
         if $E = 30 \times 10^6$ psi and $v = 0.29$.

Use *ststrain.tk* lead model to solve this problem.

3-10)   Convert the lead model *ststrain.tk* so that it can have either FPS or SI
        units.

Stress Analysis of Beams

4.1 Introduction

There are two basic methods for determining stresses in a machine element such as a beam: the experimental method and the analytical method. Nowadays, a popular, powerful analytical method is the *finite element technique*, in which a machine component, subjected to service loads, is studied by creating a so-called finite-element model. Another analytical method uses bending moment, shear, torque, and axial load diagrams to determine stresses based upon equilibrium conditions. The *elasticity approach,* based on equilibrium equations and differential equations, is also used.

In this chapter, first we shall apply the experimental techniques to measure strains and then stresses in a machine element and then analytical techniques to beams. The analytical techniques used are based on equations of equilibrium.

4.2 Introduction to the Theory of Strain Gage Analysis

Machine elements designed by using mathematical theory must be checked for actual performance. One way is to determine stresses under service loads in a machine element is by using experimental techniques. The reason for experimentally testing the machine elements is to make sure that the measured stresses under service loads are sufficiently below the allowable stresses. The experimental method involves measuring the strains first, as stresses cannot be measured easily, and then relating these strains to the stresses.

Many electrical, mechanical, and optical systems have been developed for measuring the average strain at a point on the surface of a machine element. The most popular and reliable method for measuring strains is by means of electrical strain gages in which a metal foil, consisting of a grid obtained by photoetching process is cemented between two sheets of paper or plastic

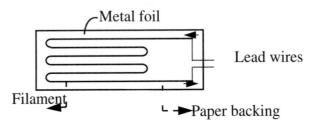

Figure 4-1 Electrical strain gage

backing, is bonded to a machine element at a point of interest. The general appearance of such a strain gage is as shown in Fig. 4–1. Upon loading the machine element, the metal foil is stretched, causing a change in the resistance of the filament. This change in resistance is measured by a special instrument called a *strain gage indicator*. It has been found experimentally that the change in resistance in such strain gages is directly proportional to the strain in the same direction ($\varepsilon \propto \Delta R$). This strain can then be related to stress.

The following equations show the relationships between strains and stresses in a homogeneous, isotopic machine element. This strain-stress relationship is based upon the Generalized Hooke's Law equations which were presented in Chapter 3. The equations are valid for isotropic and homogeneous materials. The six equations are:

$$\varepsilon_x = \frac{1}{E}(\sigma_x - \nu(\sigma_y + \sigma_z)) \tag{4–1}$$

$$\gamma_{xy} = \frac{\tau_{xy}}{G} \tag{4–2}$$

$$\varepsilon_y = \frac{1}{E}(\sigma_y - \nu(\sigma_x + \sigma_z)) \tag{4–3}$$

$$\gamma_{yz} = \frac{\tau_{yz}}{G} \tag{4–4}$$

$$\varepsilon_z = \frac{1}{E}(\sigma_z - \nu(\sigma_x + \sigma_y)) \tag{4–5}$$

$$\gamma_{zx} = \frac{\tau_{zx}}{G} \tag{4–6}$$

where E = Modulus of Elasticity in psi or MPa, G = Shear Modulus in psi or MPa, and ν = Poisson's ratio (dimensionless).

It is also possible to prove that the material constants E, G, and ν are related by the equation

$$G = \frac{E}{2(1+\nu)} \tag{4-7}$$

Thus, two independent material constants exist, (i.e., either E and ν, or G and ν, etc., that must be determined experimentally.) The foregoing strain–stress equations can be converted to the following stress–strain equations:

$$\sigma_x = \frac{E}{(1+\nu)(1-2\nu)}[(1-\nu)\varepsilon_x + \nu(\varepsilon_y + \varepsilon_z)] \tag{4-8}$$

$$\tau_{xy} = G\gamma_{xy} \tag{4-9}$$

$$\sigma_y = \frac{E}{(1+\nu)(1-2\nu)}[(1-\nu)\varepsilon_y + \nu(\varepsilon_z + \varepsilon_x)] \tag{4-10}$$

$$\tau_{yz} = G\gamma_{yz} \tag{4-11}$$

$$\sigma_z = \frac{E}{(1+\nu)(1-2\nu)}[(1-\nu)\varepsilon_z + \nu(\varepsilon_x + \varepsilon_y)] \tag{4-12}$$

$$\tau_{zx} = G\gamma_{zx} \tag{4-13}$$

These equations also hold for the relation between principal stresses and principal strains if

$$\sigma_x = \sigma_1, \sigma_y = \sigma_2, \sigma_z = \sigma_3 \qquad \varepsilon_x = \varepsilon_1, \varepsilon_y = \varepsilon_2, \varepsilon_z = \varepsilon_3 \;.$$

However, if TK Solver software is being used, then Equations (4–1) through (4–6) are sufficient to solve for the stresses if the strains and material constants are known. However, for hand calculations, Equations (4–8) through (4–13) directly give the stresses if the strains and material properties are known. Either way, such stresses can then be used to determine principal stresses and the maximum shear stress.

If the strain $\varepsilon_{a'}$ in the x' direction on the surface of a machine element is measured using a strain gage, as shown in Fig. 4–2, then it can be proved that this measured strain is related to the strains ε_x, ε_y, and γ_{xy}; that is,

$$\varepsilon_{a'} = \frac{\varepsilon_x + \varepsilon_y}{2} + \frac{\varepsilon_x - \varepsilon_y}{2}\cos 2\theta' + \frac{\gamma_{xy}}{2}\sin 2\theta'$$

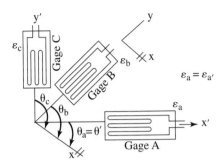

Figure 4-2   Arrangement of strain gage with respect to the x axis.

Next consider the equations

$$\varepsilon_a = \frac{\varepsilon_x + \varepsilon_y}{2} + \frac{\varepsilon_x - \varepsilon_y}{2}\cos 2\theta_a + \frac{\gamma_{xy}}{2}\sin 2\theta_a \tag{4-14}$$

$$\varepsilon_b = \frac{\varepsilon_x + \varepsilon_y}{2} + \frac{\varepsilon_x - \varepsilon_y}{2}\cos 2\theta_b + \frac{\gamma_{xy}}{2}\sin 2\theta_b \tag{4-15}$$

$$\varepsilon_c = \frac{\varepsilon_x + \varepsilon_y}{2} + \frac{\varepsilon_x - \varepsilon_y}{2}\cos 2\theta_c + \frac{\gamma_{xy}}{2}\sin 2\theta_c \tag{4-16}$$

It can be seen that in Equation (4–14), for example $\varepsilon_{a'} = \varepsilon_a$ and $\theta' = \theta_a$ are known values, as ε_a is a strain measured by a strain gage at known angle θ_a, and $\varepsilon_x, \varepsilon_y$, and γ_{xy} are three unknown variables. Therefore, one must measure three strains at three different angles to obtain three simultaneous linear equations. If the three strains are measured using three strain gages, as shown in Fig. 4–2 (for instance, $\varepsilon_a, \varepsilon_b$, and ε_c at angles θ_a, θ_b, and θ_c with respect to the x-axis), then the left side of the three equations and the orientations θ for each gage are known. The quantities $\varepsilon_x, \varepsilon_y$, and γ_{xy} can then be determined by simultaneously solving Equations (4–14) through (4–16). TK Solver simplifies the process of solving a set of simultaneous equations. The sample problems will show how to solve for three unknowns from three equations without rearranging the equations for the unknowns or using any one of the methods of solving simultaneous equations. TK Solver has a built-in equation solver for dealing with linear or nonlinear simultaneous equations.

Any machine surface, such as that of a beam, always has a two-dimensional state of stress (plane stress), as stresses perpendicular to the surface are zero, assuming that the surface is free of any external pressure or load.

Electrical strain gages are capable of measuring strains on a critical surface of a machine element. If the gages can measure strain in any randomly chosen three directions, then one can obtain ε_x, ε_y, and γ_{xy}. Then σ_x, σ_y, and τ_{xy} can be found using Equations (4–8) through (4–12), and ε_z can be obtained from Equation (4–5). With σ_x, σ_y, and τ_{xy} known, it is easy to determine the principal stresses σ_1, σ_2, and one of the maximum shear stresses, τ_{max}, in the plane of the surface using the equations

$$\sigma_1 = \frac{\sigma_x + \sigma_y}{2} + \sqrt{\left(\frac{\sigma_x - \sigma_y}{2}\right)^2 + \tau_{xy}^2} \qquad (4\text{--}17)$$

$$\sigma_2 = \frac{\sigma_x + \sigma_y}{2} - \sqrt{\left(\frac{\sigma_x - \sigma_y}{2}\right)^2 + \tau_{xy}^2} \qquad (4\text{--}18)$$

$$\tau_{max} = \sqrt{\left(\frac{\sigma_x + \sigma_y}{2}\right)^2 + \tau_{xy}^2} = \frac{\sigma_1 - \sigma_2}{2} \qquad (4\text{--}19)$$

However, it is important to realize that Equation (4–19) yields the value of the maximum shear stress in the plane containing σ_1 and σ_2, which is not necessarily *the* maximum shear stress. To obtain the maximum shear stress, first σ_1 and σ_2 should be calculated using Equations (4–17) and (4–18). Then, knowing that $\sigma_z = \sigma_3 = 0$, we can rearrange the values of σ_1, σ_2, and σ_3 such that $\sigma_1 > \sigma_2 > \sigma_3$. Then the equation for τ_{max} is given by

$$\tau_{max} = \frac{\sigma_1 - \sigma_3}{2} \quad (\text{if } \sigma_1 > \sigma_2 > \sigma_3) \qquad (4\text{--}20)$$

Note that τ_{max} is the largest possible shear stress value that a designer is interested in, because it is this shear stress that might cause failure in a machine element. This shear stress must be compared with the maximum shear stress obtained during simple test, as explained in Chapter 7.

In sum, the principal stresses and the maximum shear stress at a critical location of a machine element can be determined using electrical strain gages. Principal stresses and shear stresses are necessary quantities for the design of a machine element.

The following example applies the preceding equations and presents a practical approach to solving problems involving strain gages.

There are three basic types of strain rosettes—that is, sets of three strain gages to measure strains along three different axes.

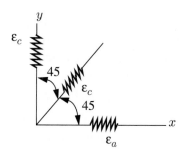

Figure 4–3   Rectangular strain rosette

1. The rectangular gage rosette, in which the angle between each consecutive gage is 45°.
2. The triangular gage rosette, in which the angle between each strain gage is 60°.
3. The star gage rosette, in which the angle between each consecutive gage is 120°.

Some authors have developed specialized equations for each type of rosette. *However, we need not do so as our lead model on strain gage analysis will solve all three rosettes.* Sketches of each rosette will be shown in the example problems given below.

Example 4-1

A rectangular strain gage rosette has been affixed at a critical location of a beam element subjected to a complex state of force. The rosette, with angles of $0^\circ, 45^\circ$, and 90° (see Fig. 4–3), measures strains in the three directions. The strain readings at maximum load are, and $\varepsilon_a = 0.0011$, $\varepsilon_b = 0.003$, and $\varepsilon_c = 0.0006$. Determine ε_x, ε_y and γ_{xy}; the principal stresses σ_1, σ_2, σ_3, and the maximum shear stress τ_{max}, if $\nu = 0.30$ and $E = 30 \times 10^6$ psi.

Solution

Since a rectangular rosette is used, we can let the x- and y-axes coincide with respect to gages measuring the strain ε_a and the strain ε_c, respectively. Then $\theta_a = 0$, $\theta_b = 45^\circ$, and $\theta_c = 90^\circ$.

Using Equations (4–14) through (4–16) and substituting known values, we obtain

$$0.0011 = \frac{\varepsilon_x + \varepsilon_y}{2} + \left(\frac{\varepsilon_x - \varepsilon_y}{2}\right)\cos 2(0) + \frac{\gamma_{xy}}{2}\sin 2(0)$$

$$0.003 = \frac{\varepsilon_x + \varepsilon_y}{2} + \left(\frac{\varepsilon_x - \varepsilon_y}{2}\right)\cos 2(45) + \frac{\gamma_{xy}}{2}\sin 2(45)$$

$$0.0006 = \frac{\varepsilon_x + \varepsilon_y}{2} + \left(\frac{\varepsilon_x - \varepsilon_y}{2}\right)\cos 2(90) + \frac{\gamma_{xy}}{2}\sin 2(90)$$

Solving these simultaneous equations reveals that $\varepsilon_x = 0.0011$, $\varepsilon_y = 0.0006$, and $\gamma_{xy} = 0.0044$.

Next, let us use an alternative approach to determine the principal stresses. From equations developed in earlier mechanics courses, and knowing ε_x, ε_y, and γ_{xy}, the principal strains ε_1, ε_2, and ε_3 can be determined. The principal strains are related to the strains in the x- and y- directions and also to the shear strain γ_{xy} by the equation

$$\varepsilon_1 = \left(\frac{\varepsilon_x + \varepsilon_y}{2}\right) + \sqrt{\left(\frac{\varepsilon_x - \varepsilon_y}{2}\right)^2 + \left(\frac{\gamma_{xy}}{2}\right)^2}$$

$$\varepsilon_1 = \left(\frac{0.0010 + (0.0006)}{2}\right) + \sqrt{\left(\frac{0.0010 - (0.0006)}{2}\right)^2 + \left(\frac{0.0044}{2}\right)^2}$$

$$\varepsilon_1 = 0.00301$$

Similarly,

$$\varepsilon_2 = \left(\frac{\varepsilon_x + \varepsilon_y}{2}\right) - \sqrt{\left(\frac{\varepsilon_x - \varepsilon_y}{2}\right)^2 + \left(\frac{\gamma_{xy}}{2}\right)^2}$$

Substituting values for ε_x, ε_y, and γ_{xy}, we have

$$\varepsilon_2 = -0.00141$$

and since, for two-dimensional stress problems (plane stress), $\sigma_3 = 0$, it follows that

$$\varepsilon_3 = \frac{-v}{1-v}(\varepsilon_1 + \varepsilon_2) \quad \text{[modified Equation (4-12)]}$$

$$\varepsilon_3 = -0.000686$$

Now modified versions of Equations (4-8) through (4-12) can be used; that is,

$$\sigma_1 = \frac{E}{(1+v)(1-2v)}[(1-v)\varepsilon_1 + v(\varepsilon_2 + \varepsilon_3)]$$

$$\sigma_2 = \frac{E}{(1+v)(1-2v)}[(1-v)\varepsilon_2 + v(\varepsilon_3 + \varepsilon_1)]$$

$$\sigma_3 = \frac{E}{(1+v)(1-2v)}[(1-v)\varepsilon_3 + v(\varepsilon_1 + \varepsilon_2)]$$

Substituting given values of E and v and calculated values of ε_1, ε_2, and ε_3, one can determine σ_1, σ_2, and σ_3 using the preceding equations. The results are

$$\sigma_1 = 85264.3 \ \ \text{psi}$$

$$\sigma_2 = -3.13e^{-12} \approx 0 \ \ \text{psi}$$

and

$$\sigma_3 = -16692.9 \ \ \text{psi}$$

Now we rearrange the principal stresses such that $\sigma_1 > \sigma_2 > \sigma_3$, to get

$$\sigma_1 = 85264.3 \qquad \sigma_2 = 0 \qquad \sigma_3 = -16692.9$$

Note that by chance, σ_1 had the largest value and σ_3 had the lowest value. However, this is not always true. Therefore, always rearrange calculated stresses such that $\sigma_1 > \sigma_2 > \sigma_3$.

Once the stresses are arranged as required, the maximum shear stress is

$$\tau_{max} = \frac{\sigma_1 - \sigma_3}{2} = \frac{85264.3 - (-16692.88)}{2} = 50978.60 \text{psi}$$

4.3 Sample Solution Using TK Solver for Strain-Gage Analysis

All sample problems in this chapter and the following chapters are solved using the UNIX format. However, they can also be solved using personal computers (PCs) with a Windows operating system and TK Solver software. Also included is the Unit Sheet that is used in every lead model.

NOTE: All lead models accept FPS (English) and SI (International Standard) units. The user must be aware of what units he or she would like to have for input or output. Input and output units can be the same or different, but must be given. Do not change units if you have a value in the input or output field, as that value will change as soon as you change units. This could result in incorrect answers. It is also advised that one check the hard copy of Units sheets to ensure that multiplication factors are correct.

RULES SHEET

The lead model called *ex4-1.tk* or *stgage.tk* is saved on the disk that accompanies this text.. Three new functions are used in this model: **Sqrt, cosd**, and **sind**. **Sqrt** is a command for square root, and **cosd** and **sind** are, respectively, cosine and sine functions in degrees. (**Note**: cos and sin are cosine and sine functions in radians.) TK Solver will automatically append variables from the Rules Sheet onto the Variables Sheet. Enlarge the display window and file the Rules Sheet and Variables Sheet using the **Tile** command from the **Windows** menu. Choose **Append Variable Name** from the **Options** menu if a variable is missing on the Variables Sheet. The equations are now ready to be solved. Notice that TK will not solve simultaneous equations automatically if there is more

VARIABLES SHEET

```
                                        stgage.tk
 File  Edit  Commands  Options  Windows  Sheets  Help
┌─────────────────────────────── Variables ──────────────────────────────┐
St Input---- Name--- Output--- Unit----- Comment-----------------------------
     .0011    ea                in/in    Normal strain of Gage a in/in or m/m
     .001     eb                in/in    Normal strain of Gage b in/in or m/m
     .0001    ec                in/in    Normal strain of Gage c in/in or m/m

     0        ta                deg      Orientation of Gage a deg or rad
     45       tb                deg      Orientation of Gage b deg or rad
     90       tc                deg      Orientation of Gage c deg or rad

              ex       .0011    in/in    Nor.strain along x axis in/in or m/m
              ey       .0001    in/in    Nor.strain along y axis in/in or m/m
              gxy      .0008    in/in    Nor.strain along z axis in/in or m/m

     3E7      E                 psi      Young's modulus psi or ksi or Pa, MPa
     .3       v                          Poisson's ratio

              e1       .00124031 in/in   Principal normal strain in/in or m/m
              e2       -4.031E-5 in/in   Principal normal strain in/in or m/m
              e3       -.0005143 in/in   Principal normal strain in/in or m/m

              S1       40490.726 psi     Prin. stress psi or ksi or Pa or MPa
              S2       10937.845 psi     Prin. stress psi or ksi or Pa or MPa
              S3       -3.13E-12 psi     Prin. stress psi or ksi or Pa or MPa

              tmax1    14776.441 psi     Max. shear st. psi or ksi or Pa or MPa
              tmax2    5468.9226 psi     Max. shear st. psi or ksi or Pa or MPa
              tmax3    -20245.36 psi     Max. shear st. psi or ksi or Pa or MPa
└─────────────────────────────────────────────────────────────────────────┘
(14) Input: .3                                                        F9 OK
```

than one unknown in a given equation. TK needs at least guessed values of the unknown variables to invoke its iterative solver, which then solves the simultaneous equations. Guessed values must be given by entering **G** in the status field of the unknown variable. (One can guess values close to roots.) Enter the known data into the input field of the Variables Sheet, and delete all other inputs, if they exist, by pressing the space bar. Solve the equations by pressing the F9 function key or by choosing **Solve** from the **Commands** menu. The answers should appear in the output field of the Variables Sheet. Verify the results by checking them against hand calculations.

Observe that the TK Solver does only the mathematical part of solving sets of simultaneous equations; users have to type equations correctly and decide the correct input and output units. Desired units may be entered in the unit field on the Variables Sheet. TK treats the units as symbols and users have to decide the correct units and conversion numbers for variables that are not available in the Unit Sheet provided. An incorrect conversion number will produce a wrong answer. Following are some guidelines for using TK Solver:

1.   Develop all necessary equations by hand before using TK Solver. TK Solver only solves equations; it does not check for errors, scientific or otherwise, in the equations. Make sure that the

equations are correct before you enter them into the Rules Sheet. *Users should not rely completely on TK Solver.* Always compare the TK solutions with hand-calculated answers. This eliminates errors due to incorrect inputs or equations. Once the equations are tested for accuracy, one can use the model with confidence.

2.   Write brief descriptions of each variable in the **Comment** column on the Variables Sheet. The descriptions will help the viewer to understand a model and will be beneficial for other users in the future.

3.   Using the lead model with file name *ex4-1.tk* or *stgage.tk* will save time in solving problems related to various types of strain rosettes. This lead model is applicable to any type of strain rosette.

One of the sample problems dealt with earlier is used on the Rules and Variables Sheets—that is, given ε_a = 0.0011, ε_b = 0.003, and ε_c = 0.0006 in/in a (plane-stress problem) and θ_a = 0°, θ_b = 45°, and θ_c = 90°, with $E = 30 \times 10^6$psi, and ν =.3, determine ε_x, ε_y γ_{xy} ε_1, ε_2, ε_3, σ_1, σ_2, σ_3, and τ, the maximum shear stress.

See the Variables Sheet for all the answers. The reason such a model is applicable to any rosette is that the difference between any two types of rosettes lies in the values of θ_a, θ_b, and θ_c. Since such angles are inputs, one can have a rosette with any orientation, as long as the exact values of θ_a, θ_b, and θ_c are known.

Once the values of σ_1, σ_2, and σ_3 are calculated, we can calculate the value of the maximum shear stress by rearranging σ_1, σ_2, and σ_3 such that $\sigma_1 > \sigma_2 > \sigma_3$. When we do so, we obtain

$$\tau_{max} = \left(\frac{\sigma_1 - \sigma_3}{2}\right)$$

Then $\gamma_{max} = \tau_{max}/G$

The *ex4-1.tk* or *stgage.tk* model was used to solve a sample problem discussed earlier. The results can be obtained in FPS or SI units. According to whether one enters, for example, the stress in FPS or SI units. This lead model is complete with a Units Sheet and has been verified to give correct answers if input data are entered correctly. You may use the model to check your hand-calculated answers or to check answers obtained by using your own model.

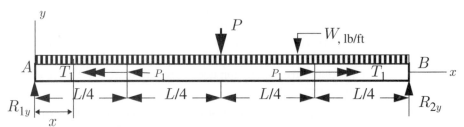

Figure 4–4   Beam showing loads applied and their locations (double
arrow indicates torque or moment)

4.4 Analytical Stress Analysis: Introduction to Shear, Bending Moment, Torque, and Axial Load Diagrams

An analytical method of determining stresses in machine elements is to develop equations to determine stresses using equations of equilibrium. With knowledge of the behavior of materials one can develop enough equations to yield analytical values for internal stresses developed by axial, bending, shear, and torque loads.

In general, machine elements, beams, or shafts are loaded in longitudinal and transverse directions, as shown in Fig. 4–4. Naturally, such loads create internal moments, shears, and axial loads in the x, y, and z directions. Let us develop the equations necessary for dealing with axial forces, bending moments, shear forces, torques, etc., so that one can plot the variation in the values of these variables as a function of the distance x, y, or z measured along a machine element. Plots of such bending moment, shear, torque, and axial loads are needed to obtain the distribution of the moments, loads, etc., along the length of a machine element or a beam and to obtain a cross section in which one or all of the values of the preceding variables are maximum in magnitude. Then such a cross section can be treated as a critical section. However, this is time consuming, and easier ways must be developed to expedite the process. It turns out that the process can be made easier by developing proper equations for the axial load, shear, bending moments, and torque as a function of the distance x along the length of a beam. These equations can then be used in software such as TK Solver to obtain all of the needed diagrams using plotting routines available with the software. It is important to concentrate only on writing correct equations for the total range of, for example, a beam. Then one can have the software solve for such equations and/or plot necessary diagrams.

The following is a simple example that illustrates how to write axial, shear, bending moment, and torque equations. Once the concept of writing correct equations for a given range is understood, it can be extended to more difficult problems that include 3-D forces.

In solving these types of problems, use the following guidelines:

1.   Make sure that all external loads are shown clearly and that the restrictions at the reaction locations are understood. For example, a roller support cannot support a load in the longitudinal direction of a shaft.

2.   When one writes equations for shear, bending moment, etc., such equations are written for a range, depending upon load, cross sectional, or material changes.

3.   To develop such equations, one needs to cut a beam at, say, distance x. Then one works only with that portion of the cut beam containing x and does not look at the other cut part of the beam.

4.   Assume unknowns (reactions, moments, etc.) as positive values; the sign of the resulting values will tell whether the unknowns are in fact negative. For example, if the $+ve$ assumed reaction results in a $-ve$ answer, then the correct direction is opposite to the assumed direction.

These rules will help you solve problems, as well as develop computer models.

All the examples that follow in this chapter use the preceding instructions.

Example 4-2

A beam of length L is subjected to a concentrated load P and a distributed load W. It is also subjected to a torque T_1 and an axial load P_1, as shown in Fig. 4–4.

Develop equations to determine the reactions at A and B, as well as equations for the reactive axial load, moment, shear, and torque for the range $0 \le x \le L$. (x is measured from station A.)

Solution

The two unknowns R_{1y} and R_{2y} can be obtained by using static equilibrium equations; that is,

$$\Sigma F_y = 0 \Rightarrow R_{1y} + R_{2y} - P - WL = 0$$

$$\Sigma M_{nA} = 0 \Rightarrow -R_{2y}L + P(L/2) + WL(L/2) = 0$$

where M_{nA} is the moment at A about an axis perpendicular to the x–y axis.

These two equations can be solved for both of the reactions, which are

$$R_{1y} = R_{2y} = W(L/2) + P/2 \text{ (due to symmetry)}$$

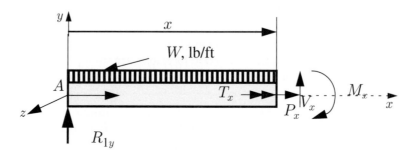

Figure 4–5   The portion of beam under equilibrium for the range $0 \leq x \leq L/4$

Now it can be seen that the equations required for the range $0 \leq x \leq L$ must be written for the four ranges $0 \leq x \leq L/4$, $L/4 \leq x \leq L/2$, $L/2 \leq x \leq 3L/4$, and $3L/4 \leq x \leq L$ as load changes take place at those four locations.

Drawing the free body diagram of the cut beam is an important step, as the diagram will yield equations for the shear, bending moment, etc., without any error. Note that when a beam is cut into two parts, only the free-body diagram of the beam containing x should be used (see Fig. 4–5). We should ignore any part of the beam that does not contain x, as including it may create confusion. The study of the cut portion of the beam shows that, for static equilibrium, one requires internal shear, moment, axial force and torque at that location. The requirement is for a system to be in equilibrium, i.e., $\Sigma F_x = 0$, $\Sigma F_y = 0$, $\Sigma M_z = 0$, and $\Sigma M_x = 0$. For example, equilibrium dictates that, for the range $0 \leq x \leq$ L/4, the following equations apply, assuming that one requires internal V_x, P_x, M_x, and T_x at the cut section to balance external forces, moments and torques:

Therefore, for $\Sigma F_y = 0$, will yield $V_x + R_{1y} - Wx = 0$

for $\Sigma M_z = 0$ will yield $M_x + R_{1y}(x) - \dfrac{Wx^2}{2} = 0$

for $\Sigma M_x = 0$ will yield $T_x = 0$

for $\Sigma F_x = 0$ will yield $P_x = 0$

In this problem there are no external forces in the z direction and hence $\Sigma F_z \equiv 0$.

It is always better to assume all the internal unknown forces like P_x, V_x, etc. as positive to the right and upward respectively. Similarly, the internal

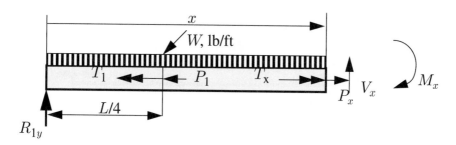

Figure 4–6 The portion of beam under equilibrium for the range
$L/4 \leq x \leq L/2$

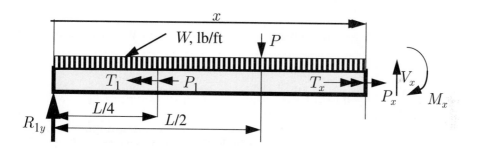

Figure 4–7 The portion of beam under equilibrium for the range $L/2 \leq x \leq 3L/4$

clockwise moment, M_x, and internal torque, T_x, if in the positive x direction, are positive. However, one can choose his/her own sign convention, but once decided, it must be used consistantly.

The preceding equations are for the range $0 \leq x \leq L/4$. Using the same procedure, shear, moment, torque, and axial force equations can be obtained for the other ranges of the beam. These are presented in Table 4–1, on the basis of Figures 4–5, 4–6, and 4–7.

It is very important to write correct equations, so that one can solve them using a calculator or computer. Now, all the foregoing equations containing external forces and moments will yield, at any distance x, the values of the axial force, shear, bending moment, and torque that must be developed internally to keep a cut beam in equilibrium.

Let us use the lead model called *ex4-2.tk* or *bstadiag.tk,* which contains all of the preceding equations in its Rules Sheet. Assume that the beam length $L =$

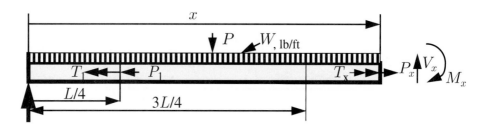

Figure 4–8 The portion of beam under equilibrium for the range $3L/4 \leq x$ $\leq L$

Table 4–1

$0 \leq x \leq L/4$	$L/4 \leq x \leq L/2$	$L/2 \leq x \leq 3/4\ L$	$3/4L \leq x \leq L$
$V_x = -R_{1y} + Wx$	$V_x = -R_{1y} + Wx$	$V_x = -R_{1y} + Wx + P$	$V_x = -R_{1y} + Wx + P$
M_x $= -R_{1y}(x) + \dfrac{Wx^2}{2}$	M_x $= -R_{1y}(x) + \dfrac{Wx^2}{2}$	$Mx = -R_{1y}(x) +$ $\dfrac{Wx^2}{2} + P\left(x - \dfrac{L}{2}\right)$	$M_x = -R_{1y}(x) +$ $\dfrac{Wx^2}{2} + P\left(x - \dfrac{L}{2}\right)$
$T_x = 0$	$T_x = T_1$	$T_x = T_1$	$T_x = 0$
$P_x = 0$	$P_x = P_1$	$P_x = P_1$	$P_x = 0$

800 inches, $W = 100$ lb/in, $P_1 = 5000$ lb, $T_1 = 5000$ in-lb, and the transverse load P $= 5000$ lb are given. We are required to determine the reactions at each support and draw the shear, bending, torque, and axial force diagrams for the range $0 \leq x$ $\leq L$.

First, we enter all of the equations into the Rules Sheet. Then we let x assume values between 0 and L in increments of, for example, $0.01L$. Next, by list filling and then list solving, we can find V_x, M_x, T_x, and P_x as functions of x. Finally, we can obtain plots of the V_x (shear) diagram, M_x (moment) diagram, etc., as a function of x along the length of the beam. (See the Rules, Variables, and Plot sheets on pages 63–64 for all of this information.)

We emphasize that writing the *correct* equations for the shear, bending moment, torque, and axial forces at any distance x is important. The computer cannot correct wrong equations. If the equations entered are correct and if plots of these values are required, it can be done very quickly using the TK Plot Sheet. Once the plots are obtained for the shear, bending moment, torque, and axial forces, the critical cross section in which the combination of plotted values is

RULES SHEET

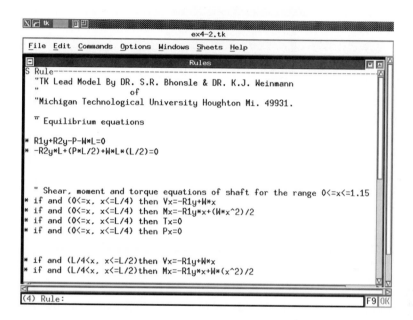

```
X ⌐ tk        ⊡ ⊟                                          ex4-2.tk
 File  Edit  Commands  Options  Windows  Sheets  Help
⊟                              Rules                          ⊡ ⊟ △
S Rule─────────────────────────────────────────────────────── △
  "TK Lead Model By DR. S.R. Bhonsle & DR. K.J. Weinmann
  "                    of
  "Michigan Technological University Houghton Mi. 49931.

  " Equilibrium equations

* R1y+R2y-P-W*L=0
* -R2y*L+(P*L/2)+W*L*(L/2)=0


  " Shear, moment and torque equations of shaft for the range 0<=x<=1.15
* if and (0<=x, x<=L/4) then Vx=-R1y+W*x
* if and (0<=x, x<=L/4) then Mx=-R1y*x+(W*x^2)/2
* if and (0<=x, x<=L/4) then Tx=0
* if and (0<=x, x<=L/4) then Px=0


* if and (L/4<x, x<=L/2)then Vx=-R1y+W*x
* if and (L/4<x, x<=L/2)then Mx=-R1y*x+W*(x^2)/2
                                                            ▽
(4) Rule:                                              F9 OK
```

```
⊟                              Rules                          ⊡ ⊟
S Rule─────────────────────────────────────────────────────── △
* if and (L/4<x, x<=L/2)then Tx=T1
* if and (L/4<x, x<=L/2)then Px=P1


* if and (L/2<x, x<=3*L/4)then Vx=-R1y+W*x+P
* if and (L/2<x, x<=3*L/4)then Mx=-R1y*x+W*(x^2)/2+P*(x-L/2)
* if and (L/2<x, x<=3*L/4)then Tx=T1
* if and (L/2<x, x<=3*L/4)then Px=P1


* if and (3*L/4<x, x<=L)then Vx=-R1y+W*x+P
* if and (3*L/4<x, x<=L)then Mx=-R1y*x+W*(x^2)/2+P*(x-L/2)
* if and (3*L/4<x, x<=L)then Tx=0
* if and (3*L/4<x, x<=L)then Px=0


* if (x>L) then Vx=0
* if (x>L) then Mx=0
* if (x>L) then Tx=0
* if (x>L) then Px=0
                                                            ▽
```

VARIABLES SHEET

```
┌──────────────────────────────────────────────────────────────────┐
│ X ⌐ tk ▭▭▭▭ ⊡⊡                                                     │
│                              ex4-2.tk                              │
│  File  Edit  Commands  Options  Windows  Sheets  Help             │
│ ┌──────────────────────────────────────────────────────────────┐ │
│ │⊟                           Variables                    ▭ ▭  │▲│
│ │St Input---- Name--- Output--- Unit----- Comment------------- │▭│
│ │             R1y      42500     lb                            │ │
│ │             R2y      42500     lb                            │ │
│ │    100      W                  lb/in                         │ │
│ │    800      L                  in                            │ │
│ │   5000      P                  lb                            │ │
│ │L   800      x                  in                            │ │
│ │L            Vx       42500     lb                            │ │
│ │L            Mx      -7.47E-9    lb-in                         │ │
│ │L            Tx       0          lb-in                         │ │
│ │L            Px       0          lb                            │ │
│ │   5000      T1                 lb-in                          │ │
│ │   5000      P1                 lb                             │ │
│ │                                                              │ │
│ │                                                              │ │
│ │  ─                                                        ▼  │▼│
│ │◄                                                          ►  │ │
│ ├──────────────────────────────────────────────────────────────┤ │
│ │(16)  Input:                                          F9│OK    │ │
└──────────────────────────────────────────────────────────────────┘
```

highest can be spotted easily. The following example illustrates how a model can be created for obtaining the required results.

Suppose we wish to create a Table Sheet (see Table 4–2) to accommodate values of x, V_x, M_x, T_x, and P_x. Then first make sure that you type "L I" in the status column of the x value row to convey to the computer that the x value is the **list input**, and type "L O" in the status field of the variables V_x, M_x, T_x, and P_x to relay to the computer that these are the **list output** values. (Remember, you will not see "I" and "O" on the monitor when you type them.) Once you are in the last subsheet of a table sheet, let the cursor be at x, and then list-fill x values, say, $x = 0$, increment 1, and final value 800. This will fill the x column with the values 0 to 800 in increments of 1. List filling in the last subsheet is important, as you will not be able to use the list solve command effectively if you try to list solve while you are in any other sheet. Also, make sure *before* you start using this file that all the List Sheets and the input column in the Variables Sheet are blank. When you list solve, the table subsheet must be filled with values of the unknowns as soon as the solving process is complete. Only then can you plot x against any other variable. Implementing this model will make you proficient in TK Solver, as you will have used most of the capabilities of the software. You will find that the computer-aided solution saves you time and allows you to learn the fundamentals, while also obtaining all the information a designer needs.

All necessary graphs are plotted in Figs. 4–1 through 4–4.

Table 4–2 Distance x vs. M_x, V_x, T_x and P_x

```
XF tk        OP
                         ex4-2.tk
  File  Edit  Commands  Options  Windows  Sheets  Help

  □              Int Table: table1                □ □
Title:
Element x--------- Vx-------- Mx-------- Tx-------- Px---
1        0         -42500     0          0          0
2        1         -42400     -42450     0          0
3        2         -42300     -84800     0          0
4        3         -42200     -127050    0          0
5        4         -42100     -169200    0          0
6        5         -42000     -211250    0          0
7        6         -41900     -253200    0          0
8        7         -41800     -295050    0          0
9        8         -41700     -336800    0          0
10       9         -41600     -378450    0          0
11       10        -41500     -420000    0          0
12       11        -41400     -461450    0          0
13       12        -41300     -502800    0          0
14       13        -41200     -544050    0          0
15       14        -41100     -585200    0          0
16       15        -41000     -626250    0          0
17       16        -40900     -667200    0          0
18       17        -40800     -708050    0          0
19       18        -40700     -748800    0          0
20       19        -40600     -789450    0          0
21       20        -40500     -830000    0          0
22       21        -40400     -870450    0          0
23       22        -40300     -910800    0          0
24       23        -40200     -951050    0          0
25       24        -40100     -991200    0          0
26       25        -40000     -1031250   0          0
27       26        -39900     -1071200   0          0
28       27        -39800     -1111050   0          0
29       28        -39700     -1150800   0          0
30       29        -39600     -1190450   0          0
31       30        -39500     -1230000   0          0
32       31        -39400     -1269450   0          0

 (1,1)  x: 0                                     F9 OK
```

Plot 4.1—SHEAR FORCE VS DISTANCE x

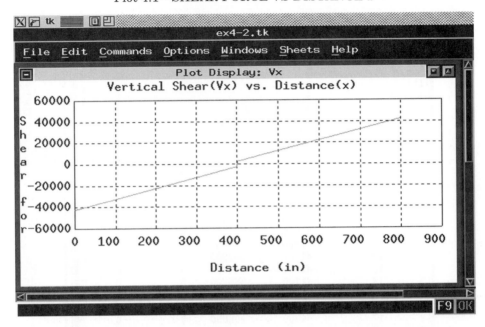

Plot 4.2—BENDING MOMENT VS DISTANCE x

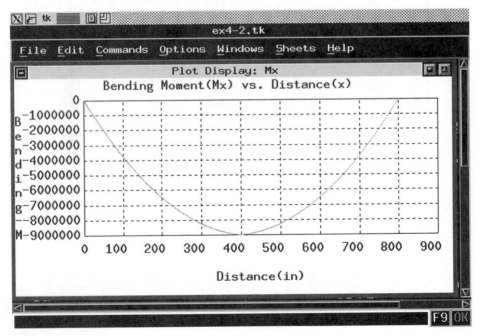

Plot 4.3—TORQUE VS DISTANCE x

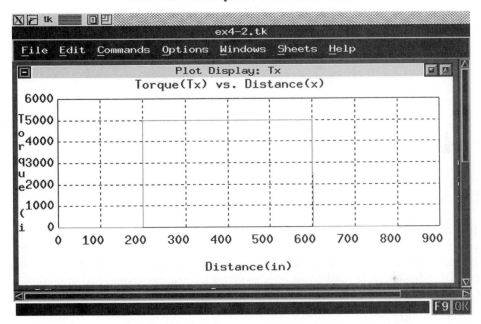

Plot 4.4—AXIAL FORCE VS DISTANCE x

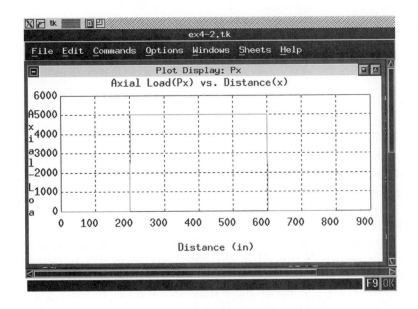

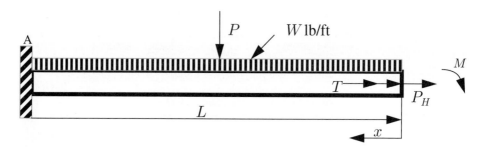

Figure 4–9   Cantilever beam subjected to external loads.

4.5 Review of Stress Equations

The following stress formulas are derived in an introductory course in the mechanics of materials. Such formulas are used in this text wherever necessary. The stress formulas are applicable to straight beams only. The formula for shear stress due to torque is applicable only to straight circular cross-section beams.

Bending Stress Formula:

$$\sigma = \frac{My}{I}$$

Shear Stress (Vertical) Formula:

$$\tau = \frac{VQ}{It}$$

Shear Stress Due to Torque T:

$$\tau = \frac{T\rho}{J}$$

In the preceding equations, M is the moment, V is the shear force, T is the torque, I is the moment of inertia, J is the polar moment of inertia, y is the distance from the neutral axis, ρ is the radial distance from the center of a shaft, Q is the moment of an area above a line of interest about the neutral axis, and t is the thickness at a line of interest.

Example 4-3

Fig. 4–9 shows a cantilever beam of circular cross section that is fixed at A and loaded as shown. Write all of the equations necessary to determine the unknown reactions and also to obtain shear, bending moment, torque, and axial load diagrams. Then plot all of these diagrams for the range $0 \leq x \leq L$ and determine the critical cross section for which the combination of axial force, shear, bending moment, and torque are maximum. Data are as follows:

RULES SHEET

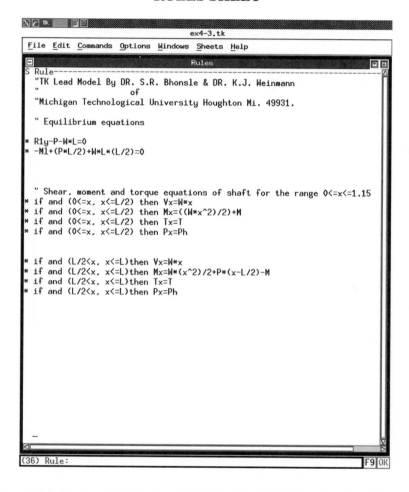

$W = 100$ lb/ft, $P = 2000$ lb, $T = 300$ ft-lb, $M = 250$ ft-lb, $P_H = 5000$ lb,

$L = 18'0"$ (vertical load P is @ $9'0"$ from the left end)

Solution

All of the required equations are contained in the model called *ex4-3.tk* or *cant.tk.* (See the TK Solver Rules Sheet.) Note how the range is expressed in the Rules Sheet. Some of the ranges are stated in a fashion that makes all diagrams close whenever necessary. All the required diagrams are presented on pages 71–72. The "diagrams to be closed" concept comes from the equilibrium requirement, which dictates that axial force, shear force, moment, and torque values must go to zero just left of left end and just right of right end of a beam.

VARIABLES SHEET

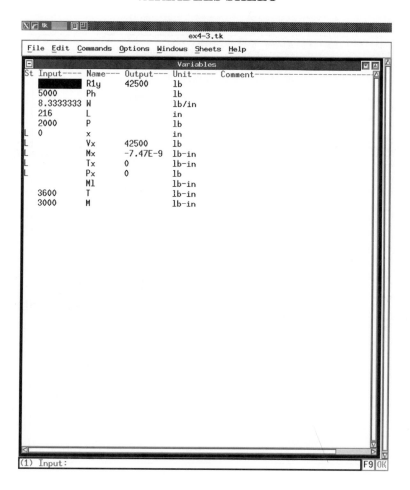

Plot 4.5—SHEAR FORCE DIAGRAM

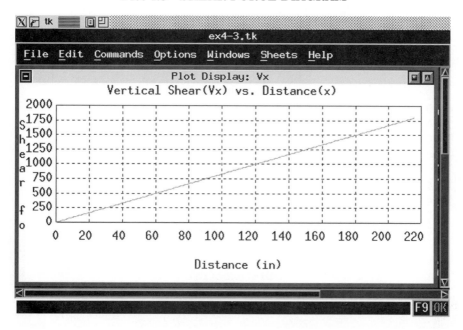

Plot 4.6—BENDING MOMENT DIAGRAM

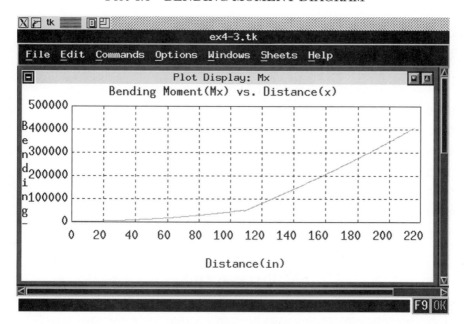

Plot 4.7—TORQUE DIAGRAM

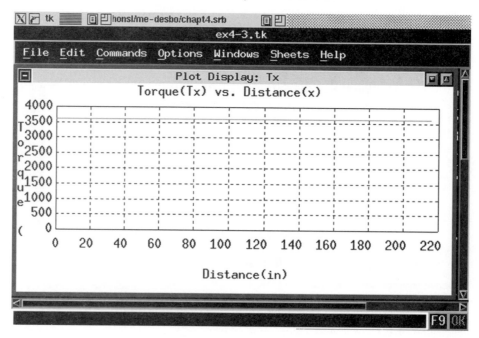

Plot 4.8—AXIAL FORCE DIAGRAM

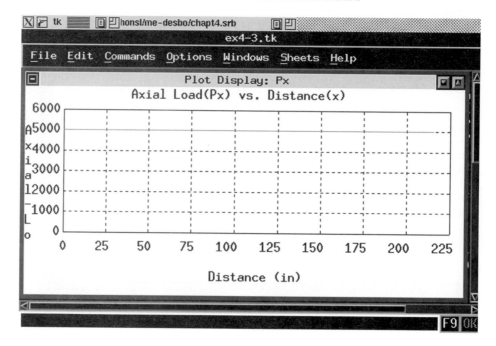

PROBLEMS

4–1) The triangular strain rosette shown in the following diagram was used to obtain strain data at a point on the free surface of a machine part made of steel.

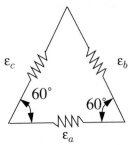

The measured strain values are as follows: $\varepsilon_a = 900\ \mu$, $\varepsilon_b = -450\ \mu$, and $\varepsilon_c = 400\ \mu$ (where $\mu = 10^{-6}$), $E = 30 \times 10^6$ psi, and ν (Poisson's ratio) = 0.3.

Assuming that this is a plane-stress problem, determine:

a. The strain components ε_x, ε_y, and γ_{xy} (Answer: 900, –333.34, 981.50, all in μ in/in)

b. The principal strains ε_1, ε_2, and ε_3. (Answer: 1071.44, –504.77, –242.86, all in μ in/in)

c. The principal stresses and maximum shear stress at the point $(\sigma_1, \sigma_2, \sigma_3, \tau_{max})$. (Answer: 3.033×10^4, 0×10^4, -6.044×10^3, 1.82×10^4 psi)

4–2) The rectangular strain rosette shown in the following diagram was used to obtain normal strain data at a point on a free surface of a machine part made of steel.

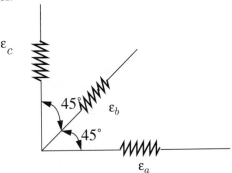

If ε_a = 1100 μ, ε_b = 1000 μ, ε_c = −100 μ, E = 30 × 10^6 psi, and ν = 0.3, determine:

a.   The strain components ε_x, ε_y and γ_{xy} (Answer: 1100, −100, 1000, all in μ)

b.   The principal strains ε_1, ε_2, and ε_3. (Answer: +1281.02, −281.02, −428.51, all in μ)

c.   The principal stresses and the maximum shearing stress at the point σ_1, σ_2, σ_3, τ_{xy}.

   (Answer for part c: σ_1 = 3.94 × 104 psi, σ_2 = 3.40 × 103 psi, σ_3 = 0, and τ_{max} = 19.7 × 103 psi)

d.   Check the answer in part a, b, and c by using the lead model for strain gage analysis.

4–3) At a point on the free surface of a machine part, the known strains are ε_x = 1000 μ, ε_y = −200 μ, and γ_{xy} = +400 μ. Given E = 10^6 psi, ν = 0.3, determine the following, assuming a plane-stress problem:

a.   The principal strains ε_1, ε_2, and ε_3.

b.   The principal stresses and maximum shearing stress at the point (σ_1, σ_2, σ_3, and τ_{max}).

4–4) A critical surface in a machine element was studied for stresses using a star strain gage rosette as shown in the following diagram:

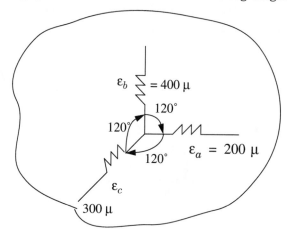

Determine the principal stresses at the point of interest if $E = 200 \times 10^9$ Pa., $v = 0.3$, and $\varepsilon_a = 200\mu$ in/in, $\varepsilon_b = 300\mu$ in/in., $\varepsilon_c = 400\mu$ in/in. (Note: E is in SI units and strains are in FPS units, but the strains in FPS and SI units are the same. Why?) Also, determine the shear modulus G and the maximum shear strain γ_{max}.

Answers (in microinches/inch): ($\varepsilon_x = 200\mu$, $\varepsilon_y = 400\mu$, $\gamma_{xy} = 115.47\mu$, $\varepsilon_1 = 415.47\mu$, $\varepsilon_2 = 184.53\mu$, $\varepsilon_3 = 257.14\mu$)

4–5) Given a triangular strain rosette's strain values as shown in the following diagram, determine the principal strains, principal stresses, and maximum shear stress.

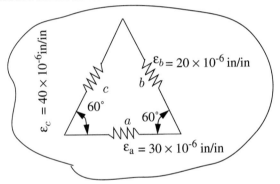

4–6) On the surface of a machine element, a triangular rosette is bonded to measure strains. If $\varepsilon_a = 100 \times 10^{-6}$ in/in, $\varepsilon_b = 500 \times 10^{-6}$ in/in, and $\varepsilon_c = 350 \times 10^{-6}$ in/in are the measured strains, determine e_x, e_y, γ_{xy}, e_1, e_2, e_3, σ_1, σ_2, and σ_3 and maximum shear stress, if $E = 30 \times 10^6$ psi and $v = 0.29$.

4–7) A strain rosette, shown in the following diagram, was bonded to a machine element:

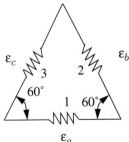

The load on the machine element created the following strains:

Gage 1 = 20×10^{-6} in/in, Gage 2 = 4000×10^{-6} in/in, Gage 3 = 100×10^{-6} in/in

Determine $\varepsilon_1, \varepsilon_2, \varepsilon_3, \sigma_1, \sigma_2, \sigma_3, \tau_{max}$ if $E = 200$ GPa and $\nu = 0.3$. (1GPa = 10^9 Pa.)

4–8) Given the triangular rosette and the strain values shown in the following diagram, if $\varepsilon_a = 200 \times 10^6$ in/in, $\varepsilon_b = 300 \times 10^6$ in/in, and $\varepsilon_c = -200 \times 10^6$ in/in, $E = 70$ GPa, and $\nu = 0.29$, assume that the problem is a plane-stress problem. Determine $\varepsilon_1, \varepsilon_2, \varepsilon_3, \sigma_1, \sigma_2, \sigma_3, \tau_{max}$ by hand and using TK.

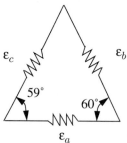

4–9) A rectangular rosette's strain measurements are $\varepsilon_a = 1000 \times 10^{-6}$ in/in, $\varepsilon_b = 2000 \times 10^{-6}$ in/in, and $\varepsilon_c = 500 \times 10^{-6}$ in/in. Determine e_x, e_y, γ_{xy} $\varepsilon_1, \varepsilon_2, \varepsilon_3, \sigma_1, \sigma_2, \sigma_3$ and the maximum shear stress, if $E = 10 \times 10^6$ psi and $\nu = 0.29$. Assume the plane-stress condition. The rosette is shown in the following diagram:

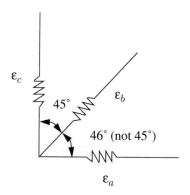

4–10) If, for plane stress, $\tau_{max} = \dfrac{\sigma_1 - \sigma_2}{2}$ and $\gamma_{max} = \dfrac{\tau_{max}}{G}$, prove that $G = E/(2(1 + v))$. (Hint $\gamma_{max} = \varepsilon_1 - \varepsilon_2$.)

4–11) For the beam shown in the following diagram, using TK Solver, draw shear and bending moment diagrams by developing equations and typing in the Rules Sheet, and also determine the location(s) of the maximum bending moment and the maximum shear force using the TK model that you will develop if $W = 1000$ lb/ft, $P = 5000$ lb, and $L = 100"$.

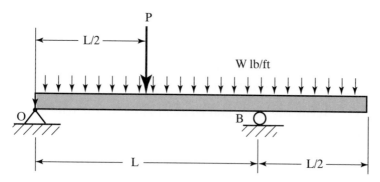

4–12) For the following figure, assume that the bearing at O takes a load in the x, y, and z directions and the bearing at B, only in the y and z directions. Note $\bar{F}_D = -.242F_D\bar{i} - .242F_D\bar{j} + .940F_D\bar{k}$,

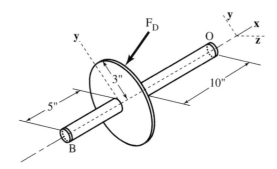

a. Find the reactions at O and B if $F_D = 1000$ lb.

b. Draw two shear diagrams, two moment diagrams, and the torque and axial force diagrams by hand by writing proper equations for the range 0 to 15 inches, measured from O.

c.   Use the TK Solver software to answer parts (a) and (b). Plot the diagrams obtained in (b).

4–13) Beam ABC is simply supported at A and B as shown in the following diagram:

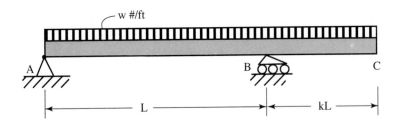

w #/ft

A   B   C

L   kL

a.   Determine reactions at A and B if $k = .3$ in terms of W and L.

b.   Write the shear and moment equations for the range A to C.

c.   Find the value of k such that the maximum moment in the range AB is nearly the same in magnitude as that in the range BC if $L = 100"$ and $W = 200$ lb/in.

d.   Create a TK model for part (a) and (b) and draw all necessary diagrams. Use the value for k as obtained in (b).

4–14) Consider the beam shown in the following diagram:

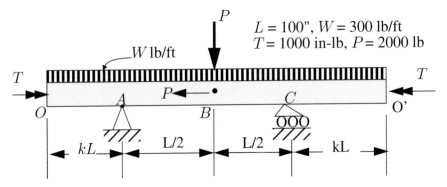

P

$L = 100"$, $W = 300$ lb/ft
$T = 1000$ in-lb, $P = 2000$ lb

W lb/ft

T T

O A P B C O'

kL   L/2   L/2   kL

a.   Determine the reactions at A and B if $k = .5$.

b.   Draw all necessary axial, shear, bending, and torque diagrams.

c.   Find the value of k such that the bending moment at A nearly equals the bending moment at B.

d.   Find the critical cross section where you think that the torque
     and the bending stress is maximum.

4–15) The following diagram shows the forces acting on a steel shaft.

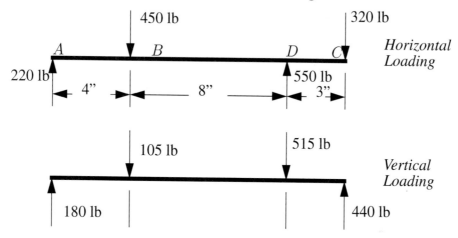

The allowable shear stress is 12,000 psi.

a.   Sketch horizontal and vertical shear and bending moment dia-
     grams; show the values at the change points.

b.   Determine the resultant bending moment and shear force.

c.   Determine the diameter of the shaft (to the nearest 0.01 inch)
     necessary for critical section, using an allowable shear stress
     equal to 40,000 psi.

4–16) A landing gear has a motor driving the pinion and the gear into the
       vertical or closing position. A bevel pinion and shaft are shown in the
       following figure.
       Bearing A takes thrust. The left end of the shaft is coupled to an electric
       motor, and the right end is free. The load components applied by the
       mating bevel gear are shown.

a.   Draw axial load, torsional load, bending load, and shear load
     diagrams.

b.   Determine the critical section, and at that section find the
     maximum shear stress due to shear force, V. (Use the equation
     $\tau = \dfrac{VQ}{It}$.)

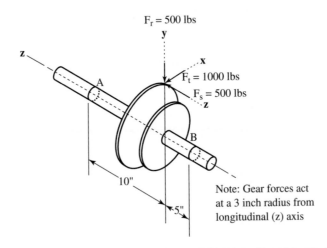

$F_r = 500$ lbs

$F_t = 1000$ lbs

$F_s = 500$ lbs

A

B

10"

5"

Note: Gear forces act
at a 3 inch radius from
longitudinal (z) axis

c.   Using maximum shear stress design criterion, find the dia–
meter of the shaft required to if $\tau_{y.p.} = 40{,}000$ psi and safety
factor is 1.6.

4–17) Write equations for the bending moment, shear, and axial force as a
function of an angle θ as shown in the following diagram for $0 \le \theta \le 180°$:

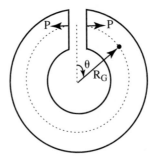

P   P

θ

R_G

The ring has a uniform cross section, and the radius to the cross-section
(not shown) center of gravity is R_G.

4–18) Write the equations necessary for determining the bending moment, shear, and axial force as a function of θ if the pressure at an angle θ is given by $p_\theta = \dfrac{P_{max}}{\sin\theta_a}\sin\theta$. (See the following diagram.)

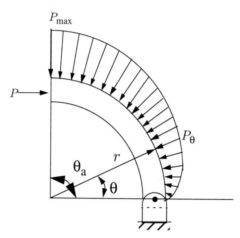

4–19) Consider the following diagram:

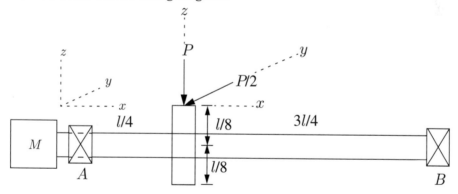

Write equations for the bending moment, shear, and torque for the range $0 \le x \le l$. AB are roller bearings. The motor M supplies necessary torque in the x direction. If $l = 80''$, and $P = 5000$ lb, draw the necessary shear, bending moment, and torque diagrams, and determine the horsepower required if H.P. $= (2\pi n T/33000)$. Assume the motor is rotating at $n = 1800$ revolutions per minute (rpm).

Deflections and Castigliano's Theorem

5.1 Introduction

Most types of machine elements, such as shafts, crane booms, airplane wings, etc., deflect due to external loads. An understanding of deflections in machine elements due to deformation is very important. When such deflections create interference with other machine elements, they must be prevented. Also, one must make certain that critical deflections are not excessive in order to avoid failure. Naturally, one must know how to find deflections. There are many ways to determine deflections due to external loads on a flexible machine element. An efficient method applicable to circular beams, triangular plates, helical compression springs, etc., is presented in this chapter.

The method of finding deflections based on strain energy stored in a machine element is superior to many other methods—even more so when it is necessary to deal with deflections in machine elements such as piston rings, coil springs, tappets, crankshafts, leaf springs, c-clamps, etc. The superiority of this method will become obvious as we study it.

5.2 Strain Energy Method

To understand the means of finding strain energy stored in a machine element, let us first study a simple machine element such as the tension specimen shown in Fig. 5–1. This machine element is subjected, let us say, to an axial stress σ_y, resulting in a corresponding tensile strain ε_y. That is, if σ_y is increased, there will be an increase in the ε_y. The proportional range of the stress–strain curve will appear as shown in Fig. 5–1(b).

The area under the triangle—let us call it u—is equal to $\sigma_y \varepsilon_y / 2$, in units of in-lb/in^3 (in the British (FPS) system), which represents energy per unit volume.

83

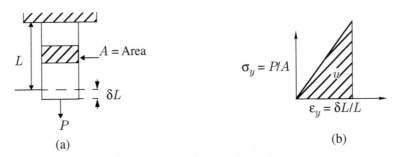

Figure 5–1   a) Tension-bar-like machine element subjected to an axial
load, b) Curve of stress vs. strain for the same element

Therefore, the total energy stored in the machine element (for a given stress and
strain), U, is given by the integral of the product udV, where $dV = Adx$ in this case.
Hence, the integral for the total energy is

$$U = \int_V u\,dV$$

However, $u = \dfrac{\sigma_y \varepsilon_y}{2}$ and consequently, because σ_y and ε_y are constant for

this case,

$$U = \frac{\sigma_y \varepsilon_y}{2} \int_V dV$$

$$\therefore U = (\sigma_y \varepsilon_y / 2)V$$

where $V = AL$ for a machine element of uniform cross section as shown in Fig. 5–
1. Now let us first find the deflection at P using Hooke's law and then, in the next
section, find the same deflection using the concept based on total energy U.

In order to determine the elongation δL, which is the deflection due to the
load P and is at P, one can use Hooke's law, which relates the modulus of elasticity E to an external stress; that is,

$$E = \frac{\sigma_y}{\varepsilon_y}$$

Since this machine element is subjected to a pure axial load, we know that

$$\sigma_y = P/A \quad \text{and} \quad \varepsilon_y = \delta L/L$$

Therefore,

$$E = \frac{(P/A)}{\delta L/L}$$

Solving for δL, we get

$$\delta L = \frac{PL}{AE}$$

But δL is the deflection (elongation) due to the load P and therefore can be written as

$$\delta_P = PL/AE \qquad (5\text{--}1)$$

where δ_p represents deflection in the direction of P at P. Thus, using Hooke's law, we can find the deflection at P in the direction of P.

5.3 Energy Approach to Determining Deflections

Now let us find the same deflection using the concept based upon the energy method. As stated earlier, the total energy in a machine element of volume V is given by

$$U = \int_V u \, dV \qquad (5\text{--}2)$$

In the foregoing case, $u = \dfrac{\sigma_y \varepsilon_y}{2}$

where $\sigma_y = \dfrac{P}{A}$, $\varepsilon_y = \dfrac{\sigma_y}{E} = \dfrac{P}{AE}$, and $dV = AdL$ (because the area of the cross section is constant in this example). It follows that

$$U = \int_o^L \frac{P^2}{2A^2E} AdL$$

$$= (P^2/2EA)\int_0^L dL$$

On integrating from 0 to L, we find that the total strain energy U in the bar is given by

$$U = \frac{P^2 L}{2EA} \qquad (5\text{--}3)$$

We can now show that if one differentiates Equation (5–3) with respect to P, one will get the deflection as in Equation (5–1), that is,

$$\frac{\partial U}{\partial P} = \frac{2PL}{2AE}$$

$$\frac{\partial U}{\partial P} = \frac{PL}{AE}$$

But $\delta_P = \dfrac{PL}{AE}$ [See Equation (5–1)]

Therefore, we conclude that the deflection in the direction of P can be found by taking the partial derivative of the total energy U with respect to P, that is,

$$\delta_p = \frac{\partial U}{\partial P} = \frac{PL}{AE} \tag{5–4}$$

Further, it can be shown that if the total energy U due to all externally applied loads, moments, and torques stored in a machine element is found, and the partial derivative of U with respect to, for example, P is calculated, then such partial yields the *deflection due to all external loads in the direction of* P *at* P. This discovery by Castigliano is presented as *Castigliano's theorem*.

The *theorem* states that when forces $F_1, F_2, ..., F_i$, torques $T_1, T_2, ..., T_i$, and moments $M_1, M_2, ..., M_i$ are applied on an elastic machine element, one can find the total energy stored in that element. Then, taking the partial derivative of the total strain energy stored with respect to, for example, the force F_i will result in the displacement (deflection) at F_i in the direction of F_i, as shown in the following sketch.:

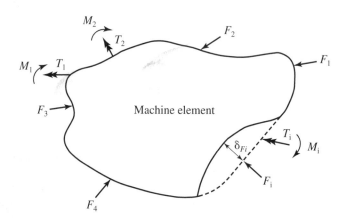

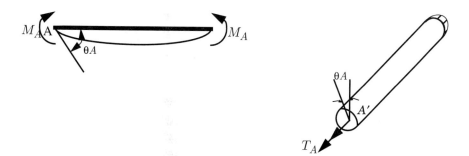

Figure 5–2   Relation of M_A or T_A to an angle θ_A for a beam and circular shaft.

Mathematically, the theorem can be stated as

$$\frac{\partial U}{\partial F_i} = \delta_{Fi} \tag{5–5}$$

Similarly, it can be shown that

$$\frac{\partial U}{\partial M_i} = \theta_A \tag{5–6}$$

and

$$\frac{\partial U}{\partial T_i} = \theta_A \tag{5–7}$$

where θ_A is the angular rotation (deflection) at the moment M_i or the torque T_i about an axis about which M_i or T_i is taking place. (See Fig. 5–2.) M_A is the moment applied on some machine element at point A, T_A is the torque applied to a circular shaft at point A', and θ_A is the angle of tangent or of rotation at A or A'.

It is not necessary that the state of stress in a machine element be one dimensional. Castigliano's theorem can be extended to machine elements subjected to a general (3-D) state of stress. Using the method of superposition, one can show that, for a machine element subjected to a 3-D state of stress the unit energy is given as

$$u = \frac{\sigma_x \varepsilon_x}{2} + \frac{\sigma_y \varepsilon_y}{2} + \frac{\sigma_z \varepsilon_z}{2} + \frac{\tau_{xy} \gamma_{xy}}{2} + \frac{\tau_{yz} \gamma_{yz}}{2} + \frac{\tau_{zx} \gamma_{zx}}{2} \tag{5–8}$$

To find the total energy U one has to integrate the unit energy over the volume of a machine element; that is,

$$U = \int_V u\,dV \qquad\qquad (5\text{--}9)$$

Finding U for a general state of stress can be a complicated process. However, most of the time there are only two or three stresses, due to the nature of the loading, which creates an internal energy. Thus, the partial derivative of U with respect to the load F can be expressed as a simple equation. The steps that follow show the procedure. We have already seen that

$$\frac{\partial U}{\partial F} = \delta_F$$

But $U = \int_V u\,dV$. Therefore,

$$\frac{\partial U}{\partial F} = = \frac{\partial}{\partial F}\int_V u\,dV \qquad\qquad (5\text{--}10)$$

If u were as shown in Equation (5–8), integration could become involved, but it can be simplified further if, for example, we assume that only a normal stress σ_x is acting and all other stresses are zero. Then the preceding equation may be simplified to

$$\frac{\partial U}{\partial F} = \delta_F = \frac{\partial}{\partial F}\int_V \frac{\sigma_x \varepsilon_x}{2}\,dV$$

because $u = \dfrac{\sigma_x \varepsilon_x}{2}$

$$\varepsilon_x = \frac{\sigma_x}{E} \text{ (in the proportional range)}$$

Consequently,

$$\delta_F = \frac{\partial}{\partial F}\int_V \frac{\sigma_x^2}{2E}\,dV \qquad\qquad (5\text{--}11)$$

This equation can be further simplified if σ_x is attributed to a bending moment M_x, since the bending stress at y is $\sigma_{xy} = \dfrac{M_x y}{I_z}$, which is really σ_x. (Note that here M_x means the moment at a distance x, but it is about the z-axis.)

Writing $(\sigma_x)^2 = \left(\dfrac{M_x y}{I_z}\right)^2$ in Equation (5–11), we get

$$\delta_F = \frac{\partial}{\partial F}\int_V \frac{M_x^2 y^2}{I_z^2 2E}\,dV$$

With $dV = dA\,dx$ (assume the general case of a nonuniform cross section), we separate proper variables to integrate and get

$$\delta_F = \frac{\partial}{\partial F}\int_0^A\int_0^L \frac{M_x^2 y^2\,dA\,dx}{I_z^2 2E}$$

Rearranging for convenience, we have

$$\delta_F = \frac{\partial}{\partial F}\left[\int_0^L \frac{M_x^2\,dx}{I_z^2 2E}\int_0^A y^2\,dA\right]$$

$$= \frac{\partial}{\partial F}\int_0^L \frac{M_x^2\,dx}{I_z^2 2E}\left[I_z\right] \quad \left(\text{since }\int_0^A y^2\,dA = I_z\right)$$

Therefore,

$$\delta_F = \frac{\partial}{\partial F}\int_0^L \frac{M_x^2\,dx}{I_z 2E}$$

Now, taking the partial derivative of M_x with respect to the force F, and assuming that I_z and E are constant throughout the length of a machine element, we get

$$\delta_F = \int_0^L \frac{\partial M_x}{\partial F}\frac{M_x}{EI_z}\,dx \tag{5–12}$$

Accordingly, to determine δ_F, the deflection due to bending stress under a load F in the direction of F, one needs to use Equation (5–12). To solve this equation, $\dfrac{\partial M_x}{\partial F}$, M_x, E, and I_z as a function of the length L must be known.

The following examples illustrate the use of Equation (5–12) to solve a beam deflection problem.

Example 5-1

A simply supported beam is subjected to a load P at the middle and has a constant EI for the total length of the beam. Find the deflection of the beam under the load P. (Neglect shear stress deflection; the shear stress deflection theory is presented later on.) The load on the beam is shown in the following diagram:

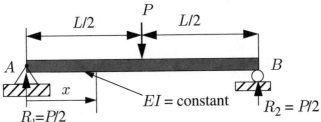

Solution

Observation shows that, since the beam is loaded symmetrically, $R_1 = R_2 = P/2$.

Now, cut the beam at a distance x from the left end. Then, for $0 \le x \le L/2$,

$$M_x = (P/2)(x)$$

Therefore,

$$\frac{\partial M_x}{\partial P} = \frac{x}{2}$$

For $L/2 \le x \le L$,

$$M_x = (P/2)(x) - P(x-L/2)$$

Hence,

$$\frac{\partial M_x}{\partial P} = \frac{x}{2} - x + \frac{L}{2} = -\frac{1}{2}(x-L)$$

The partial derivative of the total strain energy with respect to P in the first and second halves of the beam and which is deflection under load P is given by

$$\delta_p = \int_0^{\frac{L}{2}} \frac{\partial M_x}{\partial P} \frac{M_x}{EI} dx + \int_{\frac{L}{2}}^{L} \frac{\partial M_x}{\partial P} \frac{M_x}{EI} dx$$

Note that in the range 0 to $L/2$, $\dfrac{\partial M_x}{\partial P} = \dfrac{x}{2}$ must be used, and in the range

$L/2$ to L, $\dfrac{\partial M_x}{\partial P} = -\dfrac{1}{2}(x - L)$ must be used, as is shown shortly.

On substituting the corresponding values for M_x and $\partial M_x/\partial P$ for each range, δ_p (the deflection under the load P) can be obtained as

$$\delta_p = \left(\int_0^{\frac{L}{2}} \frac{(x)\left(\frac{P}{2}x\right)dx}{EI} dx + \int_{\frac{L}{2}}^{L} -\frac{1}{2}(x - L)\frac{\left[\left(\frac{P}{2}x\right) - P\left(x - \frac{L}{2}\right)\right]dx}{EI} \right)$$

Upon integrating and rearranging, we get

$$\delta_p = \frac{PL^3}{48EI}$$

This is the same value obtained by the conventional method derived in any standard book on the strength of materials.

The foregoing calculations can be simplified, because the load, bending moment, and modulus of elasticity for the beam are symmetric. If one finds the energy on half of the beam and then doubles it, one can then differentiate the energy to get the deflection under the load P; that is,

$$\delta_p = 2\int_0^{\frac{L}{2}} \frac{(x/2)(P/2(x))}{EI} dx$$

(This is possible because the bending moment, modulus of elasticity, and moment of inertia are symmetric about an axis passing through $x=L/2$.) Therefore,

$$\delta_p = \frac{2P}{4EI} \int_0^{\frac{L}{2}} x^2 dx$$

$$= \frac{P}{2EI}\left[\frac{x^3}{3}\right]_0^{\frac{L}{2}}$$

$$= \frac{P}{2EI}\left(\frac{L^3}{24}\right)$$

and

$$\delta_p = \frac{PL^3}{48EI}$$

as before.

It is always advisable to see if the beam has an axis of symmetry. It is important to note that one has a symmetric beam if and only if the beam has geometrical, mechanical, and load symmetry. This simply means that moment of inertia, modulus of elasticity (E), and external loads must be symmetric about an axis to call it an axis of symmetry. Let us further study another example where the advantage of symmetry will be utilized.

Example 5-2

A simply supported beam is subjected only to a distributed load W. Determine the deflection at the center of the beam. (Neglect the deflection due to shear loads; later we will discuss the procedure to account for this shear effect.) The load on the beam is shown in the following diagram:

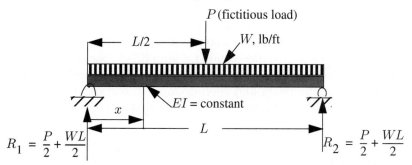

To obtain the deflection at the center, one needs to have a fictitious load P placed at middle to obtain deflection at middle. In other words, since there is no load in the middle, the fictitious load P is placed there. Now the moment equations for the ranges $0 \leq x \leq L/2$ and $L/2 \leq x \leq L$ have to be written. The left and right reactions are $R_1 = P/2 + WL/2 = R_2$, due to symmetry of loads.

The moment equation for the left half range is

for $0 \leq x \leq \dfrac{L}{2}$

$$M_x = \left(\frac{P}{2} + \frac{wL}{2}\right)x - \frac{wx^2}{2}$$

Therefore,

$$\frac{\partial M_x}{\partial P} = \frac{x}{2}$$

Taking advantage of symmetry about an axis passing through the center of the beam, we can state Castigliano's equation as

$$\delta_p = 2 \int_0^{\frac{L}{2}} \frac{\partial M_x}{\partial P} \frac{M_x}{EI} dx$$

$$\delta_p = 2 \int_0^{\frac{L}{2}} \left(\left(\frac{x}{2} \right) \left(\frac{P}{2} + \frac{wL}{2} \right) x - \frac{wx^2}{2} \right) \frac{dx}{EI}$$

Now set $P = 0$ and integrate, but note that it is important to know that once the $\dfrac{\partial M_x}{\partial P}$ is found, only then can one can set $P = 0$, and not anytime before.

Now, on integrating the preceding equation, one finds the deflection

$$\delta_p = \frac{5}{384} \frac{wL^4}{EI}$$

The solutions to the previous two examples can be checked against the solutions in any standard mechanics books.

Let us now summarize the equations used to determine the deflection due to bending energy under a load P.

$$\delta_p = \int_l \frac{\partial M_x}{\partial P} \frac{M_x}{EI} dx$$

The deflection due to the bending moment M is created under load P anywhere along the span of a beam.

Similarly, the shear stress in a beam also creates deflection, which we have been neglecting so far. It can be shown that the shear deflection for a beam subjected to a real or fictitious load P is given by:

$$\delta_p = \int_l fs \frac{\partial V_x}{\partial P} \frac{V_x}{GA_x} dx$$

where V_x is the shear force at a distance x, G is the shear modulus, A_x is the area of a cross section at a distance x, and fs is a factor depending upon given cross section of a beam. Table 5–1 gives values of fs for various cross sections. (You are requested to derive this fs value for various cross sections in a homework problem.)

Table 5–1 fs values for shear energy deflection

Type of Cross Section				
fs	6/5	A/A_{web}	10/9	2

The rotation due to a bending moment M_A (see Fig. 5–2) is given by integrating Equation (5–6); that is,

$$\theta_A = \int_l \frac{\partial M_x}{\partial M_A} \frac{M_x}{EI} dx$$

where θ_A is the rotation at location A due to bending energy.

Similarly, the angle of twist at say A due to the external torque energy (see Fig. 5–2) is given by Equation (5–7); that is,

$$\theta_A = \int_l \frac{\partial T_x}{\partial T_A} \frac{T_x}{GJ} dx$$

where θ_A is the rotation at A' due to the external torque.

5.4 Shear Deflection Using the Strain Energy Approach

Let us consider the effect of a shear force V_x on the deflection in Example 5-2. We can write the shear equation for the beam for the range $0 \le x \le L/2$ as

$$V_x = \left(\frac{P}{2} + \frac{wL}{2}\right) - wx$$

Therefore,

$$\frac{\partial V_x}{\partial P} = \frac{1}{2}$$

Note that we need not find V_x for $L/2 \le x \le L$, as the beam has an axis of symmetry about the vertical axis passing through the point $x = L/2$. (Symmetry load, as well as material and geometrical symmetry exists.)

Now, the shear deflection is given by the formula

$$\delta_P = \int_L fs \frac{\partial V_x}{\partial P} \frac{V_x}{G_x A_x} dx$$

For a beam with axis of symmetry at $L/2$, we can modify this equation to read

$$\frac{\partial U}{\partial P} = 2 \int_0^{\frac{L}{2}} \frac{\partial V_x}{\partial P} fs \frac{V_x}{G_x A_x} dx$$

If we assume a rectangular cross section for the beam and use $fs = 6/5$, as given in Table 5–1, then, substituting for $\frac{\partial V_x}{\partial P} = \frac{1}{2}$ and for V_x as shown before, we get

$$\delta_{Pv} = 2 \int_0^{\frac{L}{2}} \left(\frac{1}{2}\right)\left(\frac{6}{5}\right) \left(\frac{\left(\frac{P}{2} + \frac{wL}{2}\right) - wx}{G_x A_x}\right) dx$$

Setting $P = 0$ as we already have found $\frac{\partial V_x}{\partial P}$, (and realizing GA = a constant) we obtain

$$\frac{\partial U}{\partial P} = \delta_{Pv} = \frac{6}{5GA} \int_0^{\frac{L}{2}} \left(\frac{wL}{2} - wx\right) dx$$

$$= \frac{6}{5GA}\left[wL\frac{x}{2} - \frac{wx^2}{2}\right]_0^{\frac{L}{2}}$$

$$= \frac{6}{5GA}\left[\frac{wL^2}{4} - \frac{wL^2}{8}\right]$$

$$= \frac{6}{5GA}\left[\frac{wL^2}{8}\right]$$

Therefore,

$$\delta_{pv} = \frac{3}{20}\frac{wL^2}{GA}$$

Therefore, the total deflection due to bending moment and shear is given by

$$\delta_p = \frac{5}{384}\frac{wL^4}{EI} + \frac{3}{20}\frac{wL^2}{GA}$$

Most of the time, engineers do not calculate shear deflection, or else they neglect it, assuming that it is not significant. Let us see, however, what one must understand about such significance or insignificance. If L is short, then shear deflection (the second term in the preceding equation) is significant, but if L is large, then the shear deflection is insignificant compared to the bending deflection. It is the responsibility of a design engineer to make sure that the significance of the shear deflection is evaluated. For example, in machine design, shafts in a gearbox are very short in length, and therefore, the deflection due to shear becomes significant. However, in a very long crane boom, shear deflection can be neglected if an engineer designing the crane justifies neglecting it.

5.5 Use of Castigliano's Theorem to Solve Indeterminate Structures

A statically indeterminate structure is a structure in which equilibrium equations (i.e., $\Sigma \overline{F} = 0, \Sigma \overline{M} = 0$) are not sufficient to solve for the reaction forces and moments developed at fixed supports. Naturally, for indeterminate structures, additional equations are needed to solve for the unknowns. Castigliano's theorem provides these equations. The following example illustrates the use of the theorem to generate additional equations needed to solve for moments and/or reactions at supports and then, if necessary, deflections at a desired location in an indeterminate structure.

Example 5-3

A beam fixed at both of its ends is subjected to a distributed load as shown in the following diagram, (E, I, and A are constant):

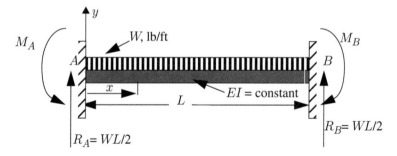

Determine M_A and M_B, the fixed-end moments.

Solution

The total number of unknowns is four, namely, the left reaction, left moment, right reaction, and right moment. We note that we have only two equilibrium equations available: $\Sigma M_z = 0$ and $\Sigma F_y = 0$. Therefore, the problem is indeterminate. Due to symmetry,

$$R_L = R_R = wL/2, \qquad\qquad M_A = M_B$$

(However, we still need to find M_A or M_B.) Now, let us write the moment equation at a distance x from the left end of the beam for the range $0 \le x \le L$:

$$M_x = \frac{wL}{2}(x) - M_A - \frac{wx^2}{2}$$

We know that $\theta_A = \int_l \frac{\partial M_x}{\partial M_A}\frac{M}{EI}dx$. To use this equation, we need

$\partial M_x / \partial M_A$; we can get this from the earlier equation for M_x as stated above

$$\frac{\partial M_x}{\partial M_A} = -1$$

However, the boundary condition at the left end of the beam dictates that $\theta_A = 0$ at the left support, as the beam is fixed at that support. Therefore,

$$0 = \int_o^L (-1)\left(\frac{wL}{2}(x) - M_A - \frac{wx^2}{2}\right)\frac{dx}{EI}$$

On integrating, we get

$$0 = \frac{wL}{2}\left[\frac{x^2}{2}\right]_o^L - M_A[x]_o^L - \frac{w}{2}\left[\frac{x^3}{3}\right]_o^L \qquad (\textit{Note: } EI \text{ cancels out.})$$

$$0 = \frac{wL^3}{4} - M_A L - \frac{w}{6}L^3$$

Solving for M_A, we obtain

$$M_A = \frac{wL^3}{4L} - \frac{wL^3}{6L}$$

$$M_A = \frac{3wL^3 - 2wL^3}{12L}$$

$$M_A = \frac{wL^2}{12}$$

We know that, due to symmetry,

$$M_A = M_B = \frac{wL^2}{12}$$

and

$$R_A = R_B = \frac{WL}{2}$$

Thus, an otherwise very difficult problem becomes simple when one makes use of the energy approach to solve for M_A.

The next section presents some more examples to determine deflections or reactions (forces or moments) in indeterminate structures.

5.6 Solutions to Problems Using Castigliano's Theorem and the Lead Model

The previous section demonstrated the use of Castigliano's theorem to solve for deflections and/or reactions in indeterminate structures. Once again, it is important to realize that the indeterminate structures are defined as those structures which have more unknowns than the number of equilibrium equations available. Therefore, for two-dimensional (2-D) indeterminate structures, it becomes necessary to use two equilibrium equations (say, $\Sigma F_y = 0$, $\Sigma M_0 = 0$) and one, two, or more equations based upon Castigliano's theorem. This means that a set of three, four, or five simultaneous nonhomogeneous equations have to be solved. Solving one or two equations for one or two unknowns is easy. However, solving three, four, or more equations becomes time consuming and complicated in some cases, and the solution can be subject to errors, especially if the simultaneous equations become long or nonlinear. TK Solver aids in solving such simultaneous, nonhomogeneous, linear or nonlinear sets of equations.

Needless to say, this is another instance in which TK Solver requires the user to understand the theory well and also requires that the problem be stated in algebraic form rather than in integration form. Since a certain number of unknowns require the same number of equations, one must develop such equations correctly.

The next example requires the user to write equilibrium equations, and then other required equations are written using Castigliano's theorem. The combined set of equations is solved simultaneously, as shown in the Rules Sheet. The correct answers can be calculated only if the equations are accurately entered into the sheet.

Examples 5-4 and 5-5 are easy to understand and require solving only three or four simultaneous equations. Example 5-6 is quite involved, but becomes simple with the help of TK Solver. Try solving it without TK, and you will notice that the symbolic answers require lengthy manipulations.

Example 5-4

A beam is fixed at its left end and is simply supported at a point B. If $L = 96$ in and $w = 1000$ lb/ft, write the two available equilibrium equations, and then use Castigliano's theorem to solve for R_A, R_B, and M_A. The beam and its load are shown in the following diagram:

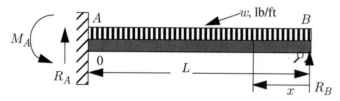

Solution

There are three unknowns in this problem: R_A, R_B and M_A. Let us first use the equilibrium equations, $\Sigma M_0 = 0$ will yield

$$M_A - wL\left(\frac{L}{2}\right) + R_B L = 0 \tag{5-13}$$

and $\Sigma F_y = 0$ will yield

$$R_A + R_B - wL = 0 \tag{5-14}$$

These are the only two static equilibrium equations that are nontrivial. To obtain the required additional equation, let us use boundary conditions and Castigliano's theorem—that is, $\delta_B = 0$ due to the support at B. (We could have used $\theta_A = 0$ as a boundary condition; try it if you want.) Now, Castigliano's theorem states that

$$\delta_B = \frac{\partial U}{\partial R_B} = \int_L \frac{M_x}{EI} \frac{\partial M_x}{\partial R_B} dx$$

The moment equation must be written for $0 \le x \le L$.

For $0 \le x \le L$ (note that x is measured from the right end of the beam),

$$M_x = \frac{wx^2}{2} - R_B x$$

Therefore,

$$\frac{\partial M_x}{\partial R_B} = -x$$

and

$$\delta_B = \frac{1}{EI} \int_0^L (-x)\left(\frac{wx^2}{2} - R_b x\right) dx$$

But $\delta_B = 0$ because the support at B is rigid, even though the beam is simply supported. Thus,

$$0 = \int_0^L \left(\frac{-wx^3}{2} + R_B x^2 \right) dx$$

$$0 = \left(\frac{wx^4}{8} - R_B \frac{x^3}{3} \right) \Bigg|_0^L$$

$$0 = \frac{-wL^4}{8} + \frac{R_B L^3}{3}$$

Solving for R_B, we obtain

$$R_B = \frac{3}{8} wL$$

Now,

$$R_A = wL - \frac{3}{8} wL$$

$$R_A = \frac{5}{8} wL$$

To find M_A, we use Equation (5–13):

$$M_A = \frac{wL^2}{2} - R_B L = \frac{wL^2}{2} - \frac{3}{8} wL^2$$

$$M_A = \frac{wL^2}{8}$$

The foregoing equations are entered into the lead model *ex5-4.tk* (not shown here). If input values for W and L are given, then one can solve for R_B, as well as for the other unknowns using two static equations (Equations (5–13) and (5–14)).

Example 5-5

Let a beam be fixed at both ends and subjected to triangular loading. If $L = 96$ in and $w = 10,000$ lb/ft, determine R_A, R_B, M_A, and M_B by writing and solving four equations. (Use Castigliano's theorem to obtain additional necessary equations.) The beam and its load are shown in the following diagram:

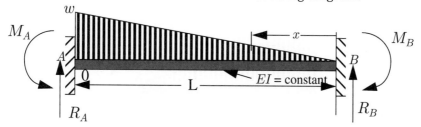

Solution

In this problem, one has no obvious answer that relates R_A or R_B to load W and length L.

Therefore, using the static equilibrium equation for moment about O at R_A, we get

$$\Sigma M_o = 0$$

and taking counterclockwise as positive will yield

$$M_A - M_B + R_B L - \frac{wL^2}{6} = 0 \tag{5-15}$$

Using

$$\Sigma F_y = 0$$

will yield

$$R_A + R_B - \frac{wL}{2} = 0 \tag{5-16}$$

Since there are four unknowns and only two equilibrium equations, this problem is a statically indeterminate problem of second degree. Therefore, two additional equations are required. Now, we can use the boundary conditions to solve the problem; that is,

$$\theta_B = 0 \text{ and } \delta_B = 0$$

But

$$\theta_B = \int_L \frac{M_x}{EI} \frac{\partial M_x}{\partial M_B} dx$$

and

$$\delta_B = \frac{\partial U}{\partial R_B} = \int_L \frac{M_x}{EI} \frac{\partial M_x}{\partial R_B} dx \qquad \text{(Castigliano's theorem)}$$

Now let us write moment equations for the range $0 \leq x \leq L$. The free-body diagram for a beam of length x is as follows:

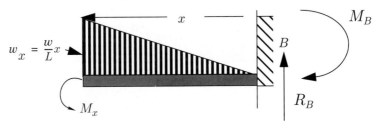

The moment equilibrium equation is

$$M_x - M_B + R_B x - \frac{w_x x^2}{6} = 0$$

but

$$\frac{w}{L} = \frac{w_x}{x}$$

Therefore,

$$M_x - M_B + R_B x - \frac{w x^3}{6L} = 0 \text{ as } w_x = \frac{wx}{L}$$

Now take partial of above equation with respect to R_B and M_B, we get

$$\frac{\partial M_x}{\partial R_B} = -x \qquad \text{and} \qquad \frac{\partial M_x}{\partial M_B} = 1$$

Now,

$$\delta_B = 0 = \int_0^L \frac{1}{EI}\left(M_B - R_B x + \frac{w x^3}{6L}\right)(-x)dx$$

$$0 = \frac{1}{EI}\int_0^L \left(-M_B x + R_B x^2 - \frac{w x^4}{6L}\right)dx$$

Upon integrating, one gets

$$0 = -\frac{M_B x^2}{2} + \frac{R_B x^3}{3} - \frac{w x^5}{30L}\Bigg|_0^L \qquad\qquad (5\text{--}17)$$

$$= A_1 - A_2$$

where A_1 is Equation (5–17) with $x = x_1 = L$ and A_2 is Equation (5–17) with $x = x_2 = 0$. (See the Rules Sheet for lead model *ex5-5.tk* to better understand A_1 and A_2.)

Now, we know that

$$\theta_B = \frac{\partial U}{\partial M_B} = \int_0^L \frac{M_x}{EI} \frac{\partial M_x}{\partial M_B} dx$$

But $\theta_B = 0$; therefore,

$$0 = \int_0^L \frac{1}{EI}(1)\left(M_B - R_B x + \frac{wx^3}{6L}\right) dx$$

$$0 = \frac{1}{EI}\left[\left(M_B x - R_B \frac{x^2}{2} + \frac{wx^4}{24L}\right)\right]_0^L \qquad (5\text{–}18)$$

$$0 = A_3 - A_4$$

where A_3 is Equation (5–18) with $x = x_3 = L$ and A_4 is Equation (5–18) with $x = x_4 = 0$.

The lead model *ex5-5.tk*, or *castind.tk*, is developed to solve Example 5-5. Study this model (which is presented on pages 104 and 105) carefully as it shows how to write A_1, A_2, A_3, and A_4 equations when one enters Equations (5–15), (5–16), (5–17), and (5–18). By substituting the given values and making guesses for M_A, M_B, R_A, and R_B, one finds that M_A = 384 kip in, M_B = 256 kip in, R_A = 28,000 lb, and R_B = 12,000 lb.

In sum, one can become proficient in using Castigliano's theorem to solve for deflections in determinate structures and deflections and reaction forces or moments in indeterminate structures. This approach will also offer further experience, understanding and confidence in solving difficult problems.

There is always a question of whether one should find the total energy U by integration of unit energy first and then differentiate to find the deflection or reaction, or differentiate first—that is, find $\dfrac{\partial U}{\partial P}$ first and then integrate $\int_l \dfrac{\partial M}{\partial P} \dfrac{M}{EI} dx$ second. It does not matter which path one takes. It is, however, true that if you *differentiate first* and then *integrate second*, it will become easier to manage a difficult problem.

RULES SHEET

```
ex5-5.tk

 File  Edit  Commands  Options  Windows  Sheets  Help

                              Rules

S Rule------------------------------------------------------

  "Lead Model Developed by DR.S.R.Bhonsle & DR.K.J.Weinmann"


  _

  "Moment Equilibrium Equation"
* Ma-Mb+Rb*L-(w*L^2)/6=0

  "Force Equation"

* Ra+Rb-w*L/2=0


  "Castigliano's Equations"

* A1=-((Mb*x1^2)/2)+((Rb*x1^3)/3)-(w*x1^5)/(30*L)
* A2=-((Mb*x2^2)/2)+((Rb*x2^3)/3)-(w*x2^5)/(30*L)
* A3=(Mb*x3)-((Rb*x3^2)/2)+(w*x3^4)/(24*L)
* A4=(Mb*x4)-((Rb*x4^2)/2)+(w*x4^4)/(24*L)
* A1-A2=0
* A3-A4=0



(4) Rule:                                                  F9 OK
```

VARIABLES SHEET

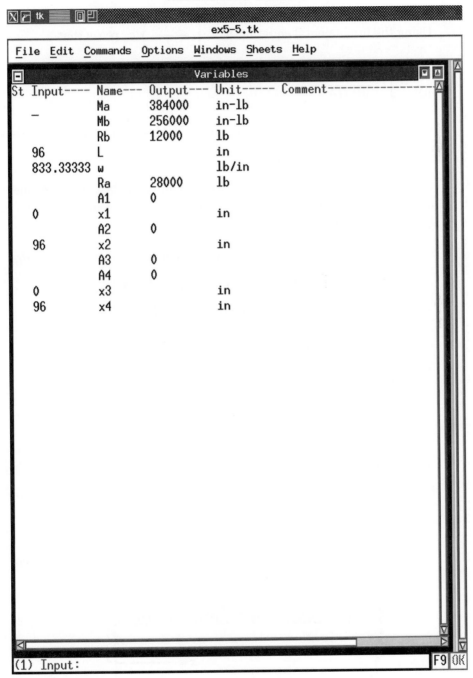

```
X  tk        

                        ex5-5.tk

 File  Edit  Commands  Options  Windows  Sheets  Help

                        Variables

St Input----  Name---  Output---  Unit-----  Comment----------------
               Ma       384000    in-lb
   —           Mb       256000    in-lb
               Rb       12000     lb
   96          L                  in
   833.33333   w                  lb/in
               Ra       28000     lb
               A1       0
   0           x1                 in
               A2       0
   96          x2                 in
               A3       0
               A4       0
   0           x3                 in
   96          x4                 in
```

(1) Input: F9 OK

PROBLEMS

5–1) Using the fundamental equation $u = \int_V \left(\frac{\tau_{xy}\gamma_{xy}}{2}\right)dv$, prove that

$$\frac{\partial u}{\partial P} = \int f s \frac{\partial V_x}{\partial P} \frac{V_x}{GA}dx$$

where $\tau_{xy} = (V_x Q)/(It)$ and all other variables are as described earlier in this chapter.

5–2) Given a circular ring of radius R_G with modulus of elasticity E, moment of intertia I, area of cross section A, and subjected to a load P as shown in the following diagram, determine:

   a.   The moment at $A - A$ if $A - A$ makes an angle θ having radius R_G.

   b.   The moment at B if B is at an angle $\pi/2$ and $2R_G$ from P.

   c.   The deflection in the direction of P at P.

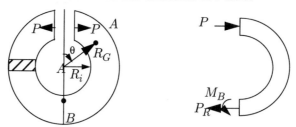

(*Hint*: Take the free-body diagram of the half ring, as shown, to find M_B. Realize that $\frac{\partial U}{\partial M_B} = 0$ at B. Neglect shear and axial effects. First deter-mine M at some angle θ ($M_\theta = P(R_G - R_G \cos\theta)$)).

5–3) Consider the following diagram of a beam subjected to the loads shown:

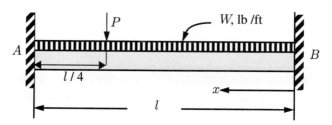

a. Determine the moments and reactions of the beam at A and B.

b. Determine the deflection under the load P

Use $P = 1000$ lb, $W = 1000$ lb/ft, $l = 100$ in, $E = 30 \times 10^6$ psi, and $I = 200$ in^4. (*Hint*: First write moment equations. for $0 \le x \le 3/4$ and then for $3/4 \le x \le l$, with x measured as shown. Find the deflection in terms of P, W, L, E, I, and then substitute the given values.)

5–4) Using Castigliano's theorem, determine, for the diagram that follows:

a. The value of the load P if the deflection under P is 0.04". Take $E = 30 \times 10^6$ psi.

b. The maximum bending stress in the beam and its location ($\sigma = MC/I$).

c. The deflection at B. (*Hint*: Transfer the load P to B and the moment due to P ($M_B = 10P$), write the moment equation from B to A, and continue.)

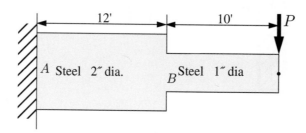

5–5) Determine the deflection under the load P for the triangular beam of width b and thickness t as shown in the following diagram.

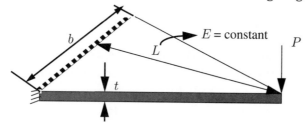

Then find the spring constant $k = P/\delta_p$.

5–6) Given a piston ring of the size shown in the following diagram, and assuming that $\sigma = My/I$ is a valid relationship, determine r_G the radius to the CG of beam and deflection at load P if the ring has an internal radius r_i = 3" and a modulus of elasticity E (use Castigliano's theorem).:

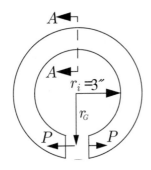

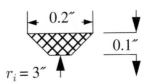

Cross Section @ A–A
load P is acting at C.G. of cross section shown above

5–7)   Find the horizontal deflection at B (positive to the right) due to the application of the vertical and horizontal forces F and P at point A in the following bracket (use Castigliano's method and the coordinates shown):

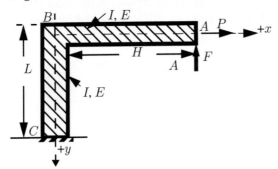

5–8)   Consider the following diagram of a beam subjected to loads as shown.

   a.   Determine the reactions at A, B, and C.

   b.   Determine the deflection under load P_1 at point D.

   c.   If l = 36", $P = P_1 = P_2$ = 15,000 lbs, and $E = 30 \times 10^6$ psi, determine the required moment of inertia if the deflection at D is 0.4".

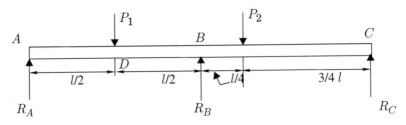

Analysis and Design of Columns

6.1 Introduction

Columns are long, straight, prismatic bars subjected to a compressive axial load. As the load increases, it causes the column to deflect laterally, and the load eventually will buckle and ultimately fracture the column at a stress level below the proportional limit (or below the yield-point stress in compression). Obviously, the stress formulas derived so far cannot be used to evaluate the load-carrying capacity of a column or to determine the required size of a column to support a given load. Therefore, it is the purpose of this chapter to analyze columns in order to determine the relationship between compressive load capacity and the physical and mechanical properties of a column. A safe compressive load can then be determined by applying the appropriate safety factor.

As we derive the column design equations (a mathematical model) it can be seen that the behavior of a column depends primarily upon the length of the column, the type of loading, the size of the column, and the mechanical properties of the material out of which the column is made.

The columns can be divided into four types:

1. Long columns with concentric loading

2. Intermediate-length columns with concentric loading

3. Columns with eccentric loading

4. Struts, or short columns, with eccentric loading

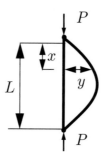

Figure 6–1   Long and hinged column subjected to concentric load P

6.2 Long Column with Concentric Loading

Consider a column as shown in Fig. 6–1 which is assumed to be hinged at both ends and is subjected to a concentric load P. The load P is a compression load, and the column is long and has a tendency to buckle. Let us develop the necessary mathematical model to understand design of columns.

If one cuts a column at a distance x, and takes a moment of the external load, one obtains

$$M_x = -Py \qquad\qquad (6\text{–}1)$$

where M_x is the moment at distance x, P is the external concentric load, and y is the deflection at a distance x.

However, in basic mechanics, the relationship between the deflection and bending moment is

$$EI\frac{d^2y}{dx^2} = M_x$$

Rearranging, and upon substituting Equation (6–1) for M_x, we get

$$EI\frac{d^2y}{dx^2} + Py = 0 \qquad\qquad (6\text{–}2)$$

which is the well-known ordinary differential equation (ODE) for simple harmonic motion.

The solution of this ODE is

$$y = A_1 \sin\sqrt{\frac{P}{EI}}\,x + B_1 \cos\sqrt{\frac{P}{EI}}\,x \qquad\qquad (6\text{–}3)$$

where A_1 and B_1 are constants to be determined on the basis of the boundary conditions. One of the boundary conditions is at $x = 0$, $y = 0$, as can be seen in Fig. 6–1. Therefore, letting $x = 0$ and $y = 0$ in Equation (6–3), we get

$$B_1 = 0$$

Similarly, another boundary condition is at $x = L$, where, again, $y = 0$. Therefore, letting $x = L$ and $y = 0$ in Equation (6–3), we get

$$0 = A_1 \sin \sqrt{\frac{P}{EI}} L$$

for a nontrival solution $A_1 \neq 0$; therefore,

$$\sin \sqrt{\frac{P}{EI}} L = 0$$

which is true when $\sqrt{\dfrac{P}{EI}} L = n\pi$. Now for $n = 1$, (i.e., the first harmonic mode) and substituting for $P = P_{cr}$, where P_{cr} is the critical load we get on rearranging and squaring:

$$P_{cr} = \frac{\pi^2 EI}{L^2} \tag{6–4}$$

Equation (6–4) was derived by Euler; hence, it is called the *Euler Column Formula*. If we divide this equation by the cross-sectional area A of a column, we get

$$\frac{P_{cr}}{A} = \frac{\pi^2 E}{(L/\rho)^2} \tag{6–5}$$

(Note that $(I/A) = \rho^2$ where ρ is called the *radius of gyration* of the column and (L/ρ) is called the slenderness ratio.)

Equation (6–5) has one major fault, which can be seen by plotting P_{cr}/A vs. $(\pi^2 E) / (L/\rho)^2$. Plotting such an equation (see Fig. 6–2) shows that P_{cr}/A approaches infinity if L/ρ approaches zero. But this is a physical impossibility. To avoid the problem, the following approach is taken. Select a point T along the $S_{yp}/2$ line as shown in Fig. 6–2. This limiting value of $S_{yp}/2$ restricts the value of P_{cr}/A from being larger than $S_{yp}/2$ or, in other words, the Euler Formula is made applicable for $0 \leq \dfrac{P_{cr}}{A} \leq \dfrac{S_{yp}}{2}$, where S_{yp} = yield point strength of the column's material.

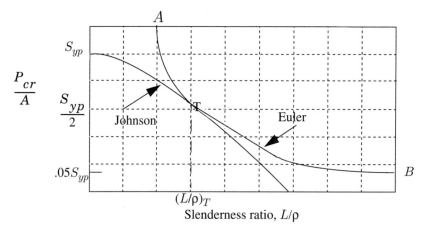

Figure 6-2 Plot of P_{cr}/A vs. (L/ρ), the slenderness ratio.

However, we need to determine the value of slenderness ratio (L/ρ) at T for further study. This can be easily done by realizing that at T the stress is $S_{yp}/2$, that is,

$$\frac{P_{cr}}{A} = \frac{S_{yp}}{2} = \frac{\pi^2 E}{(L/\rho)_T^2}$$

$$\therefore \left(\frac{L}{\rho}\right)_T = \left(\frac{2\pi^2 E}{S_{yp}}\right)^{\frac{1}{2}} \tag{6-6}$$

It is obvious that the limiting value of slenderness ratio $(L/\rho)_T$ depends upon the material properties, such as E and S_{yp}. Equation (6–6) can be used as the lower limiting value of L/ρ, beyond which the Euler formula can be used. This means that if L/ρ for a column is known to be greater than $(L/\rho)_T$, the Euler formula is applicable.

If one plots the Euler equation as a function of (L/ρ), the curve will look like ATB in Fig. 6–2 for the range $0 < (L/\rho) \leq \infty$. As suggested, however, the limiting value of the Euler formula should be $P_{cr}/A = S_{yp}/2$. Thus, the Euler Column formula is:

$$\frac{P_{cr}}{A} = \frac{\pi^2 E}{(L/\rho)^2} \qquad \text{if } \left(\frac{L}{\rho}\right) > \left(\frac{L}{\rho}\right)_T$$

Table 6–1 Equivalent Length (L_e) values for various end conditions

Column End	L_e (Recommended by ASCE)
Fixed – Fixed	$L_e = 0.5\,L$
Fixed – Hinged (not fixed hinge)	$L_e = 2.0\,L$
Fixed – Hinged (fixed hinge)	$L_e = 0.80\,L$
Hinged – Hinged	$L_e = L$

In sum, columns whose slenderness ratio (L/ρ) value is greater than $(L/\rho)_T$ are considered long columns and are designed using Euler's Column Formula.

The previous derivation was for a column that was hinged at both ends. However, columns can be fixed at either one or both ends. To accommodate such boundary conditions, the preceding equation is modified by setting $L = L_e$, the equivalent length which depends upon the boundary conditions. Table 6–1 gives values of L_e for various end conditions.

We now will modify Euler's Column formula which is to be used if $(L_e/\rho) > (L_e/\rho)_T$ as:

$$\frac{P_{cr}}{A} = \frac{\pi^2 EI}{L_e^2 A}$$

where L_e is the equivalent length of the column as given in Table 6–1.

Since $I = A\rho^2$, the preceding equation can be rearranged to yield

$$\frac{P_{cr}}{A} = \frac{\pi^2 E}{(L_e/\rho)^2} \tag{6–7}$$

Clearly, the higher the modulus of elasticity, the greater is the load-carrying capacity and similarly, the lower the value of L_e/ρ, the higher the load-carrying capacity. Now the question remains as to which equation is to be used if $(L_e/\rho) < (L_e/\rho)_T$. In this regard, one should first determine (L_e/ρ) for a given column. If (L_e/ρ) for a given column is less than $(L_e/\rho)_T$, then such a column will be called an intermediate column or intermediate-length column.

6.3 Intermediate Column with Central Loading

Intermediate-length columns are, by definition, columns whose L/ρ is less than $(L/\rho)_T$ as given by Equation (6–6). The intermediate column theory can be derived by drawing a parabolic curve passing through S_{yp} at $L_e/\rho = 0$ and $S_{yp}/2$, at $(L_e/\rho) = (L_e/\rho)_T$, as shown in Fig. 6–2. Let us select T as discussed earlier and shown in that figure. T has the coordinates $S_{yp}/2$, $(L_e/\rho)_T$ and a parabolic curve is assumed to fit two points, one at S_{yp}, and the other at $(L_e/\rho)_T$.

Now let us use the general equation for a parabola that has the form $y = a - b(x)^2$. From Fig. 6–2, it is obvious that the y axis is really the (P_{cr}/A) axis and the x axis is the (L_e/ρ) axis for our case. Therefore, the parabolic equation can be written as

$$\frac{P_{cr}}{A} = a - b(L_e/\rho)^2 \qquad\qquad (6\text{–}8)$$

The constants a and b can be evaluated using two boundary conditions, i.e. when $(L_e/\rho) = 0$, $P_{cr}/A = S_{yp}$

Substituting into Equation (6–8), we get

$$a = S_{yp}$$

Similarly, when $L_e/\rho = (L_e/\rho)_T$, $P_{cr}/A = S_{yp}/2$. Substituting into Equation (6-8) and letting $a = S_{yp}$, we get

$$\frac{S_{yp}}{2} = S_{yp} - b(L_e/\rho)_T^2$$

but $(L_e/\rho)_T = \left(\dfrac{2\pi^2 E}{S_{yp}}\right)^{1/2}$ [see Equation (6–6)]

When rearranged, this equation will yield $b = (S_{yp}/2\pi)^2 \, (1/E)$.

Now we can rewrite Equation (6–8) as

$$\frac{P_{cr}}{A} = S_{yp} - (S_{yp}/2\pi)^2 (L_e/\rho)^2 (1/E) \qquad\qquad (6\text{–}9)$$

Equation (6–9), which was first derived by J.B. Johnson, is called *Johnson's Formula*. This formula is the design formula for intermediate columns—that is, for columns having

$$0 \le (L_e/\rho) \le (L_e/\rho)_T.$$

The Johnson Formula is widely used by civil and mechanical engineers. We will study it in some sample problems later on.

We reiterate that the Euler and Johnson formulas are valid only for *concentric loads.*

Columns subjected to nonconcentric (eccentric) loads are designed using the theory developed in the next section. Two types of columns subjected to eccentric loads are long columns and short columns.

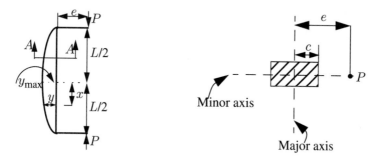

Figure 6–3   Long column subjected to eccentric loading

6.4 Long Column with Eccentric Loading

A column subjected to an eccentric load is shown in Fig. 6–3. The load P is applied at a distance e measured from the major axis of the column. It is obvious that it makes design sense to have an eccentric load buckling a column about a major axis instead of a minor axis, as such eccentricity will enable the column to carry a greater load.   Therefore, one should always ensure that the eccentricity distance is perpendicular to the major axis and not to the minor axis, of the transverse cross section of a column. Let the column be cut as shown in Fig. 6–3 at a distance x measured from the center of the column. The bending moment at a distance x due to load P is

$$M_x = -P \ (y + e)$$

As stated earlier, we know that $EI\dfrac{d^2 y}{dx^2} = M_x$

$$\therefore EI\frac{d^2 y}{dx^2} + P(y + e) = 0 \tag{6–10}$$

Rearranging, one gets

$$EI\frac{d^2 y}{dx^2} + Py = -Pe$$

This is a nonhomogeneous differential equation having a general and a particular solution. The total solution is:

$$y = C_1 \sin\left(\sqrt{\frac{P}{EI}}\right)x + C_2 \cos\left(\sqrt{\frac{P}{EI}}\right)x - e \tag{6–11}$$

As shown earlier, the constants C_1 and C_2 can be evaluated using boundary conditions; that is, to find C_1 and C_2, we realize that $dy/dx = 0$ when $x = 0$, and also $y = 0$ when $x = L/2$.

Therefore, using the first condition, we can show that

$$C_1 = 0$$

and using the second condition yields

$$C_2 = \frac{e}{\cos\left[\sqrt{\frac{P}{EI}\left(\frac{L}{2}\right)^2}\right]}$$

Substituting the values of C_1 and C_2 into Equation (6–11), we get

$$y = \frac{e}{\cos\left[\sqrt{\frac{PL^2}{4EI}}\right]} \cos\left(\sqrt{\frac{P}{EI}}\right) x - e \tag{6–12}$$

The deflection y is maximum at $x = 0$. (See Fig. 6–3.) Hence,

$$\therefore y_{max} = \frac{e}{\cos\sqrt{\frac{PL^2}{4EI}}} - e \tag{6–13}$$

We still have not related the load P to the geometrical and mechanical properties of a column. To achieve such relation, however, we know that when the load P is at a distance e, the stress at an extreme fiber is maximum, due to the axial and bending loads. The elementary formula for normal stress due to axial and bending loads is

$$\sigma = P/A + Mc/I$$

where c is the distance from the major (neutral) axis to the extreme fiber (see Fig. 6–3), I is the moment of inertia of the cross section of the column about the major axis, and M = the moment about the major axis. This stress should not exceed the yield stress of the material, that is,

$$\sigma_{max} = \frac{P_{cr}}{A} + \frac{Mc}{I} = S_{yp}$$

where S_{yp} is compressive yield stress.

Note that we now have P as P_{cr} as it will be the maximum load for a given maximum allowable stress S_{yp}. Realizing that moment M is maximum at $x = 0$ (i.e., $M = P_{cr}(y_{max} + e)$ and taking $I = \rho^2 A$, we obtain

$$\sigma = S_{yp} = \frac{P_{cr}}{A} + \frac{(P_{cr}(y_{max} + e))c}{I}$$

where y_{max} is given by Equation (6–13). Substituting for y_{max}, solving, rearranging, and designating P_{cr}/A as S_{cr}, the critical stress, results in

$$S_{cr} = \frac{P_{cr}}{A} = \frac{S_{yp}}{1 + \left(\dfrac{ec}{\rho^2}\right)\sec\left[\dfrac{L_e}{\rho}\right]\sqrt{\dfrac{P_{cr}}{4AE}}} \qquad \text{for } e > 0 \qquad (6\text{–}14)$$

where $(ec)/\rho^2$ is called the *eccentricity ratio* and L_e is the equivalent length as given in Table 6–1.

Equation (6–14) is a nonlinear equation that requires special techniques to solve. However, TK Solver can solve such equations very easily. The equation is known as the *secant formula* and is applicable to columns subjected to an *eccentric* load.

At times, when P is large and e is small, it is necessary to check for buckling about the minor axis by assuming that P is a concentric load. In this situation, one must realize that the lesser of the two loads is to be selected, as such a load will cause buckling in a column about the minor axis first. Of course, one can stiffen the column to prevent buckling about a weak axis.

6.5 Short Columns (Struts)

For very short columns loaded eccentrically, one has to make sure that the combined axial and bending stress at a critical location does not exceed the allowable yield stress in tension or compression, that is,

$$\sigma_{yp} = \frac{P_{cr}}{A} + \frac{P_{cr}c}{I}$$

where A is the area of cross section of the strut, e is the eccentricity, c is the extreme fiber distance, I is the moment of inertia about the bending axis, and P_{cr} is the critical eccentric load. It now becomes easy to find the value of P_{cr} if σ_{yp} and the column's geometrical properties are known.

Thus, we have covered all possible types of columns engineers will encounter. It is, however, important that if P_{cr} is obtained, then the design load is given by the equation

$$P = P_{cr}/SF$$

where SF is a safety factor (with recommended value 1.6 to 2.5).

The next section summarizes all the possible types of columns and applicable formulas. It is presented as a summary, and you will not have to go through the lengthy derivations of the preceding formulas again.

6.6 Review of Column Equations

Long columns with central loading are best described by the *Euler Column Formula*, which assumes an ideal, straight column, a precise axial load, and stresses within the proportional limit. The critical buckling load is given by

$$P_{cr} = \frac{\pi^2 EI}{L_e^2} \tag{6–15}$$

where E is the modulus of elasticity, I is the moment of inertia of the column about the minor axis and such axis is always passing through the center of gravity, and L_e is the equivalent length of a column. (See Table 6–1 for values of L_e.) Dividing Equation (6–15) by cross-sectional area, we get

$$S_{cr} = \frac{P_{cr}}{A} = \frac{\pi^2 E}{(L_e/\rho)^2}, \text{ For } \left(\frac{L_e}{\rho}\right) \geq \left(\frac{L_e}{\rho}\right)_T$$

The corresponding critical stress is the minimum stress needed to place the column in an unstable equilibrium state; that is,

Note: $$\left(\frac{L_e}{\rho}\right)_T = \left(\frac{2\pi^2 E}{S_{yp}}\right)^{\frac{1}{2}}$$

The Euler Formula fails to compare well with experimental results for intermediate-length columns (in other words, when $L_e/\rho < (L_e/\rho)_T$. Therefore, the *Johnson*, or *parabolic*, *formula* is used. The equation is

$$S_{cr} = \frac{P_{cr}}{A} = S_{yp} - \frac{S_{yp}^2}{4\pi^2 E}\left(\frac{L_e}{\rho}\right)^2$$

The parabolic curve is always tangent to the Euler curve at the point where the slenderness ratio and critical stress are

$$\left(\frac{L_e}{\rho}\right)_T = \sqrt{\frac{2\pi^2 E}{S_{yp}}}$$

and

$$\frac{P_{cr}}{A} = \frac{S_{yp}}{2}$$

This point is called the *tangent point*.

Columns with a slenderness ratio greater than the tangent point are considered to be long columns; otherwise they are considered to be intermediate columns. The parabolic curve is used for intermediate-length columns whereas the Euler curve is used for long columns. (See Fig. 6–2.)

The *secant column formula* differs from both the Euler and Johnson formulas in that it quantifies stress due to *eccentric* loads. It can be written in the form

$$S_{cr} = \frac{P_{cr}}{A} = \frac{S_{yp}}{1 + (ec/\rho^2)\sec[(L_e/\rho)\sqrt{P_{cr}/4AE}]}$$

where ec/ρ^2 is the eccentricity ratio of the column and c is the distance from the neutral axis to the extreme fiber measured perpendicular to the major axis. (Note that this is a nonlinear equation, with the dependent variable occurring on both sides of the equation.)

As the slenderness ratio increases, the critical stress from the secant formula approaches the critical stress from the Euler formula. Notice that the secant formula depends on an eccentricity ratio to determine the level of critical stress and that greater eccentricity reduces the load a column can carry. Let us next study some examples in which the theory we have presented is used to analyze or design various types of columns.

Example 6-1

A column with the cross section shown in Fig. 6–4 is subject to concentric load (not shown):

      a.   Determine the critical concentric load.

      b.   Determine the critical nonconcentric load if $ec/\rho^2 = 1$.

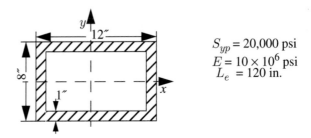

$$S_{yp} = 20,000 \text{ psi}$$
$$E = 10 \times 10^6 \text{ psi}$$
$$L_e = 120 \text{ in.}$$

Figure 6–4   Cross-section of a column

Solution

Let us first determine the geometrical properties for the column subjected to a concentric load. We know that, for a rectangular section, the moment of inertia is given by the equation

$$I = \frac{bd^3}{12}$$

$$\therefore I_x = \frac{12(8)^3}{12} - \frac{10(6)^3}{12} = 512 - 180 = 332 in^4$$

$$I_y = \frac{8(12)^3}{12} - \frac{6(10)^3}{12} = 512 - 180 = 652 in^4$$

$$A = bd - b_1 d_1$$

$$= 12(18) - 10(6)$$

$$A = 96 - 60 = 36 in^2$$

Hence,

$$\rho_x = \sqrt{\frac{I_x}{A}}$$

$$\rho_x = \sqrt{\frac{332}{36}} = 3.037 \text{ in}$$

Similarly,

$$P_y = \sqrt{\frac{J_v}{A}}$$

$$\rho_y = \sqrt{\frac{652}{36}} = 4.25 \text{ in}$$

$$\therefore L_e/\rho_x = \frac{120}{3.03} = 39.51$$

(*Note: L_e/ρ* is found about the weak axis as the column will buckle about this axis.)

Now find $(L_e/\rho)_T$ i.e.

$$(L_e/\rho)_T = \sqrt{\frac{2E\pi^2}{S_{yp}}}$$

$$= \left[\frac{2 \times 10 \times 10^6 (\pi)^2}{20,000}\right]^{1/2}$$

$$= 99.92$$

This shows that

$$(L_e/\rho_x) < (L/\rho)_T$$

i.e. $39.51 < 99.92$

Therefore, we use Johnson's formula, which is

$$S_{cr} = S_{yp} - \frac{S_{yp}^2}{4\pi^2 E}(L_e/\rho)^2$$

$$= 20,000 - \frac{20,000^2}{4\pi^2 (10(10^6))}(39.51)$$

$$= 20,000 - 1582$$

$$S_{cr} = 18,418 \text{ psi}$$

but $S_{cr} = P_{cr}/A$

$$\therefore P_{cr} = (S_{cr})A$$

$$P_{cr} = (18,148)(36)$$

$$P_{cr} \approx 663,000 \text{ lb}$$

Knowing the critical value, one can determine the safe load if the safety factor is known as $P_{safe} = \frac{P_{cr}}{SF}$ where SF is the safety factor.

For the second part, we have $ec/\rho^2 = 1$ and $L_e/\rho_y = 120/4.25 = 28.23$. (Note that L_e/ρ_x is not used. Why?) Using the *secant column formula* one gets $S_{cr} =$

9510 psi. This stress value is less than the value 18,418 psi obtained using Johnson's formula, as we now have a column that is eccentrically loaded.

We can determine the critical load as follows:

$$P_{cr} = (S_{cr})(A) = (9510)(36) = 342,360 \text{ lb}$$

A safe load can be obtained from this critical load if a safety factor is assumed or given.

6.7 Sample Solution Using Lead Model

The lead model with the file name *coldes.tk* or *ex6-1.tk* has the appropriate equations in the Rules Sheet as per Table 6–1. It is important to note that two new functions are being introduced in the model. The function ELT () returns the number of that instance (i.e., the number of the element currently being processed during list solving). For example, function PLACE ('P, ELT () + 1) = P, puts the current value of P into the next element of list P. The input value will be used by the Iterative Solver as a guess value during the next list-solving instance. There are a few lines in the Rules Sheet beginning with double quotes ("). TK regards all text appearing to the right of the quotation marks as comments.

At this point, all the variables used should have appeared in the Variables Sheet. Enter the known variables into the input field. Enter G into the status field of P to indicate that guessed values are being provided for solving the non-linear secant formula. Now click the **Option** menu at the top of the screen, and choose **Solver Control**. Change the maximum iteration count in the dialog box to 1000, and then close the dialog box and switch to the Variables Sheet. Try to solve the secant formula when the slenderness ratio is equal to zero. Use the output as an initial guess for later list solving. List the appropriate variables by typing L into the status field. Enter I and O into the status fields of all input and output variables, respectively.

Create a table and open the Table Subsheet. List the slenderness ratio and the critical stresses, and then open the subsheet and fill in the slenderness ratios from 0 to 250 with increments of 0.5. (*Note*: A larger step size can be used, but the equations may not converge to the desired output.) **List solve** the equations, and use the data points to plot graphs. *Notice that some lines may not show on the screen even though they have been plotted.* Don't give up!!! Try to print the graph, because the missing lines may be there.

(*Note*: Make sure to clear the tables (all but the list-fill section) before re-solving the problem with different data, in order to reduce the chance of error.)

RULES SHEET

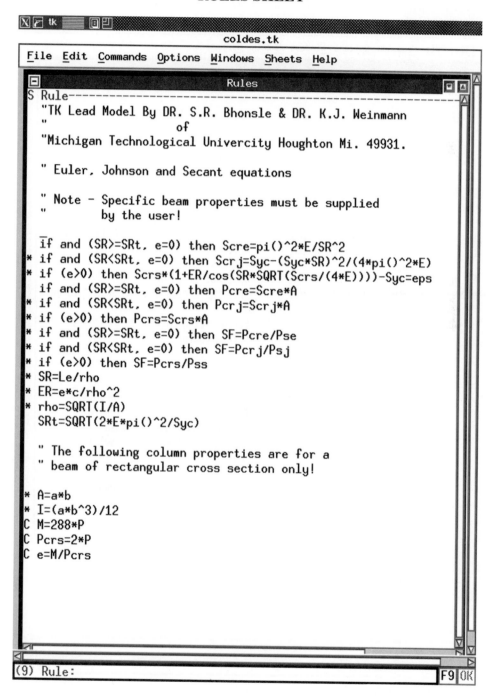

```
coldes.tk
```

File   Edit   Commands   Options   Windows   Sheets   Help

```
                              Rules
S Rule----------------------------------------------------------------
  "TK Lead Model By DR. S.R. Bhonsle & DR. K.J. Weinmann
  "                     of
  "Michigan Technological Univercity Houghton Mi. 49931.

  " Euler, Johnson and Secant equations

  " Note - Specific beam properties must be supplied
  "          by the user!

  if and (SR>=SRt, e=0) then Scre=pi()^2*E/SR^2
* if and (SR<SRt, e=0) then Scrj=Syc-(Syc*SR)^2/(4*pi()^2*E)
* if (e>0) then Scrs*(1+ER/cos(SR*SQRT(Scrs/(4*E))))-Syc=eps
  if and (SR>=SRt, e=0) then Pcre=Scre*A
* if and (SR<SRt, e=0) then Pcrj=Scrj*A
* if (e>0) then Pcrs=Scrs*A
* if and (SR>=SRt, e=0) then SF=Pcre/Pse
* if and (SR<SRt, e=0) then SF=Pcrj/Psj
* if (e>0) then SF=Pcrs/Pss
* SR=Le/rho
* ER=e*c/rho^2
* rho=SQRT(I/A)
  SRt=SQRT(2*E*pi()^2/Syc)

  " The following column properties are for a
  " beam of rectangular cross section only!

* A=a*b
* I=(a*b^3)/12
C M=288*P
C Pcrs=2*P
C e=M/Pcrs
```

(9) Rule: F9 OK

VARIABLES SHEET

X🔁 tk ▨▨ 🔲🔳

coldes.tk

File Edit Commands Options Windows Sheets Help

▣	Variables	▣▣

St	Input----	Name---	Output---	Unit-----	Comment-------------------------------
	1E7	E		psi	Young´s modulus (psi,ksi or Pa or MPa)
		c		in	Distance from NA to extreem fiber
	0̄	e		in	Eccentricity
L	88.626292	SR			Slenderness ratio
		SRt	99.345883		Slenderness ratio at the tangential point of Euler and Johnson curve
L		Scre	0	psi	Critical stress base on Euler formula
L		Scrj	12041.606	psi	Critical stress base on Johnson formula
	20000	Syc		psi	Critical yield stress
		Scrs		psi	Critical stress base on Secant formula
		ER	0		Eccentricity ratio
		SF			Safety factor
	16	A		in^2	Cross sectional area
		Pcre		lbf	Critical load (Euler formula)
		Pcrj	192665.7	lbf	Critical load (Johnson formula)
		Pcrs		lbf	Critical load (Secant formula)
		Pse		lbf	Safety load (Euler formula)
		Psj	192665.7	lbf	Safety load (Johnson formula)
		Pss		lbf	Safety load (Secant formula)
		Le		in	Equivalent length of column
		rho		in	Radius of gyration
		I	29.333056	in^4	Second moment of inertia
		a		in	Square cross section dimension
		M		lb-in	
L		P		lbf	
L		eps			
		b			

(4) Input: F9 OK

PLOT 6.1 - CRITICAL STRESS VS. SLENDERNESS

PROBLEMS

6–1)   For the column cross section shown in the following diagram, determine
        the critical load (P_{cr}) if $E = 30 \times 10^6$ psi, $L_e = 80$" and $S_{yp} = 30,000$ psi for:

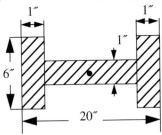

a.   e = 0

b.   e = 6". (Assume the effective length of the column is 70 inches.)
        Solve the above problem using hand calculations, and then use
        the lead model provided to check your answers.

6–2)   For a column with the square cross section shown in the following
        diagram, use lead model *ex6-1.tk* to obtain:

a.   Secant curves with eccentricity ratios of 0.1, 0.3, 0.6, and 1

b.   The Euler curve

c.   The Johnson curve

        if $E = 10 \times 10^6$psi and $\sigma_{yp} = 20,000$ psi.

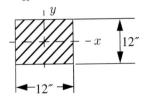

Use slenderness ratios from 0 to 250 for your plots. Now, using those plots,

d.   Determine P_{cr} if $L_e/\rho = 100$. Find P_{safe} if $SF = 2.5$. Also, find the
        effective length of the column.

e.   Determine what value of P_{safe} is possible if $L_e = 10$ ft and SF =
        2.0.

f.   What value of P_{safe} is possible if $ec/\rho = 1$, and SF = 1.6.

6–3)   For the column shown in the following diagram, determine:

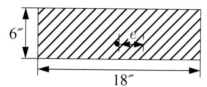

a.   The critical load (P_{cr}) if $E = 30 \times 10^6$ psi, $L_e = 10',0"$ and $e = 0$

b.   The critical load if $e = 6"$, (check both axes)

c.   The safe load for parts (a) and (b)if the safety factor = 1.56.

6–4)   The column illustrated in the following figure has the cross section as shown.

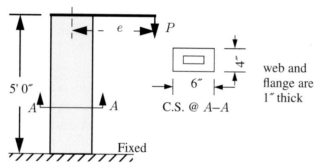

If $E = 10 \times 10^6$ psi and $S_{yp} = 20$ ksi"

a.   Determine the critical load if $e = 10"$

b.   Determine the critical load if $e = 0$.

6–5)   Consider the following cross section:

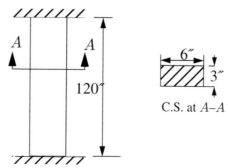

a.   Find the safe load if the safety factor is 1.6 and $e = 0$.

b.   Find the safe load if the safety factor is 2 and $\dfrac{ec}{\rho^2} = 0.6$. What is the value of e?

c.   What must be the length of the column so that either the Euler or Johnson equation can be used, assuming the end conditions shown in the figure and $e = 0$?

d.   What is the value of L/ρ (maximum) if $P_{cr}/A = 30{,}000$ psi and $e = 0$? Note that L is the effective length of the column.

6–6)   An SAE-formula car has a solid round tie rod 15" long and 1/2" in diameter, and may be considered to have pinned ends. Determine the compressive load required to buckle the rod, assuming that it is made of cold drawn steel with $S_{ut} = 80$ ksi and $S_{yp} = 54$ ksi, and that it is centrally loaded.

6–7)   For the column shown in the following diagram, (a) determine the critical load (P_{cr}) if $E = 30 \times 10^6$ psi, $L_e = 10',0"$ and $e = 0$; (b) If the eccentricity e equals 6", determine the critical load (check both the axes); (c) If the safety factor is 1.56, determine the safe load. ($S_{yp} = 30{,}000$ psi.)

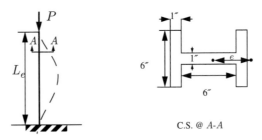

C.S. @ A-A

6–8)   For the column cross section shown in the following diagram, determine the critical load (P_{cr})

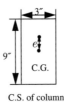

C.S. of column

a.   if S_{yp} = 30,000 psi and $E = 30 \times 10^6$ psi for $e = 0$, $L_e = 120$ inches.

b.   if $e = 6''$. (Assume that the effective length of the column is 90 inches.) For part (b) write only the necessary equation and substitute known values but do not solve for P_{cr}.

6–9)   A steel connecting rod having the diameter of $\frac{1}{2}''$ and length L_e is subjected to a maximum compressive load of 1000 lb. Determine the equivalent length L_e if either Johnson or Euler equation is applicable. (S_{yp} = 40,000 psi)

6–10)   A steel connecting rod has 12" equivalent length. Determine the diameter of the rod if the yield point stress of the rod material is 60,000 psi and modulus of elasticity is 30×10^6 psi. Assume the Euler Formula is applicable but check to see if the solution is correct. If not, then try Johnson's Equation.

Theories of Failure (Ductile and Brittle Materials)

7.1 Introduction

*T*heories of failure are mathematical predictions of the failure of a machine element subjected to a complex state of stress. Experimental evidence and past experience have shown that predicting failure, on the basis one of the theories of failure applicable to a given material is reasonably accurate, thus allowing one to design a structure without really testing it for its strength. The reader is cautioned, however, that prototype testing must be conducted in order to achieve a sound understanding of the failure possibilities of a machine component.

A structure is said to fail if it ceases to perform as intended. For example, an automobile suspension coil spring may not have broken into two pieces, but it may have an excessive permanent set, thus making it a candidate for replacement. Another type of failure could be excessive wear. In sum, failure can be due to:

    a.   Breaking

    b.   Excessive deflections

    c.   Excessive wear

    d.   Change in shape

    e.   Excessive creep

Each of these types of failures could be caused by excessive static, dynamic, or fatigue loads.

The modes of failure can be categorized as time independent or time dependent, as follows:

Time-Independent Failures

a. Failure due to yielding

b. Failure due to low stiffness

c. Failure due to excessive buckling

d. Failure due to excessive shear (shear yielding)

Time-Dependent Failures

a. Creep under constant strain or constant stress

b. Cyclic or fatigue failure

7.2 Theory of Failure

A theory of failure is used to predict whether a certain state of stress will produce plastic action in a normally ductile material or produce fracture in a normally brittle material. Theories of failure are not derivable laws. They are, rather, unifying empirical rules, usually of simple form, that provide a good fit to an experimental (uniaxial or biaxial) test results. We will first cover theories of failures for *time independent* failures. However, even before that, let us discuss when such theories are used.

7.3 How Theories of Failure Are Used

We will use subscript "$_{ME}$" to stand for a *machine element*, subscript "$_{STT}$" for a *simple tension test*, and "$_{SCT}$" for *simple compressive test*. All time-independent theories of failure require basic simple tension or compression test results, or they require knowledge of the tensile yield stress σ_{yp} and yield-point compressive stress σ_{yc} of the machine-element material. Also required are the properties called the modulus of elasticity (E) and Poisson's ratio (v). In addition, one may require a multiple number of tests to come up with statistically analyzable properties.

The next few sections present a number of theories that can be used to predict failures in machine elements.

7.4 Ranklin's Theory (Maximum Principal Stress Theory)

Ranklin's theory predicts that if the magnitude of the maximum principal stress σ_1 or $\sigma_3)_{ME}$ is equal to the yield-point stress $\sigma_{yp})_{STT}$ or $\sigma_{yp})_{SCT}$ of the same material, then the failure is impending; that is,

$$(\sigma_1)_{ME} = (\sigma_{yp})_{STT} \text{ or}$$

$$\left|\sigma_3\right|_{ME} = \left|\sigma_{yp}\right|_{SCT} \text{ if } [\sigma_3]_{ME} \text{ is compression} \qquad (7\text{–}1)$$

where $\left.\sigma_{yp}\right|_{STT}$ — yield-point stress in tension test

$\left.\sigma_{yp}\right|_{SCT}$ — yield-point stress in compression test

It is difficult to pinpoint the yield-point stress during a tension or compression test. Therefore, a 0.2% offset method to obtain the yield-point stress is used. Yield-point stress values can be obtained from a book on the properties of materials.

7.5 Coulomb Theory (Maximum-Shear-Stress Theory)

The maximum-shear-stress theory is based on the assumption that the maximum shear stress that existed at the time of failure in a simple tension test can be used as a failure criterion. The maximum shear stress in a simple tension test has the value $\sigma_{yp}/2$ for a material that has reached the yield point. (To see this, draw Mohr's circle for an element obtained from a simple tension test specimen.) Such a maximum shear stress is compared with the maximum shear stress in a machine element. In other words, the maximum-shear-stress theory states that if $\tau_{max)ME}$ is equal to $\tau_{yp)STT}$, then the failure is impending; that is,

$$(\tau_{max})_{ME} = (\tau_{yp})_{STT} = \left(\frac{\sigma_{yp}}{2}\right)_{STT}$$

but τ_{max} in a machine element subjected to 3-D stress is equal to $(\sigma_1-\sigma_3)/2$ if

$$\sigma_1 > \sigma_2 > \sigma_3 \quad \therefore \left(\frac{\sigma_1 - \sigma_3}{2}\right)_{ME} = \left(\frac{\sigma_{yp}}{2}\right)_{STT} \quad \text{or}$$

$$(\sigma_1 - \sigma_3)_{ME} = (\sigma_{yp})_{STT} \qquad (7\text{–}2)$$

7.6 St. Venant Theory (Maximum-Principal-Strain Theory)

The maximum-principal-strain theory is based on the assumption that the failure of a machine element is impending when the maximum normal strain in the element is equal to the maximum normal strain that existed during failure of the tension or compression test. Thus, the maximum normal strain in a simple tension test can be used as a failure criterion. The failure in a machine element is impending if

$$(\varepsilon_1)_{ME} = (\varepsilon_{yp})_{STT}$$

But

$$(\varepsilon_1)_{ME} = \frac{\sigma_1}{E} - \frac{v}{E}(\sigma_2 + \sigma_3) \quad \text{(Note that } \sigma_1 > \sigma_2 > \sigma_3\text{)}$$

and

$$(\varepsilon_{yp})_{STTorSCT} = \left(\frac{\sigma_{yp}}{E}\right)_{STT \text{ or } SCT}$$

$$\therefore \sigma_1 - \nu(\sigma_2 + \sigma_3) = \sigma'_{yp} \tag{7-3}$$

or

$$\left|\sigma_3 - \nu(\sigma_1 + \sigma_2)\right| = \left|\sigma''_{yp}\right| \tag{7-4}$$

where σ'_{yp} is the yield stress in tension for a specimen undergoing simple tension and σ''_{yp} is the yield stress in compression for a specimen undergoing simple compression.

7.7 Energy Theories

Material engineers' initial thoughts on failure in a machine element were based on the assumption that a machine element's failure is impending if the energy per unit volume due to applied principal stresses reaches the energy per unit volume that existed in the simple tension or compression test at the time of yielding; that is,

$$\left(\frac{\sigma_1\varepsilon_1}{2} + \frac{\sigma_2\varepsilon_2}{2} + \frac{\sigma_3\varepsilon_3}{2}\right)_{ME} = \left(\frac{\sigma_{yp}\varepsilon_{yp}}{2}\right)_{STT}$$

where σ_1, σ_2, and σ_3 are the principal stresses at a critical location in the element. σ_{yp} is the yield point stress in a simple tension test. In such a test,

$$\varepsilon_{yp} = \frac{\sigma_{yp}}{E}$$

and the machine-element strains are

$$\varepsilon_1 = \frac{\sigma_1}{E} - \frac{\nu}{E}(\sigma_2 + \sigma_3)$$

$$\varepsilon_2 = \frac{\sigma_2}{E} - \frac{\nu}{E}(\sigma_1 + \sigma_3)$$

$$\varepsilon_3 = \frac{\sigma_3}{E} - \frac{\nu}{E}(\sigma_1 + \sigma_2)$$

Substituting these values into the first equation of this section and rearranging, one gets

$$\left(\frac{\sigma_{yp}^2}{2E}\right)_{STT} = \frac{1}{2E}(\sigma_1^2 + \sigma_2^2 + \sigma_3^2 - 2\nu(\sigma_1\sigma_2 + \sigma_2\sigma_3 + \sigma_3\sigma_1))_{ME} \tag{7-5}$$

Solving for σ_{yp}, we obtain

$$\sigma_{yp} = (\sigma_1^2 + \sigma_2^2 + \sigma_3^2 - 2v(\sigma_1\sigma_2 + \sigma_2\sigma_3 + \sigma_3\sigma_1))^{1/2}$$

However, this theory is no longer favored, because when a machine element undergoes stresses in such a way that $\sigma_1 = \sigma_2 = \sigma_3$ (i.e., the element is in hydrostatic tension or compression), the element seems to store a lot more energy before failure, compared with the unit energy stored in a simple tension or compression test. The hydrostatic effect (which is not a cause of failure) can be explained using a rock as an example. If a rock is thrown into an ocean and settles at the bottom, which is, say, 1 mile down, then it is subjected to very high compressive stresses σ_1, σ_2, and σ_3 (where $\sigma_1 = \sigma_2 = \sigma_3$) due to the weight of water above it. Naturally the rock has stored a large amount of energy. But it neither breaks nor fails in any other way. This phenomenon suggests that volumetric distortion does not create failure; therefore, it should be eliminated from the stress matrix and only that part of the stress matrix which causes angular distortion should be considered. Therefore, it was postulated that yielding of a machine element is not a simple hydrostatic tension or compressive phenomenon, but is somehow related to stresses other than hydrostatic tension or compression. Such stresses are called *distortion* or *deviation* stresses, as they cause an angular distortion or deviation in the element. This is done by using the following approach. If the hydrostatic stress is considered to be the average of the three principal stresses, i.e.,

$$\sigma_{av} = \frac{\sigma_1 + \sigma_2 + \sigma_3}{3}$$

then, for a critical element in a machine subjected to σ_1, σ_2, and σ_3 the stress state can be resolved into a stress state that causes angular distortion and a stress state that causes volumetric distortion, as shown in Fig. 7–1.

The left and right side, i.e. the matrix equation for such an approach is

$$\begin{bmatrix} \sigma_1 & 0 & 0 \\ 0 & \sigma_2 & 0 \\ 0 & 0 & \sigma_3 \end{bmatrix} = \begin{bmatrix} \sigma_1 - \sigma_{av} & 0 & 0 \\ 0 & \sigma_2 - \sigma_{av} & 0 \\ 0 & 0 & \sigma_3 - \sigma_{av} \end{bmatrix} + \begin{bmatrix} \sigma_{av} & 0 & 0 \\ 0 & \sigma_{av} & 0 \\ 0 & 0 & \sigma_{av} \end{bmatrix}$$

The reason for separating elements on the right side of the equal sign is to come up with a simple method of determining the energy stored in element 7-b, which is really the distortion energy. The energy stored in element 7-a is the sum of the energies stored in elements 7-b and 7-c, that is,

$$u_T = u_d + u_v$$

where u_T is the total strain energy for a unit volume in element 7-a in Fig. 7–1, u_d is the total strain energy for a unit volume (distortion) in element 7-b, and u_v is the hydrostatic strain energy for a unit volume (volumetric) in element 7-c.

Note: $\sigma_{av} = \dfrac{(\sigma_1 + \sigma_2 + \sigma_3)}{3})$

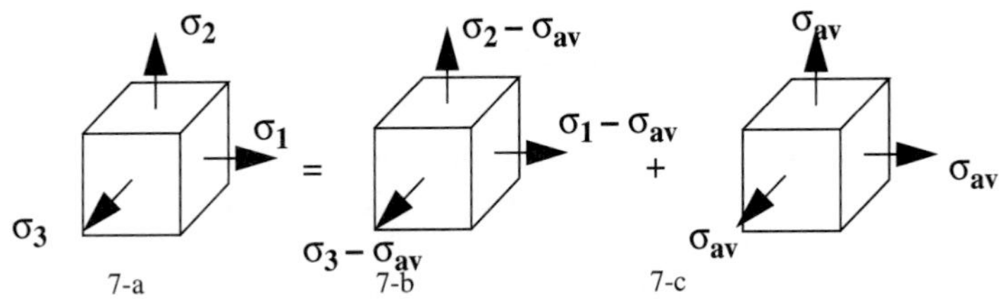

Figure 7-1   Element 7-a on the left side of the equal sign is equivalent to the two elements, 7-b and 7-c, shown on the right side.

$\therefore u_d = u_T - u_v$

One also knows that u_T is given by (see Equation (7–5))

$$u_T = \frac{1}{2E}[\sigma_1^2 + \sigma_2^2 + \sigma_3^2 - [2v[\sigma_1\sigma_2 + \sigma_2\sigma_3 + \sigma_3\sigma_1]]] \tag{7–6}$$

One can easily find u_v using equation Equation (7–6) but substituting $\sigma_1 = \sigma_2 = \sigma_3 = \sigma_{av}$; therefore,

$$u_v = \frac{1}{2E}[\sigma_{av}^2 + \sigma_{av}^2 + \sigma_{av}^2 - 2v[\sigma_{av}\sigma_{av} + \sigma_{av}\sigma_{av} + \sigma_{av}\sigma_{av}]]$$

$$u_v = \frac{1}{2E}[3\sigma_{av}^2 - 2v[3\sigma_{av}^2]]$$

$$u_v = \left[\frac{3\sigma_{av}^2}{2E}(1-2v)\right] \tag{7–7}$$

But $\sigma_{av} = \dfrac{\sigma_1 + \sigma_2 + \sigma_3}{3}$

Substituting for σ_{av} in Equation (7–7), we get

$$u_v = \left[\frac{3}{2E}\left(\frac{\sigma_1 + \sigma_2 + \sigma_3}{3}\right)^2(1-2v)\right]$$

On simplification, we obtain

$$u_v = \frac{1-2v}{6E}[\sigma_1^2 + \sigma_2^2 + \sigma_3^2 + 2\sigma_1\sigma_2 + 2\sigma_2\sigma_3 + 2\sigma_3\sigma_1] \tag{7–8}$$

Now the distortion energy can be obtained as

$$u_d = u_T - u_v$$

Substituting values for u_T from Equation (7–6), and for u_v from Equation (7–8), and then simplifying, we have

$$u_d = \frac{1+v}{3E}\left[\frac{(\sigma_1-\sigma_2)^2+(\sigma_2-\sigma_3)^2+(\sigma_3-\sigma_1)^2}{2}\right] \qquad (7\text{–}9)$$

It is now easy to understand why the rock one mile below the ocean surface does not break: the stone is subjected to hydrostatic pressure $\sigma_1 = \sigma_2 = \sigma_3 = p$ and the distortion energy in such a case is

$$u_d = 0$$

Therefore, one can postulate that the rock does not fail.

The distortion energy stored in a simple tension specimen at the time of yielding is

$$u_d)_{STT} = \frac{1+v}{3E}[\sigma_{yp}^2] \qquad (7\text{–}10)$$

since $\sigma_1 = \sigma_{yp} \qquad \sigma_2 = \sigma_3 = 0$

The distortion energy theory states that if the distortion energy in a machine element is equal to the distortion energy in simple tension at yield stress σ_{yp}, then failure is impending. Equivalently, if

$$u_d)_{ME} = u_d)_{STT}$$

then failure is impending. Substituting Equation (7–9) for a machine element energy and Equation (7–10) for simple tension yields

$$\frac{1+v}{3E}\left[\frac{(\sigma_1-\sigma_2)^2+(\sigma_2-\sigma_3)^2+(\sigma_3-\sigma_1)^2}{2}\right] = \left(\frac{1+v}{3E}(\sigma_{yp}^2)\right) \qquad (7\text{–}11)$$

On simplification, we obtain

$$\left[\frac{(\sigma_1-\sigma_2)^2+(\sigma_2-\sigma_3)^2+(\sigma_3-\sigma_1)^2}{2}\right] = \sigma_{yp}^2 \qquad (7\text{–}12)$$

Von Mises modified the foregoing equation by taking the square root of both sides, that is,

$$\left[\frac{(\sigma_1-\sigma_2)^2+(\sigma_2-\sigma_3)^2+(\sigma_3-\sigma_1)^2}{2}\right]_{ME}^{\frac{1}{2}} = (\sigma_{yp})_{STT} \qquad (7\text{–}13)$$

This theory is therefore called the *Von Mises theory*. It states that failure is impending if the yield stress σ_{yp} is equal to the machine-element equivalent stress, defined as

$$\sigma_e = \left[\frac{(\sigma_1 - \sigma_2)^2 + (\sigma_2 - \sigma_3)^2 + (\sigma_3 - \sigma_1)^2}{2}\right]^{\frac{1}{2}}$$

σ_e is also called the *Von Mises* stress. It follows that

$$(\sigma_e)_{ME} = (\sigma_{yp})_{STT}$$

for a failure to be impending.

The distortion energy theory is also called the *Shear-Energy* or *Von Mises–Hencky theory* or the *octahedral-shear-stress theory*. The theory is for ductile materials and predicts only the beginning of yielding.

Since the distortion energy theory represents angular distortion of the stressed element and agrees well with experimental evidence, it is often used by engineers. The mathematical manipulations involved in the development of the theory should not obscure the real value and usefulness of the results. One has to come up with principal stresses σ_1, σ_2, and σ_3 from a given or known $[\sigma_{ij}]$ matrix. This is easy as one already has the stress 3.tk lead model, which can solve for σ_1, σ_2, and σ_3 if the stress matrix $[\sigma_{ij}]$ is known.

In sum, for ductile materials, the maximum-shear-stress theory and distortion energy theories are preferred in design, as these theories correlate well with experimental results. The theories are applicable to materials for which $(\sigma_{yp})_{compression}$ equals $(\sigma_{yp})_{tension}$.

The reader is cautioned to be aware of failure of ductile materials subjected to very low temperatures, because these materials tend to become brittle and weak in tension as the temperature goes below the so-called transition temperature. Such brittle materials can fail at a much lower stress level. The yield theories presented here apply to ductile materials only and not to brittle materials.

Some examples will illustrate the application of the preceding theories.

Example 7-1

A pressure cylinder is subjected to the state of stress $[\sigma_{ij}]$ given by

$$[\sigma_{ij}] = \begin{bmatrix} 10000 & 5000 & 0 \\ 5000 & 20000 & 0 \\ 0 & 0 & -10000 \end{bmatrix}$$

Find the safety factor, based on maximum-shear-stress theory, normal stress theory, and distortion energy theory if the yield-point stress of the material is 40,000 psi.

Solution

Using *stress3.tk*, we can determine σ_1, σ_2, and σ_3:

$$\sigma_1 = 22{,}071.07 \text{ psi}, \ \sigma_2 = 7928.93 \text{ psi}, \text{ and } \sigma_3 = -10{,}000 \text{ psi}$$

Similarly,

$$\tau_{max} = \frac{\sigma_1 - \sigma_3}{2} = \frac{22071 - (-10000)}{2} = 16035.5 \text{psi}$$

and the Von Mises stress is

$$\sigma_e = \left[\frac{(\sigma_1 - \sigma_2)^2 + (\sigma_2 - \sigma_3)^2 + (\sigma_3 - \sigma_1)^2}{2} \right]^{\frac{1}{2}}$$

$$\sigma_e = \left[\frac{(22071 - 7929)^2 + (7929 - (-10000))^2 + ((-10000) - 22071)^2}{2} \right]^{\frac{1}{2}}$$

$$\therefore \sigma_e = 27838.76$$

Therefore,

$$SF_{ns} = \sigma_{yp}/\sigma_1$$

$$SF_{ns} = \frac{40000}{22071} = 1.812$$

$$SF_{ss} = (\sigma_{yp}/2)/\tau_{max}$$

$$SF_{ss} = \frac{20000}{16035} = 1.247$$

$$SF_{De} = \sigma_{yp}/\sigma_e$$

$$SF_{De} = \frac{40000}{27838.76} = 1.4368$$

where "ns" stands for normal stress, "ss" stands for shear stress, and "De" stands for distortion energy.

The tk model *ex7-1.tk,* or *thfail.tk,* shows the results obtained in Example 7-1 with the aid of the Rules Sheet and Variables Sheet.

RULES SHEET

```
X ⌐ tk      ▣ ⧉
                              ex7-1.tk

 File  Edit  Commands  Options  Windows  Sheets  Help

  ▣                           Rules                        ▣ ▲  ▲
S Rule---------------------------------------------------------
  " Characterestic Equations "

* Sp^3-I1*Sp^2+I2*Sp-I3=0
  I1=Sx+Sy+Sz
  I2=Sx*Sy+Sx*Sz+Sy*Sz-Txy^2-Tyz^2-Txz^2
  I3=Sx*Sy*Sz+2*Txy*Tyz*Txz-Sx*Tyz^2-Sy*Txz^2-Sz*Txy^2
  S=(1/3*R)^.5
  l=acosd(-Q/(2*T))
  R=1/3*I1^2-I2
  Q=1/3*I1*I2-I3-(2/27)*I1^3
  T=((1/27)*R^3)^.5


  S̄a=2*S*((cosd(l/3)))+1/3*I1
  Sb=2*S*((cosd((l/3+120))))+1/3*I1
  Sc=2*S*(((cosd(l/3+240))))+1/3*I1


  if and (Sa>Sb, Sa>Sc) then S1=Sa
* if (Sb>Sc) then S2=Sa
* if (Sc<Sb) then S3=Sc

* if and (Sb>Sa, Sb>Sc) then S1=Sb
* if (Sc>Sa) then S2=Sc
* if (Sa<Sc) then S3=Sa

* if and (Sc>Sb, Sc>Sa) then S1=Sc
* if (Sb>Sa) then S2=Sb
* if (Sa<Sb) then S3=Sa

  "Theories of failures
  S1-Syp/SFns=err3
  (S1-S3)-Syp/SFss=err2
  Syp= (((((S1-S2)^2)+((S2-S3)^2)+((S3-S1)^2))/2)^.5)*SFen

 ◁                                                        ▷ ▽ ▽
(14) Rule:                                               F9 OK
```

VARIABLES SHEET

```
 X  tk      
                                ex7-1.tk
 File  Edit  Commands  Options  Windows  Sheets  Help
┌─────────────────────────── Variables ───────────────────────┐
St Input---- Name--- Output--- Unit----- Comment------------------------------
             Sp                          Variable in characteristic equation

   10000     Sx                 psi      Normal stress along x-axis
   20000     Sy                 psi      Normal stress along y-axis
   -10000    Sz                 psi      1
   5000      Txy                psi      Shear stress on plane x,dir-y
   0         Tyz                psi      Shear stress on plane y,dir-z
   0         Txz                psi      Shear stress on plane x,dir-z

             I1      137895146           Constant for given sress matrix
             I2      -5.942E15           Constant for given sress matrix
             I3      -5.736E23           Constant for given sress matrix
             S       63980640            Constant for given sress matrix
             l       101.69958           Constant for given sress matrix
             R       1.2281E16           Constant for given sress matrix
             Q       1.0622E23           Constant for given sress matrix
             T       2.6191E23           Constant for given sress matrix

             Sa      22071.068 psi       Principle stress
             Sb      -10000    psi       Principle stress
             Sc      7928.9322 psi       Principle stress

             S1      22071.068 psi       Principle stress
             S2      7928.9322 psi       Principle stress
             S3      -10000    psi       Principle stress


   40000     Syp               psi       Yield point stress
             SFns    1.8123274           Safety factor normal stress theory
             SFss    1.2472301           Safety factor shear stress theory
             SFen    1.4368424           Safety factor equivilent stress theory
L            Sig_r             psi       Radial stress
L            Sig_l             psi       Longitudinal stress
L            Sig_t             psi       Tangential stress
             A                           Constant for given pressures and radi
             B                           Constant for given pressures and radi
└──────────────────────────────────────────────────────────────┘
 (46) Unit: lb-in                                              F9 OK
```

7.8 Theories of Failure for Brittle Materials

Mohr's Theory of Failure

If a material has a tension yield strength different from its compression yield strength, then none of the preceding theories can be used. Brittle materials do not have just such a property: therefore a different approach must be used to predict failure in brittle materials. Mohr established a criterion for these materials. The criterion is based upon the drawing of three Mohr circles for three different tests: simple tension, simple compression, and simple shear. Such test data will result in a set of Mohr circles, as shown in Fig. 7–2(a). (Note that the tension circle is smaller than the compression circle, indicating that the specimen could be a brittle material, as these materials, by definition, have less tension strength than compression strength.) Now if one draws the tangents ABC and DEF to these circles, then such tangents are called the Mohr envelope. Mohr's criterion states that if any machine element has a critical biaxial principal state of stress that falls within such an envelope, then failure is not impending. If the state of stress is on the border, failure is impending, and if it is outside of the envelope, failure has taken place.

Mohr's theory is also applicable to ductile materials with significantly different compressive strength compared to tension strength.

Modified Mohr Theory

A slightly different concept due to Mohr can be used for a machine element having stresses σ_1, σ_2, and σ_3 at some critical point. This so-called *modified Mohr envelope* is plotted as follows, with reference to Fig. 7–2(b).

STEP 1   Draw σ_1, σ_2 axes.

STEP 2   Plot point A with coordinates $(\sigma'_{yp}, 0)$.

Plot point B with coordinates $(0, \sigma'_{yp})$.

Plot point C with coordinates $(\sigma'_{yp}, -\sigma'_{yp})$.

Plot point D with coordinates $(0, -\sigma''_{yp})$.

Plot point E with coordinates $(-\sigma''_{yp}, 0)$.

Plot point F with coordinates $(\sigma'_{yp}, \sigma'_{yp})$.

Plot point G with coordinates $(-\sigma''_{yp}, \sigma'_{yp})$.

Plot point H with coordinates $(-\sigma''_{yp}, -\sigma''_{yp})$

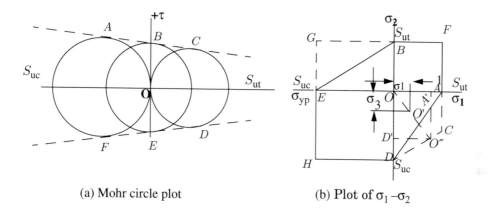

(a) Mohr circle plot (b) Plot of $\sigma_1 - \sigma_2$

Figure 7-2 Mohr envelope for a material with different tension and compression yield-point values

STEP 3 Join points A, F, B, G, E, H, D and C.

STEP 4 If the machine element has a state of stress with principal stresses σ_1, σ_2, and σ_3 (where $\sigma_1 > \sigma_2 > \sigma_3$), then plot point 0' with coordinates σ_1, σ_3 (assuming that σ_3 is compressive stress).

STEP 5 Then find the coordinates of points where the σ_1, σ_3 line originated at (0,0), extending the line from its origin and passing through 0' until it hits point 0″.

STEP 6 The safety factor is determined by taking the ratio of either $0D'/\sigma_3$ or $0A'/\sigma_1$; that is, SF $= \dfrac{0D'}{|\sigma_3|} = \dfrac{0A'}{\sigma_1}$

Measurements or calculations requiring the magnitude of $0D'$ or $0A'$ are done graphically or using laws of similar triangles. The following example illustrates such a procedure.

Example 7-2

If the material of a machine has a yield-point tension strength of 100 ksi and a yield-point compression strength of 200 ksi, determine the safety factor if the machine element has $\sigma_1 = 55$ ksi and $\sigma_3 = -110$ ksi.

Solution

Knowing the tensile yield strength and the compression yield strength, we can plot the modified Mohr values in the fourth quadrant, since the state of stress is in that quadrant of Fig. 7–2(b). The point 0' has coordinates (55, –110), as shown in the following diagram (not to scale):

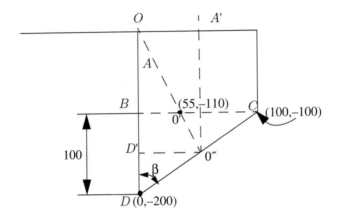

We see that $\tan\beta = \dfrac{BC}{BD} = \dfrac{100}{100} = 1$ but $\tan\beta = \dfrac{0''D'}{D'D}$ shows that $0''D' = D'D = 200 - 0D'$. Now,

$$\frac{0''D'}{0D'} = \frac{55}{110} = 0.5$$

Therefore

$$\frac{200 - 0D'}{0D'} = 0.5$$

Therefore

$$1.50D' = 200$$

$$0D' = 200/1.5$$

$$= 133.33$$

Therefore

$$SF = 0D'/0B = \frac{133.33}{100} = 1.33$$

PROBLEMS

7–1)   A machine element made of ductile steel is loaded so that the principal stresses are $\sigma_1 = 40$ ksi, $\sigma_2 = 0$, and $\sigma_3 = -20$ ksi. The material has a yield strength of 80 ksi in both tension and compression. Find the safety factor for each of the following failure theories:

   a.   maximum normal stress theory

   b.   maximum shear stress theory

   c.   distortion energy theory

7–2)   A pressure cylinder with an internal radius of 6″ and a thickness of 1″ is subjected to an internal pressure of 5000 psi, a twisting torque of 50,000 in-lb, and a bending moment of 50,000 in-lb as shown in the following diagram:

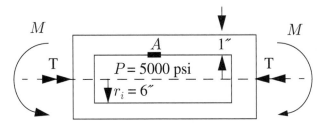

If the transverse stress due to internal pressure is $\sigma_t = 32{,}692$ psi, the longitudinal stress due to internal pressure is $\sigma_l = 13{,}846$ psi, and the radial stress due to internal pressure is $\sigma_r = 5000$ psi at $r = r_i = 6″$ (at an element near A), then:

   a.   Determine the bending and torsional stress at $r = 6″$.

   b.   Superimpose the preceding stresses on pressure stresses, and determine, from such stresses, the principal stresses σ_1, σ_2, and σ_3 and the maximum shear stress.

   c.   Determine the safety factor using normal stress theory, maximum-shear-stress theory, and distortion energy theory if $\sigma_{yp} = 40{,}000$ psi.

7–3)   Assume that the cylinder of Problem 7–2 has a 7.2″ internal radius, and a yield-point stress in tension of 40,000 psi, and a yield-point stress in compression of 80,000 psi. Using modified Mohr theory, determine the safety factor if the principal stresses due to M, T, and the pressure P are $\sigma_1 = 33{,}000$ psi and $\sigma_2 = -50{,}000$ psi as resultant stresses.

7–4)   Use the modified Mohr theory to determine the safety factor in a machine subjected to $\sigma_1 = 30$ ksi, $\sigma_2 = 0$, and $\sigma_3 = -80$ ksi and if ASTM class-40 gray cast iron with a tensile strength of 42.5 ksi and a compressive strength of 140 ksi is used as a material for the machine element.

Statistical Analysis

8.1 Introduction

Statistics and probability have become working tools of engineers. Engineering design requires one to use data that have low precision; however, the proper application of statistical analysis can assist greatly with design. This chapter deals with statistical tools required for designing experiments. Such experiments are required in understanding the behavior of machine elements. When a large number of observations are made from a random sample of a machine element, a method is needed to arrange these observations into a number of equally valued class intervals and to determine the frequency of observations falling within each interval. Thus, in the practice of machine design or fatigue analysis, one is concerned with frequency of occurrence of certain events, for example, failure versus life of an element. One can sufficiently increase the number of observations to obtain a frequency curve that represents the frequency distribution of the sample and that can be shown as a density function. A density function can be defined as a function which yields the probability that the random variable takes on any one of the admissible values.

8.2 Experimental Testing

Designing machine elements on the basis of theory alone is not enough, especially for critical machine components. Some kind of field, laboratory, or prototype testing is required. Naturally, a good test is one that provides the required information with the minimum amount of time and effort. Anything else means wasting someone's money.

To achieve a good experimental test, tests must be designed and conducted according to a certain procedure. Specifically, the following steps must be taken:

STEP 1   During design, if in doubt, redesign it. If still in doubt decide whether a test is required.

STEP 2   Define the scope of the test:

    a.   Is the test destructive or nondestructive?

    b.   Will testing be carried out to completion or will tests just be run to failure of the element?

    c.   Should testing duplicate operating conditions?

    d.   Should testing be accelerated?

    e.   Will the tests sample representative parts?

STEP 3   Design the test so that statistically significant information is obtained.

STEP 4   The test must be related to a real problem and not to a problem that you *think* exists.

STEP 5   All the test data should be properly systemized and analyzed. With current statistical tools, we can now do this job more effectively than in the past.

(*Note*: If the test data are bad, even the best statistician will not be able to do much with them.)

8.3 Statistical Tools

Statistical tools allow us to evaluate and present experimental results in a fashion that will make sense to the user. They also allow us to improve upon certain parameters that enhance the performance of the machine elements being tested.

Let us consider a couple of examples in order to understand test results. Suppose we ran a test on bearings for their life to failure from a given population. The results are tabulated as follows:

Sample #	Number of bearings tested	Average life to failure (10^6 cycles)
1	10	2.9
2	10	8.1
3	10	0.7
4	10	0.9
5	10	10.0
6	10	4.5

If plotted on a graph paper, these results exhibit considerable scatter. The question, then, is which average is the average of the population? We don't have the answer yet, but we will soon find ways to analyze the data.

Another example of testing is the following (each test involved 10 gears):

Number of gears tested	Number of hours to failure	
	Gear design A	Gear design B
10	120	110
10	150	160
10	210	350
10	250	400
10	260	410
10	270	430
10	510	500
10	870	520
10	980	530
10	1140	570
Average	476	398

Gear design A appears to be better than gear design B if one uses only common sense and not mathematical reasoning. However, this may not in fact be so. As we shall see, the average strength, life, etc., is not necessarily a good indication (index) of the quality of the two sets of data, and therefore, to analyze the data, we need to have a systematic (statistical) approach to arrive at wise conclusions.

8.4 The Concept of Random Variables

Because the samples tested are always selected randomly, values of the average life obtained from testing can be considered random. Random variables can be classified into two categories: continuous and discrete. A random variable associated with a bearing is a continuous random variable, since it can assume any value from 0 to ∞.

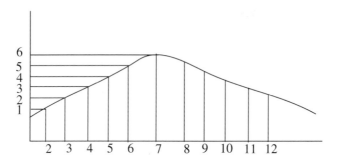

Figure 8–1   Sum of two dice vs. number of ways of reaching that sum

To understand a discrete random variable, consider a pair of dice. When the dice are rolled, there are 36 possible combinations, which can be expressed in matrix form as

$$
\begin{bmatrix}
1,1 & 1,2 & 1,3 & 1,4 & 1,5 & 1,6 \\
2,1 & 2,2 & 2,3 & 2,4 & 2,5 & 2,6 \\
3,1 & 3,2 & 3,3 & 3,4 & 3,5 & 3,6 \\
4,1 & 4,2 & 4,3 & 4,4 & 4,5 & 4,6 \\
5,1 & 5,2 & 5,3 & 5,4 & 5,5 & 5,6 \\
6,1 & 6,2 & 6,3 & 6,4 & 6,5 & 6,6
\end{bmatrix}
$$

Dice have discrete values. Thus, a random variable associated with dice is a discrete random variable, since it can attain only one of the six discrete values, 1, 2, 3, 4, 5, and 6.

8.5 Probability of Occurrence

Fig. 8–1, based on the preceding matrix, shows that, out of 36 possible combinations on each roll of the dice,

| 1 + 1 | = 2 happens only once

the sum 3 may happen in two ways

the sum 4 may happen in three ways

the sum 5 may happen in four ways

the sum 6 may happen in five ways

the sum 7 may happen in six ways

the sum 8 may happen in five ways

the sum 9 may happen in four ways

the sum 10 may happen in three ways

the sum 11 may happen in two ways

the sum 12 may happen in one way

Therefore, the probability of occurrence $= \dfrac{\text{\# of successful occurrences}}{\text{total \# of trials}}$, or, to put it another way, is equal to the number of ways of reaching a given combination divided by the total number of combinations. If we let x be the given combination or sum and $P(x)$ the probability of its occurrence, then the following results are obtained:

x	2	3	4	5	6	7	8	9	10	11	12
$P(x)$	$\dfrac{1}{36}$	$\dfrac{2}{36}$	$\dfrac{3}{36}$	$\dfrac{4}{36}$	$\dfrac{5}{36}$	$\dfrac{6}{36}$	$\dfrac{5}{36}$	$\dfrac{4}{36}$	$\dfrac{3}{36}$	$\dfrac{2}{36}$	$\dfrac{1}{36}$

A plot of these values is shown in Fig. 8–2. We see that the probability of occurrence of an event is a positive number (between 0 and 1).

8.6 Cumulative Probability Distribution Function (CPDF)

Since by definition CPDF can be expressed as $F(x)$, i.e.,

$$F(x) = \sum_i P(x_i)$$

then the foregoing table can be modified to obtain a plot of $F(xi)$ vs. x using the following values:

x	2	3	4	5	6	7	8	9	10	11	12
$F(x)$	$\dfrac{1}{36}$	$\dfrac{3}{36}$	$\dfrac{6}{36}$	$\dfrac{10}{36}$	$\dfrac{15}{36}$	$\dfrac{21}{36}$	$\dfrac{26}{36}$	$\dfrac{30}{36}$	$\dfrac{33}{36}$	$\dfrac{35}{36}$	$\dfrac{36}{36}$

The plot will be as shown in Fig. 8–3.

To understand the density function better, we define the following three measures of central tendency.

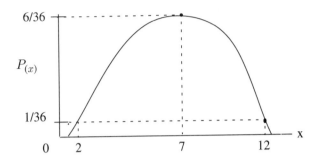

Figure 8–2 Frequency Distribution

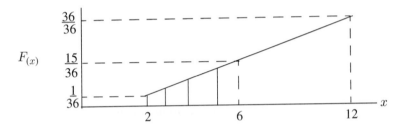

Figure 8–3 Cumulative distribution function

The Sample Mean ($\bar{x}$)

The *sample mean* or *arithmetic mean* is defined as

$$\bar{x} = \frac{\left(\displaystyle\sum_{i=1}^{n} x_i \right)}{n}$$

where x_i is the value of sample i and n is the number of occurrences.

The Sample Median (μ)

The *sample median* is the number in the middle when all test data are ranked in order of magnitude, as long as we have an odd number of samples. For even samples, we take the average of the two values in the middle of the data.

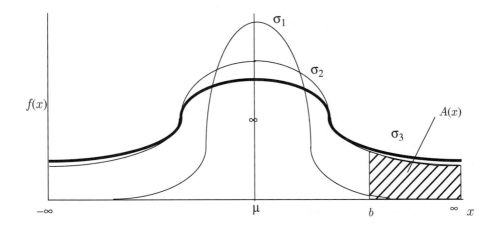

Figure 8–4   Normal distribution and effect of changing σ with μ constant

The Sample Mode

The *sample mode* is the value of the variable corresponding to the maximum probability of its occurrence. In the case of dice, the sum seven happens six times.

The Standard Deviation (σ)

To measure the variability of the sample, the standard deviation is calculated, using the formula

$$\sigma = \sqrt{\dfrac{\displaystyle\sum_{i=1}^{n} (x_i - \bar{x})^2}{n-1}}$$

where x_i is the value of sample i, $\bar{x}$ is the mean value, and n is the number of samples.

The Normal Distribution

It is common practice to analyze test results in terms of the frequency of failures and the corresponding life as shown in Fig. 8–4 for various standard deviations. After plotting the given results, we can draw a smooth curve through the data points. Many experiments will yield such a curve and, therefore, one can come up with a "ready-made" distribution called the normal distribution, discovered by Gauss, and given by an equation of the form

$$f(x) = \frac{1}{\hat{\sigma}\sqrt{2\pi}} exp(-(x-\mu)^2/(2\hat{\sigma}^2)) \text{ for the range: } -\infty \le x \le \infty$$

where $\hat{\sigma}$ is the standard deviation of the population, x is a random variable (strength, life, etc.), μ is the mean, median or mode of the population, $f(x)$ is the Gaussian frequency function, and the total area under the curve runs from $x = -\infty$ to $\infty = 1$.

Now consider the problem of determining the probabilities of occurrence of a certain event. For example, suppose the normal distribution is to be plotted as frequency of occurrence vs. life (x), and the probability of x greater than b is to be found. This probability is given by the shaded area $A(x)$ in Fig. 8–4. From the figure,

$$P(x > b) = \int_b^\infty f(x)dx$$

$$P(x > b) = \int_b^\infty \frac{1}{\hat{\sigma}\sqrt{2\pi}} exp(-(x-\mu)^2/(2\hat{\sigma}^2))dx \qquad (8–1)$$

Such an integration is difficult to carry out in standard form; however, a simple transformation can help to yield the probability of occurrence of $x > b$.

Let $z = \frac{x-\mu}{\hat{\sigma}}$, then $x - \mu = \hat{\sigma}z$

Differentiating, we have $dx = \hat{\sigma}dz$. Now substitute $\frac{x-\mu}{\hat{\sigma}} = z$ and $dx = \hat{\sigma}dz$ in Equation (8–1). Then

$$P(x > b) = \int_{\left(\frac{b-\mu}{\hat{\sigma}}\right)}^\infty \frac{1}{\hat{\sigma}\sqrt{2\pi}} exp\left(\frac{-z^2}{2}\right)(\hat{\sigma}dz)$$

But $P(x > b) = P\left(z > \frac{b-\mu}{\hat{\sigma}}\right)$

$$= \int_{\left(\frac{b-\mu}{\hat{\sigma}}\right)}^\infty \frac{1}{\sqrt{2\pi}} exp\left(\frac{-z^2}{2}\right)dz$$

Let $f(z) = \frac{1}{\sqrt{2\pi}} exp\left(\frac{-z^2}{2}\right)$

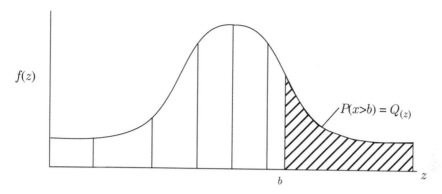

$f(z)$

$P(x>b) = Q_{(z)}$

b

z

Figure 8–5   Curve of $f(z)$ vs. z.

Then

$$P(x > b) = \int f(z)dz = Q(z)$$

where $Q(z)$ is the shaded area in Fig. 8–5 given by

$$Q(z) = f(z)(b_1 t + b_2 t^2 + b_3 t^3 + b_4 t^4 + b_5 t^5) + \varepsilon(z) \qquad (8\text{–}2)$$

where

$$|\varepsilon(z)| < 7.5 \times 10^{-8} \text{ (almost zero)}$$

and $t = \dfrac{1}{1 + (|z|)P_1}$, with $P_1 = 0.2316419$, $b_1 = 0.319381530$, $b_2 = -0.356563782$, $b_3 = 1.781477937$, $b_4 = -1.821255978$, and $b_5 = 1.330274429$.

(*Note:* $P(x > b) = P\left(z > \dfrac{b - \mu}{\hat{\sigma}}\right) = Q(z)$)

The following example will help us understand the application of probabilities of occurrence to real situations.

Example 8-1

Fifty connecting rods were tested for their strength in ksi. The mean strength of the rods is 50 ksi and the standard deviation $\hat{\sigma} = 5$ ksi. Assume that the normal distribution is applicable and answer the following questions

    a.   How many crankshafts have strength less than 40 ksi?

    b.   How many crankshafts have strength between 40 ksi and 60 ksi?

Solution

a.   With the normal distribution function assumed,

$$z_{40} = \frac{x - \mu}{\hat{\sigma}} = \frac{40 - 50}{5} = -2.0$$

Now, we can evaluate t, knowing that $z = -2$, and $P_1 = 0.2316419$. Similarly, we can evaluate $f(z)$, that is,

$$f(z) = \frac{1}{\sqrt{2\pi}} exp\left(\frac{-z_{40}^2}{2}\right) = 0.053991$$

Now, knowing $f(z)$, t, and the values of $b_1, ..., b_5$, we can calculate $Q(z)$:

$$Q(Z_{40}) = 0.02274827$$

$$\therefore P(x > 40) = 2.2\% \text{ have strength less than 40 ksi}$$

Since the shaded area on the right side is equal to that on the left side (because the standard distribution is symmetric), 2.274827% of the crankshafts have strength less than 40 ksi, i.e.,

50(2.274827%) = 1.13 (or, approximately 1) crankshaft has strength less than 40 ksi.

b.   For z_{60}, using the same procedure as in part a, we can show that $Q(z) = .02274827$.

We also know that the area under the curve of $f(z)$ vs. z is equal to 1, and therefore, the number of crankshafts having strength between 40 and 60 ksi is

$$[1 - [0.02274827 + 0.02274827]]50 = 47$$

8.7 The Weibull Distribution

Experimental data used in engineering may not follow the normal Gaussian distribution. Thus, in such cases, to assume that the experimental results follow the normal distribution will create serious errors of extrapolation or even interpolation. To deal with this problem, Weibull formulated a distribution function that follows experimental data well and yields better results than the normal distribution. The Weibull distribution is widely used in engineering practice because of its versatility. The Weibull distribution frequency function is

$$f(x) = \left\{ \frac{b}{\theta - x_o} \left(\frac{x - x_o}{\theta - x_o} \right)^{b-1} \right\} \left\{ exp\left[-\left(\frac{x - x_o}{\theta - x_o} \right)^b \right] \right\} \tag{8-3}$$

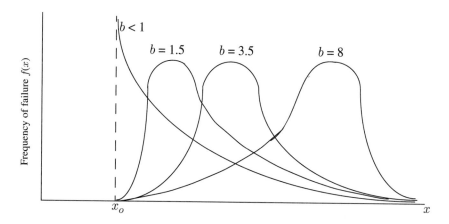

Figure 8–6 Weibull distribution for various values of b.

where x is the life of the sample, x_o is the expected minimum life of the object represented by the sample (we will see how to determine this shortly), θ is the so-called characteristics number (yet to be explained), and b is the Weibull slope (yet to be explained, and we will see why it is called a slope). x_o, θ, and b are called *Weibull parameters*. Their values depend upon the distribution and values of the experimental data.

A Weibull distribution for different values of b is shown in Fig. 8–6. The cumulative frequency function is defined as

$$F(x) = \int_{-\infty}^{\infty} f(x)dx = \int_{x_o}^{x} f(x)dx = 1 \qquad (8-4)$$

(In real engineering, of course, data do not exist for values of infinity, therefore the minimum value, x_0, is a positive number.)

We substitute Equation (8–3) for $f(x)$ into Equation (8–4) to obtain

$$F(x) = \int_{x_o}^{x} \frac{b}{\theta - x_o} \left(\frac{x - x_o}{\theta - x_o} \right)^{b-1} exp\left[-\left(\frac{x - x_o}{\theta - x_o} \right)^b \right] dx \ .$$

Let

$$\left(\frac{x - x_o}{\theta - x_o} \right)^b = y$$

Then

$$\left[b\left(\frac{x-x_o}{\theta-x_o}\right)^{b-1}\frac{1}{\theta-x_o}\right]dx = dy$$

Now substitute the values of y and dx in Equation (8–4), then

$$F(x) = \int_{y_o}^{y}e^{-y}dy$$

Integrating yields

$$F(x) = -e^{-y}\Big|_{y_o}^{y}$$

$$F(x) = -\exp-\left\{\frac{x-x_o}{\theta-x_o}\right\}^{b}\Bigg|_{x_o}^{x}$$

On expanding for $x = x$ and $x = x_o$, the following equation is obtained:

$$F(x) = 1-\exp\left\{-\left(\frac{x-x_o}{\theta-x_o}\right)^{b}\right\} \tag{8–5}$$

$F(x)$ is called the Weibull three-parameter *cumulative frequency function*. The three parameters are x_o, θ, and b. Starting with Equation (8–5), and rearranging the equation for $F(x)$, we see that, in fact, it is an equation of a straight line, viz.,

$$1-F(x) = \exp\left[-\left(\frac{x-x_o}{\theta-x_o}\right)^{b}\right]$$

Now, taking the inverse of each side of this equation, we obtain

$$\frac{1}{1-F(x)} = \exp\left[\left(\frac{x-x_o}{\theta-x_o}\right)^{b}\right]$$

Taking logarithms of left side and right side twice yields

$$\ln\left(\ln\left(\frac{1}{1-F(x)}\right)\right) = b\ln(x-x_o)-b\ln(\theta-x_o)$$

Clearly, this is an equation of a straight line, that is,

$$Y = bX \quad +c$$

where

$$Y = \ln\left(\ln\left(\frac{1}{1 - F(x)}\right)\right)$$

$$X = \ln(x - x_0)$$

$$b = \text{slope}$$

$$c = -b\ln(\theta - x_0), \quad \therefore \theta = e^{-c/b} + x_0$$

By means of regression analysis and using Cramer's rule, we can determine c, and b if there are N values of X and Y, that is,

$$b = \frac{N\Sigma X_i Y_i - \Sigma X_i \Sigma Y_i}{N\Sigma(X_i^2) - (\Sigma X_i)^2} \tag{8-6}$$

where $i = 1, ..., N$, in which N is the number of samples in a test.

Similarly, one can also determine the constant c:

$$c = \frac{\Sigma(X_i)^2 \Sigma Y_i - \Sigma X_i \Sigma X_i Y_i}{N\,\Sigma(X_i)^2 - (\Sigma X_i)^2} \tag{8-7}$$

Once c and b are determined, θ can be obtained using the equation

$$c = -b\ln(\theta - x_0) \tag{8-8}$$

or $\theta = e^{-c/b} + x_0$ (where x_0 could be a value from zero to a minimum life).

Notice that b and c can be determined using Equations (8–6) and (8–7), provided that one has sufficient data for a number of samples and provided that x_0 is also known. However, the question is how to find the value of x_0 (the value that guarantees 100% reliability, or the x_0 value at $F(x) = 0$). The approach taken is to assume x_0 values (from zero to a value a little less than the first value of failure) in some incremental fashion and find the correlation coefficient r (more on r later) for each increment of x_0. The best r value (close to the value 1) will yield the desired x_0 value. Once the x_0 value is obtained, finding b and θ is possible with the aid of Equations (8–6) and (8–7). (The equation for r is given in Example 8-2). Once the Weibull parameters are obtained, one can use a Weibull function to interpolate or extrapolate the results, provided that the Weibull distribution is a good frequency (cumulative) distribution. Always remember that extrapolation is dangerous and risky and is to be done only by an experienced engineer.

Let us consider an example of the use of the Weibull distribution function. First, we solve the stated problem in the conventional way. Then we solve the same problem using the lead model called *weibul.tk,* or *ex8-2.tk.*

Example 8-2

If the fatigue data for life to failure is given for nine bearings as 4.0, 4.3, 5.5, 6.6, 7.8, 9.9, 11.5, 12.5, and 13.5 million cycles, determine the three Weibull parameters b, θ, and x_0.

Solution

To obtain Weibull parameters, the percent of failed bearings versus the life of a bearing is needed. The percent failed is calculated using the following formula, given in any advanced text on statistical analysis:

$$y = \frac{i - 0.3}{N + 0.4} \tag{8–9}$$

For example, if $i = 1$ we are testing the first sample, then since $N = 9$ is given, the first percent failure is

$$y = \frac{i - 0.3}{9 + 0.4} = \frac{0.7}{9.4} = 0.0744681 = 7.4\,\%$$

instead of $(1/9)\,(100)\% = 11.11\%$.

The overall table used to determine the Weibull parameters—that is, b, θ and x_0, is as follows ($x_o = 0$ is assumed):

No.	x life	y $F(x)$	X $\ln(x-x_0)$ $(x_0 = 0)$	Y $\ln\ln(1/(1-F(x)))$	X^2	Y^2	XY
1	4,000,000	0.07447	15.2018	−2.5589	231.0949	6.5482	-38.9005
2	4,300,000	0.18085	15.2741	−1.6120	233.2989	2.5985	-24.6218
3	5,500,000	0.28723	15.5203	−1.0829	240.8784	1.1727	-16.8073
4	6,600,000	0.39361	15.7026	−0.69266	246.571	0.47978	-10.8766
5	7,800,000	0.5	15.8696	−0.36651	251.8453	0.13433	-5.81643
6	9,900,000	0.60638	16.1081	−0.070018	259.4691	0.00490	-1.12786
7	11,500,000	0.71276	16.2579	0.2211078	264.3179	0.04889	3.594739
8	12,500,000	0.81915	16.3412	0.536541	267.0361	0.28788	8.767745
9	13,500,000	0.92553	16.4182	0.954505	269.5573	0.91108	15.67125
	Determine all values on right:		ΣX	ΣY	ΣX^2	ΣY^2	ΣXY

Now it is easy to determine the summation values. Once these are calculated, we find b and θ using Equations (8–6), (8–7), and (8–8), provided that x_0 is known. The answers for b, θ, and c, for $x_0 = 0$ are $b = 2.345$, $\theta = 9.585 \times 10^6$, $c = -37.71044$. Notice that it was better to calculate b and c corresponding to $x_0 = 0$ because no x_0 value was available. The lead model called *ex8-2.tk* solves for these equations. Now we want to find some other value of x_0 than 0. One way to do this is to assume x_0 (i.e., from zero to the first experimental value minus say 1, or a fraction in some random increment of Δx_0) and determine the corresponding correlation. We use the x_0 value that gives the best correlation coefficient.

The preceding table can be created using TK Solver. All you have to do is enter values of the experimental data and solve for b, c, etc. using the sheet titled Table 2 on page 162.

To finish the problem in the conventional manner, derive summation values of x (i.e., ΣX) etc. using hand calculations, and place the values in the foregoing table. Then use the lead model to check the sums. Next determine b and c values if $x_0 = 0$. For $x_0 \neq 0$, follow the instructions given in the description of the lead model.

Lead Model

Now let us learn to use the lead model called *weibul.tk,* or *ex8-3.tk,* (to determine the Weibull parameters to sum the quantities shown in the preceding table, write a program using a "Procedure function" sheet. A copy of this function program is given in Table 2. Such a program is called from the Rule Sheet by inputting the value of x_0 into the Variable sheet. Since this is the first time we are inputting x_0, we use "call program," which requires that we "comment" Equation 2 and 3 in the Rule Sheet. Now, given that $x_0 = 0$, we determine b, θ, and c by pressing F9. The output will then show b, θ and c values for $x_0 = 0$: $b = 2.34$, $c = -37.7$, $\theta = 9.59 \times 10^6$ for $r = 0.97$. (Remember to clear the Table Sheet data, except for the "x" values.) Now we will do some list solving. Let x_0, for example, go from 1×10^6 to 4×10^6 in increments of 0.1×10^6 by list-filling. This is done by opening the list-fill sheet for x_0 and typing in these values in the appropriate places.

The equation for the correlation coefficient is

$$r = \frac{N\Sigma XY - \Sigma X\Sigma Y}{\sqrt{(N\Sigma X^2 - \Sigma X\Sigma X)(N\Sigma Y^2 - \Sigma Y\Sigma Y)}} \qquad (8–10)$$

Since x_0 is a set of list values, x_0 will be list input. List output will be r.

Now, upon list solving, we will get the r values. We then plot x_0 vs. r using plot 1. To do so, just go into plot 1 and click on **display** from the command menu. Next, get a print, and find at what correlating value of x_0 r is maximum. The

TABLE 1

```
┌─────────────────────────────────────────────────────────────────────┐
│ X ⊠ tk      ⊡⊡                                                        │
│                          ex8-2.tk                                     │
│  File  Edit  Commands  Options  Windows  Sheets  Help                 │
│ ┌───────────────────────────────────────────────────────────────┐    │
│ │ □              Int Table: output1                         ⊡⊡   │    │
```

Title:	Value of x, y, X, Y, X^2, Y^2, XY					
Element x-------	y-------	X-------	Y-------	Xs------	Ys------	XY------
1		.0744681	15.20179	-2.55894	231.0943	6.548178	-38.9005
2	4300000	.1808511	15.27411	-1.61199	233.2984	2.598526	-24.6218
3	5500000	.287234	15.52024	-1.08293	240.878	1.172736	-16.8073
4	6600000	.393617	15.70257	-.69266	246.5707	.4797782	-10.8765
5	7800000	.5	15.86962	-.366513	251.845	.1343317	-5.81642
6	9900000	.606383	16.10804	-.070018	259.4689	.0049025	-1.12786
7	11500000	.712766	16.25785	.2211078	264.3177	.0488887	3.594738
8	12500000	.8191489	16.34123	.536541	267.0359	.2878762	8.767741
9	13500000	.9255319	16.41819	.954505	269.5571	.9110798	15.67125

```
│ (1,1) x:                                                 │F9│OK│
```

graph may not show a good x_0 value corresponding to the maximum r. However, if one prints x_0 or values, it can be seen that r, the correlation coefficient, is maximum when $x_0 = 3.3 \times 10^6$ cycles.

To plot Weibull curves for $x_0 = 0$ and $x_0 = 3.3 \times 10^6$, we proceed as follows:

"Comment" Equation 3 in Rule Sheet

We type $x_0 = 0$ in the Variable Sheet and solve for b and c. (Note that b and c correspond to $x_0 = 0$). Next, we let $x_0 = 0$, $b_0 = b$, and $c_0 = c$, by typing b and c values into the input field of b_0 and c_0 respectively in the variables sheet also, typing L and I for the input corresponding to Y, and L and 0 for X_0. Finally, we list solve, go to plot 2, and display the plot (see Plot 2).

To obtain the third graph (Plot 3) and the Weibull curve for $x_0 = 3.3 \times 10^6$, we first type in $x_0 = 3.3 \times 10^6$ into the Variable Sheet and solve for b and c. Now we "comment" Equation 2, and "uncomment" Equation 3 in the Rule Sheet, and we give $x_0 = 3.3 \times 10^6$, $b_{op} = b$, $c_{op} = c$, Li (list input) for Y, and Lo (list output) for X_{op} (list solve).

On solving, this will give information for the second graph (plot 3) as shown. Thus, the Weibull function is now plotted separately on two graphs, one for $x_0 = 0$, and the other for $x_0 = 3.3 \times 10^6$, along with experimental data. Naturally, whenever the correlation coefficient r is positive maximum, one has a better fit to the experimental data.

TABLE 2

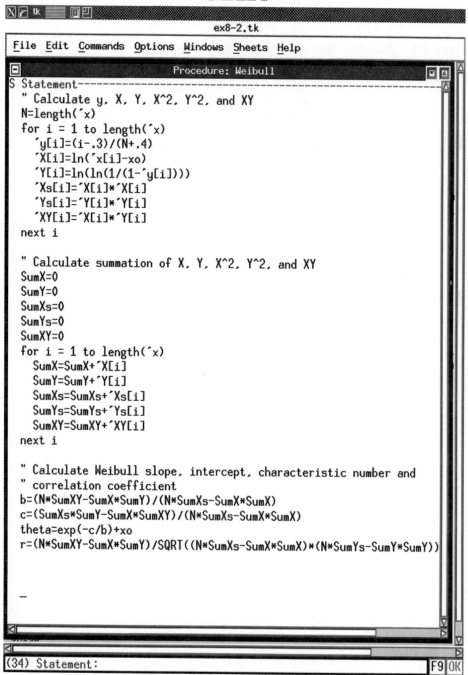

```
X r tk      
                          ex8-2.tk

 File  Edit  Commands  Options  Windows  Sheets  Help

                    Procedure: Weibull
S Statement----------------------------------------------
 " Calculate y, X, Y, X^2, Y^2, and XY
 N=length('x)
 for i = 1 to length('x)
   'y[i]=(i-.3)/(N+.4)
   'X[i]=ln('x[i]-xo)
   'Y[i]=ln(ln(1/(1-'y[i])))
   'Xs[i]='X[i]*'X[i]
   'Ys[i]='Y[i]*'Y[i]
   'XY[i]='X[i]*'Y[i]
 next i

 " Calculate summation of X, Y, X^2, Y^2, and XY
 SumX=0
 SumY=0
 SumXs=0
 SumYs=0
 SumXY=0
 for i = 1 to length('x)
   SumX=SumX+'X[i]
   SumY=SumY+'Y[i]
   SumXs=SumXs+'Xs[i]
   SumYs=SumYs+'Ys[i]
   SumXY=SumXY+'XY[i]
 next i

 " Calculate Weibull slope, intercept, characteristic number and
 " correlation coefficient
 b=(N*SumXY-SumX*SumY)/(N*SumXs-SumX*SumX)
 c=(SumXs*SumY-SumX*SumXY)/(N*SumXs-SumX*SumX)
 theta=exp(-c/b)+xo
 r=(N*SumXY-SumX*SumY)/SQRT((N*SumXs-SumX*SumX)*(N*SumYs-SumY*SumY))


 _

(34) Statement:                                          F9 OK
```

VARIABLES SHEET

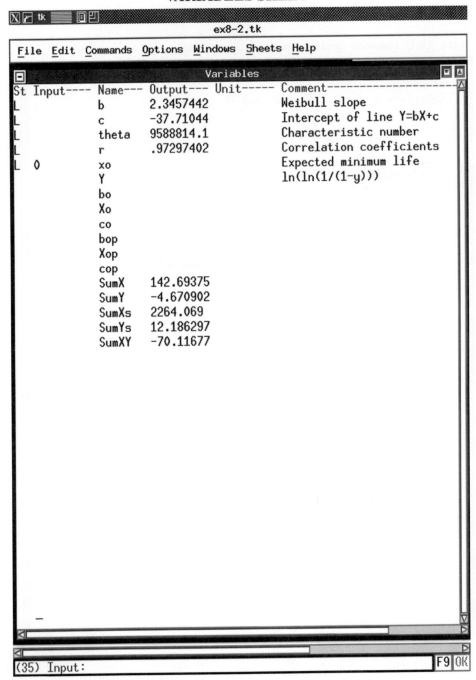

```
X [ tk      ▦  ▢▣
                              ex8-2.tk

  File  Edit  Commands  Options  Windows  Sheets  Help

 □                            Variables                              ▣ ▲
St Input---- Name--- Output--- Unit----- Comment------------------ ▲
L                b      2.3457442         Weibull slope
L                c      -37.71044         Intercept of line Y=bX+c
L              theta    9588814.1         Characteristic number
L                r      .97297402         Correlation coefficients
L      0         xo                       Expected minimum life
                 Y                        ln(ln(1/(1-y)))
                 bo
                 Xo
                 co
                 bop
                 Xop
                 cop
               SumX     142.69375
               SumY     -4.670902
               SumXs    2264.069
               SumYs    12.186297
               SumXY    -70.11677
```

```
 (35)  Input:                                                    F9 OK
```

PLOT 1 CORRELATION COEFFICIENT (r) VS. EXPECTED MINIMUM LIFE (x_o).

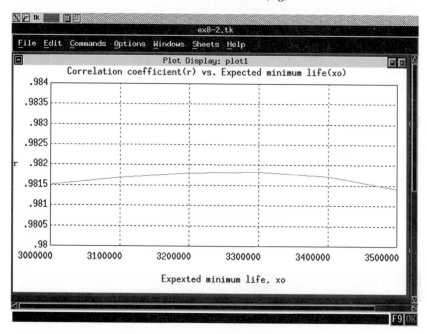

PLOT 2: Y VS. X WHEN $x_o = 0$

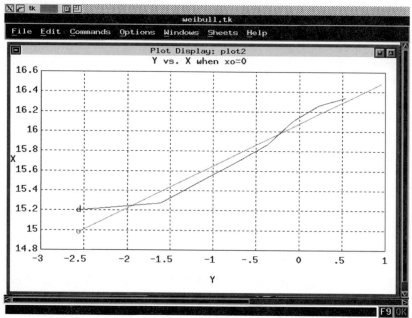

PLOT 3: Y VS. X WHEN $x_o = x_{opt}$

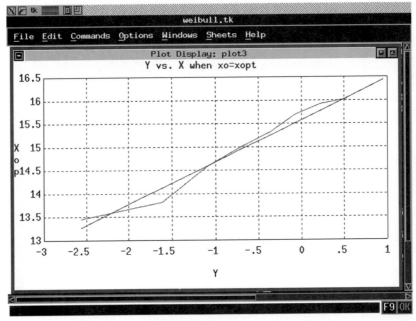

PROBLEMS

8–1)   Gearboxes are manufactured using very tightly controlled processing
        conditions and tolerances. The test results for the life of a gearbox, under
        the given load, show that the average life μ = 5 years, the standard
        deviation $\hat{\sigma}$ =.08 years, and the distribution is normal. (A total of 1000
        gears are manufactured.)

       a.   Find the number of gears that will have a life of 6 years or more.

       b.   Find the number of gears that will have a life greater than 3
           years.

       c.   Find the number of gears that will have a life between 2.5 and
           6 years.

       d.   What must be the guaranteed life if 1 failure in 1000 is accept-
           able?

                                        (ans. part (d): 2.53 years)

8–2)   Given an average life of five years and a standard deviation of one year
        for 1000 bearings tested, determine the number of bearings having a life
        more than four years, and also the number having a life more than six

years. If two percent of the bearings are allowed to fail during the warranty period, what maximum life must a company offer as the guaranteed life. If each bearing cost $5.00 without a warranty, what must be the cost of a bearing with a warranty to recover the loss due to a two percent replacement under the warranty. (Assume that a normal Gaussian bell-shaped distribution is applicable.)

8–3)   One thousand gearboxes were tested for life. The average life of a gearbox is five years and the standard deviation $\hat{\sigma}$ is 0.8 year. If no more than two failures in 1000 gearboxes is desired, find the maximum life that the company should guarantee.

8–4)   If it costs (for labor plus a new gear) twice the cost of a gearbox to replace the gearbox in Problem 8–1 and four failures in 1000 are allowed, what percent increase in the sale price of the gear must be made to recover the expected failure replacement cost up front? What guarantee (in terms of years) can one then offer?

(Part: 2.88 years)

8–5)   Take six data values from the following table and determine, for $x_0 = 0$, b and θ by hand.

Hours to failure				
70	177	242	395	652
100	182	252	425	688
125	193	285	456	748
130	195	305	526	837
152	212	325	548	880
162	217	365	596	910

Now, analyze the above data for the two-parameters and three-parameters Weibull distribution using the TK model, $weibul.tk$ (i.e., determine b and θ for $x_0 = 0$), and then find r_{max} and the corresponding x_0. Finally, determine b, c, and θ corresponding to r_{max}.

   a.   Using your calculator, calculate Weibull parameters and the Weibull equation for $x_0 = 30$. Find the one percent failure rate.

b. Using the TK model, find the one percent failure rate for $x_0 = 0$ and $x_0 = x_{optimized}$ (for all data points).

c. Do you see any correlation between parts (a) and (b)?

d. Find the life in hours if one failure in 1000 is stipulated (the 0.1% probability of failure).

Fundamentals of Metal Fatigue Analysis

9.1 Introduction

When a machine element—for example, a coil spring—is subjected to cyclic loading, it will fail at some finite lifetime even if it is loaded (stressed) well below its yield strength. Such a phenomenon (i.e., premature failure or damage) is called *fatigue failure* and manifests itself in three stages: crack initiation, crack propagation, and ultimate fracture.

In 1882, Love, the author of the famous book on elasticity, remarked in his text that "the conditions of rupture are but vaguely understood." Even today, we have only a partial understanding of fracture. However, fatigue failure, which is easily visible but difficult to describe accurately, must be studied, and an understanding of it must be used in the design of machine components subjected to cyclic loads. Every day, machine elements fail despite our awareness of cyclic loading. One of the reasons for this is that fatigue technology is in the hands of "experts," rather than people who design and build the components. The purpose of this chapter is to give both students and engineers insight into fatigue failure.

Three primary fatigue analysis methods are currently used in the design of machine elements subjected to cyclic loads. Table 9–1 presents some basic facts about these methods. Depending upon its suitability, any of the three life estimation methods can be used to study fatigue.

In the real world, the load as a function of time is neither a sine, cosine, a curve, nor even a combination of the two. However, using a mathematical technique, one can closely represent a random load curve as a sine curve, which then can have so-called mean load and amplitude load components. (We define these later.)

Table 9–1

METHOD	STRESS–LIFE APPROACH (S–N)	STRAIN–LIFE APPROACH (ε–N)	LINEAR ELASTIC FRACTURE MECHANICS (LEFM) APPROACH
AGE	Approximately 150 years old	Approximately 90 years old	(70 + Years)
FIRST RESEARCHER	August Wohher	Unknown	Griffith
ACTUAL YEARS IN USE	More than 100 years	More than 40 years	More than 30 years

9.2 The S-N Approach

The stress–life (stress vs. number of cycles to failure or S–N) method is generally used in the study of designs involving long-life fatigue or high-cycle fatigue (HCF). The method is first described using constant amplitude loading with mean load equal to zero as a function of time. It is taught at the undergraduate level and is known to all engineers. It is also easy to understand, and it is the least expensive of the three methods to conduct and study experimentally. The method, however, accounts only for elastic stress behavior and therefore has limitations in terms of the understanding of fatigue behavior in the plastic strain range.

9.3 The ε-N Approach

Some structures loaded in the elastic range create plastic strains at or around the stress concentration zone. In such cases, the strain life (ε–N) approach is used. This method provides the life for *crack initiation*. Naturally, plastic strains mean high loads and a low cycle life; therefore, the ε–N method is also called the low-cycle fatigue (LCF) method. It can be used to understand fatigue and creep interaction of machine elements in a high-temperature environment.

9.4 The LEFM Approach

The linear elastic fracture mechanics (LEFM) approach is the only method that accounts for crack propagation life. It is still in the research stage, but is considered to be complementary to the ε–N approach, which can predict crack initiation life. Together, the two methods can predict the total fatigue life—that is, the crack initiation life plus the crack propagation life.

All design situations are different and one of the three methods could be better suited for a particular situation than the other two. Each method has its advantages, disadvantages, and certain limitations. A better selection of a method can be made if the user has a good understanding of the material, the load and its history, the service environment, the component and its geometry, and the consequences of component failure. The answers we get by using one of the methods do not assure 100 percent reliability, but the situation is definitely better than designing a machine component on the basis that it is subjected only to monotonic (static) loading.

A detailed description of the *S–N* method is given in this chapter. The ε–*N* and LEFM methods are not described any further.

9.5 The *S–N* Approach

Previously, we stated that metal fatigue is a process that causes premature failure or damage of a machine component subjected to a cyclic load. As mentioned earlier, three primary fatigue analysis methods are employed to understand premature fatigue failures. The *S–N*, or stress–life, approach is used to understand the fatigue behavior of machine elements subjected to amplitude and mean load as a function of time. The fundamental test conducted to understand this behavior is the *Moore test.*

A Moore test is conducted on a tension-like specimen, as shown in Fig. 9–1. In such a test, an electric motor rotates the tension-like specimen subject to a bending load W. The load causes, in an element at a critical location, a variable load that is a sine curve having a maximum stress equal to σ_{max} and a minimum stress equal to σ_{min}. Both stresses have the same magnitude, but opposite sign; that is, if σ_{max} is tension, then σ_{min} is compression. We define the amplitude and mean stress as

$$\sigma_a = \frac{\sigma_{max} - \sigma_{min}}{2}$$

$$\sigma_m = \frac{\sigma_{max} + \sigma_{min}}{2}$$

respectively, where σ_a is the amplitude stress and σ_m is the mean stress. In this test (see Fig. 9–1), we have $\sigma_{max} = -\sigma_{min}$; therefore, $\sigma_a = \sigma_{max}$ and $\sigma_m = 0$, where

$$\sigma_{max} = \frac{MC}{I}$$

$$M = \frac{WL}{2} \qquad \text{(see Fig. 9–1)}.$$

$$C = d\,/\,2$$

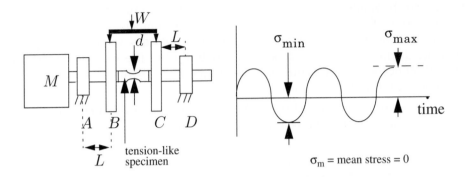

$ABCD$ are bearings

BC is the tension-like fatigue specimen

M is the electric motor

W is the variable bending load

Figure 9–1 Moore test set-up showing test fixture and load curve for rotating tension-like specimen

$$I = \frac{\pi d^4}{64}$$

and d is the diameter of the tension-like specimen at a critical location (the neck).

By conducting the Moore test on a large number of samples under different load intensities, it is possible to obtain sufficient data to plot strength vs. number of cycles to failure, called an $S–N$ curve. The typical $S–N$ curve for 120 Bhn steel is shown in Fig. 9–2. This curve is obtained by loading a test specimen with a load W, which will create a stress S_f, and the corresponding life to failure measured in terms of the number of cycles. Such a curve shows the fatigue strength (S_f) associated with a certain number of cycles (N). The process is repeated for different values of W to obtain the so-called $S–N$ curve. Most $S–N$ curves are obtained by testing at least ten specimens for a given load, and the *average life* (50 percent failure) is taken as the failure life for a given stress level.

The endurance limit S_e', which is defined as the stress level below which a machine element has more than 10^6 cycles of life, is obtained if an $S–N$ curve has a kink, as shown in Fig. 9–2. Such a kink is possible in ductile steel. At the kink and beyond, the stress level remains constant. It is however cautioned that for higher levels of reliability, the $S–N$ curve must be modified.

For a given life, the corresponding failure stress is called the *fatigue strength, S_f*. The endurance limit, S_e', is a special case of fatigue strength. It is obvious now that a particular fatigue strength has a corresponding life

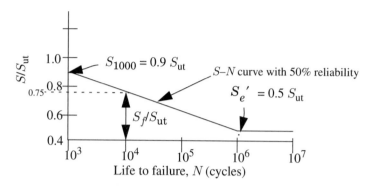

Figure 9–2   Typical stress versus life to failure (*S–N*) curve for steel.

associated with it. For example, the fatigue strength at 10^4 cycles is shown in Fig. 9–2.

In the absence of a Moore test, the value of S_f at 10^3 cycles is taken to be $0.9S_{ut}$, where S_{ut} is the ultimate tensile strength in a simple tension test, and the endurance limit is taken to be $S_{e'} = .5S_{ut}$ for steel. However, caution is advised in the use of such an empirical relationship in the absence of test data.

The Brinell hardness test measures the indentations that can be made in a metal. If the Brinell hardness number (BHN) is known and tension test data are unavailable, another empirical relationship, viz., $S_{e'} = .25$BHN in ksi or 1.73BHN in MPa, for BHN < 400 for steel, can be used. For BHN > 400, we use $S_{e'} = 100$ ksi. Again, these relationships may hold for certain steels, but may not work for other metals. Experimental evidence must support them.

If S_f at 10^3 cycles and $S_{e'}$ at 10^6 cycles are known, and the *S–N* curve between these two limits is assumed to be a straight line on $\log_{10}$–$\log_{10}$ graph paper, then the equation of a straight line can be derived, viz.,

$$\log S = b \log N + c$$

or

$$S = 10^c N^b \qquad \text{(for } 10^3 \leq N \leq 10^6 \text{)} \tag{9–1}$$

However, we know that

$$S = S_{1000} \ @ \ N = 10^3 \text{ cycles}$$

and

$$S = S_{e'} @ N = 10^6 \text{ cycles}$$

Using these two conditions, we can evaluate the two constants, b and c:

$$b = -(1/3) \log_{10} \frac{S_{1000}}{S_{e'}} \quad \text{and} \quad c = \log_{10} \frac{(S_{1000})^2}{S_{e'}}$$

The constants can further be simplified if one takes $S_{1000} = 0.9 S_{ut}$ and $S_e' = 0.5 S_{ut}$. However, this is true only for steel; thus, one should use Equation (9–1) as a general equation with b and c values calculated using the foregoing equations.

The equation $S = 10^c N^b$ can be utilized to understand the relationship between fatigue strength and the corresponding number of cycles. Hence, given the number of cycles N, one can find the corresponding fatigue strength, or vice versa, by using Equation (9–1).

Sometimes the endurance limit for a tension specimen subjected to an axial stress or shear stress is required without going through actual tension or shear fatigue testing. The Moore test can be used to understand such type of loadings. It is observed experimentally that if $S_{e'}$ is the bending fatigue endurance limit for steel, then

S_{ea} = axial fatigue endurance limit = 0.7 to $0.9 S_{e'}$

and

S_{es} = shear fatigue endurance limit = 0.5 to $0.577 S_{e'}$

Let us now study the conditions under which the Moore test is conducted and the resulting limitations of such a test. A Moore test is conducted on a tension-like test specimen that

1. is well polished and therefore has a reasonably crack-free surface

2. is subjected to bending loads only and therefore experiences a bending (uniaxial) state of stress

3. is 0.25 to 0.3 inch in diameter

4. is subjected to mean stress σ_m that is zero, that is, the specimen is subjected to pure amplitude load

5. has a ratio of $\sigma_{minimum}$ to $\sigma_{maximum}$ = -1

6. is tested at room temperature.

We see, then, that $S_{e'}$ must be modified if we deal with different loadings, surface conditions, surrounding temperatures, sizes, stress gradients, etc., in a machine element.

To account for the type of load, surface condition, and stress gradient, we have the equation

$$S_e = S_{e'}C_L C_S C_G \tag{9-2}$$

where S_e is the corrected endurance limit for a machine element, C_L is the load factor, C_S is the surface factor, and C_G is the stress gradient factor. Naturally, if one has to modify the endurance limit to account for size, temperature, etc., then further modification of Equation (9–2) is necessary.

Before that, however, we need to know how to obtain the C values. Table 9–2 gives the C_L and C_G values for steel having different conditions.

Table 9–2 Values of C_L and *CG*.

Type of Load	C_L	C_{G*}	Remarks
Axial	0.9 to 1	0.7 to 0.9	* C_G values are for $d < 10$ mm or 0.4 in.
Bending	1	1	For $d > 1$ in or 25 mm, reduce C_G by
Torsion	0.5 to 0.578	1	10%.

The surface factor C_S can be obtained using a certain equation and the values of the constants a_1, a_2, and a_3, as given in the table for various types of surfaces. The C_S values for steel machine elements are approximate and are a function of the ultimate strength of the steel. The equation is

$$C_S = a_1 + a_2 S_{ut} + a_3 S_{ut}^2$$

where a_1, a_2, a_3 are as given in Table 9–3 and S_{ut} is the ultimate strength of the steel in megapsi, or 10^6 psi. The equation for C_s is limited to steel, but similar values are available for other metals in any advanced book on properties of materials.

Table 9–3 Constants to determine C_s

Steel	Mirror Polished	Machined	Hot Rolled	Forged	Corroded in Water	Corroded in Salt Water	Fine Ground Commercial
a_1	1	0.8284	0.9414	0.7600	1.0130	0.7095	0.8295
a_2	0	−0.4296	−4.0645	−4.0092	−7.0562	−4.6957	1.5438
a_3	0	−3.0886	5.2156	6.2648	14.9767	9.1200	−7.7506

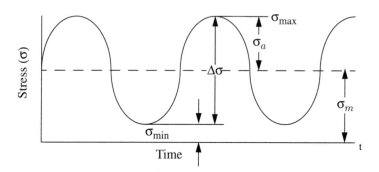

Figure 9–3   Cyclic loading with mean tensile stress.

9.6 Fatigue Strength Under Cyclic Loads Having σ_m Greater than Zero Mean Stress

The Moore test is conducted with the mean stress equal to zero. However, if the load curve shown in Fig. 9–3 applies to any given design, then the theory developed so far is not applicable, as in this case the mean stress is not zero. Let us, then, study the theory for the design of machine elements subjected to a mean stress different from zero. Toward that end, we present the equations that follow:

From Fig. 9–3, we define the stress range as

$$\Delta\sigma = \sigma_{max} - \sigma_{min} \qquad\qquad (9\text{--}3)$$

We have already defined

$$\sigma_{mean} = \sigma_m = \frac{\sigma_{max} + \sigma_{min}}{2} \qquad\qquad (9\text{--}4)$$

$$\sigma_{amplitude} = \sigma_a = \frac{\sigma_{max} - \sigma_{min}}{2} \qquad\qquad (9\text{--}5)$$

and we will define further

$$R = \frac{\sigma_{min}}{\sigma_{max}} = \text{ratio of minimum load to maximum load} \qquad\qquad (9\text{--}6)$$

$$r = \sigma_a/\sigma_m = \text{ratio of amplitude to mean stress} \qquad\qquad (9\text{--}7)$$

However, it is observed experimentally that if a machine element is subjected to a non-zero σ_{mean} *tension stress*, then the allowable fatigue strength for the element decreases. Fig. 9–4, called the *Haigh diagram*, shows the experimental data that clearly demonstrate the effect of mean stress on the

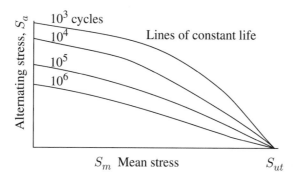

Figure 9–4   Effect of mean (tension) stress on alternating (amplitude) stress; Haigh diagram for constant life.

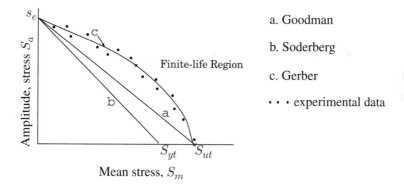

Figure 9–5   Fatigue diagrams: a) Goodman, b) Soderberg, c) Gerber.

endurance limit of a machine element. In the figure, it can be observed that S_a, the amplitude stress, and S_m, the mean stress, are maintained constant, and the corresponding cycles to failure are measured. Since we know that $S_a = SF\sigma_a$ (where SF denotes the safety factor) and $S_m = SF\sigma_m$, we can interchange, for example, S_a with σ_a, and S_m with σ_m (such relationship is possible for a proportional loading).

The data indicate that a positive σ_m reduces the endurance limit. Several ways of simplifying the curves of the Haigh diagram have been introduced over the years and are still in use. The most widely employed curves are shown in Fig. 9–5. For example, a conservative diagram that can be used in the absence of actual $S_a - S_m$ fatigue test data was presented by Goodman in 1899 in England. The Goodman curve is obtained by joining a straight line between S_e and S_{ut}, as shown in Fig. 9–5. Note S_e here is a corrected endurance limit for a given machine element and S_{ut} is the ultimate tension stress in a simple tension test.

The equation for the Goodman line is of the form $S_a = bS_m + c$, but can be changed to another form if we observe that when $S_m = 0$, $S_a = S_e$, and when $S_m = S_{ut}$, $S_a = 0$. Such conditions will determine the constants b and c, and the *Goodman equation* will assume the form

$$\frac{S_m}{S_{ut}} + \frac{S_a}{S_e} = 1 \tag{9–8}$$

where S_m is the maximum mean stress in a machine element ($S_m = $ SF (σ_m), SF is the safety factor, σ_m is the mean stress in the machine element, S_a is the maximum amplitude stress in the same machine element ($S_a = $ SF (σ_a)), σ_a is the amplitude stress in the machine element, S_{ut} is the ultimate tensile stress of the material used for the machine element in a simple tension test, and S_e is the (corrected) endurance limit for the machine element.

(*Note*: in using the safety factor SF, it is assumed that σ_a and σ_m are proportional to an external load.)

If $\sigma_a/\sigma_m = r$ is known, then $\sigma_a/\sigma_m = S_a/S_m = r$ (for proportional loading). Also, it can be shown that the following equation will yield the allowable S_m if one knows r, S_e, and S_{ut}:

$$S_m = \frac{S_e}{r + \dfrac{S_e}{S_{ut}}} \tag{9–9}$$

Once the allowable S_m is found, it follows that

$$\text{SF} = S_m/\sigma_m \tag{9–10}$$

It is obvious that the safety factor is based on the Goodman equation and also can be obtained by taking the ratio of S_a and σ_a. i.e., SF $= S_a/\sigma_a$.

The equations defining the other curves shown in Fig. 9–5 are as follows:

The Gerber Parabolic equation (Germany, 1874):

$$\frac{S_a}{S_e} + \left(\frac{S_m}{S_{ut}}\right)^2 = 1 \tag{9–11}$$

The Gerber equation is a better fit to experimental data and hence is preferred over the Goodman equation.

The Soderberg equation (United States, 1930):

$$\frac{S_a}{S_e} + \frac{S_m}{S_{yt}} = 1 \tag{9–12}$$

The Soderberg equation gives better results if the tensile yield stress S_{yt} is a failure criterion and if σ_m is close to S_{yt}. (Otherwise, the results are too conservative.) The Soderberg equation requires the yield strength S_{yt}. If S_{yt} is not available for steel, then one takes $S_{yt} = .85 \, S_{ut}$.

One additional equation introduced fairly recently is the following:

The quadratic equation:

$$\left(\frac{S_a}{S_e}\right)^2 + \left(\frac{S_m}{S_{ut}}\right)^2 = 1 \tag{9–13}$$

This equation is overfit to the experimental data and therefore is non-conservative.

A couple of examples will illustrate the application of different theories.

Example 9-1

A machine component is subjected to a cyclic load (stress) with a maximum value of 120 ksi and a minimum value of 20 ksi. (The maximum stress accounts for the safety factor.) The component is made from steel with $S_{ut} = 160$ ksi and a corrected endurance limit $S_e = 70$ ksi with fully reversed $S_{1000} = 120$ ksi. Using the Goodman, Gerber, and Soderberg criteria, determine the life of the component.

Solution

Now SF = 1 as the maximum and minimum stress account for safety factor SF = 1 means that $\sigma_a = S_a$ and $\sigma_m = S_m$. We now find

$$S_a = \frac{S_{max} - S_{min}}{2} = \frac{120 - 20}{2} = 50 \, \text{ksi}$$

$$S_m = \frac{S_{max} + S_{min}}{2} = \frac{120 + 20}{2} = 70 \, \text{ksi}$$

Next, let us calculate the fatigue strength that is required, as the endurance limit S_e may not be adequate. We will define S_f as the required fatigue strength; then the Goodman equation is

$$\frac{S_a}{S_f} + \frac{S_m}{S_{ut}} = 1$$

Substitute known values:

$$\frac{50}{S_f} + \frac{70}{160} = 1$$

Solving for S_f, we get

$$S_f = 88.9 \text{ ksi}$$

This is the required fatigue strength. Since it is greater than the endurance limit S_e (70 ksi), the machine element will have a shorter life than 10^6 cycles.

Now we use the $S\text{–}N$ diagram to get the number of cycles, or else we use Equation (9–1) rearranged as

$$N = 10^{-c/b} S_f^{1/b}$$

where

$$b = -(1/3) \log (S_{1000}/S_e)$$

$$c = \log ((S_{100})^2/S_e)$$

Knowing that $S_{1000} = 120$ ksi, and $S_e = 70$ ksi, we get $b = -0.0780$, and $c = 2.31$. Then if we solve for N, we get $N = 46,811$ cycles, or approximately 46,800 cycles of life to failure. In sum, the machine element under a given load (which includes SF) will have 46,800 cycles of life. This can be verified by the TK solution shown after the next example. Now let us determine the life to failure using the other criteria.

Using the Gerber equation, viz.,

$$\frac{S_a}{S_f} + \left(\frac{S_m}{S_{ut}}\right)^2 = 1$$

and substituting the known values, i.e.,

$$\frac{50}{S_f} + \left(\frac{70}{160}\right)^2 = 1$$

yields $S_f = 61.84$ ksi, which is less than S_e (70 ksi). Therefore, the machine element has at least 10^6 cycles of life.

For the Soderberg equation, we have $\sigma_{yp} = S_{yt} = .85 S_{ut}$, and

$$\frac{S_a}{S_{er}} + \frac{S_m}{S_{yt}} = 1$$

Substituting S_{yt}, S_a, and S_m, we get $S_f = 103.03$, and the corresponding life N can be obtained using Equation (9–1), which yields $N = 7057$ cycles.

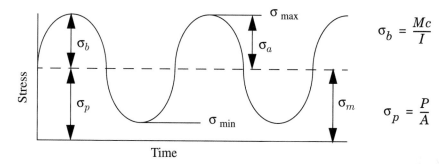

Figure 9–6   Critical stress variation as a function of time.

Thus, different values of life (in cycles) are obtained, depending upon the theory used. It is recommended that the Goodman equation be utilized for reasonably conservative results, and the Gerber equation be employed when one wants to remain close to experimental results.

Example 9-2

A solid steel shaft of diameter d and having S_{ut} = 110 kpsi, S_y = 85 kpsi, and S_e = 40 kpsi (after correcting for load, surface, and gradient) is rotating and has a bending moment of 5,000 in-lb and an axial load of 50,000 lb tension. Using the various theories presented in this section, determine the diameter of the shaft. Take the safety factor to be 2.0.

Solution

Calculate the bending and axial stress at the extreme fiber, where both stresses are critical. That is, calculate

$$\sigma_b = \frac{MC}{I} = \frac{5,000(d/2)}{\frac{\pi d^4}{64}}$$

and

$$\sigma_p = P/A = \frac{50,000}{(\pi/4)d^2}$$

It can be seen that the axial stress σ_p is independent of time even if the shaft is rotating, no matter where the element is selected in a cross section. However, the bending stress changes from tension to compression as the shaft rotates. Fig. 9–6 shows both stresses as a function of time.

Now we must find the diameter using the different theories that account for the mean stress. First, we find the amplitude and mean stress. We have

$$\sigma_{\max} = \sigma_p + \sigma_b$$

$$\sigma_{\min} = \sigma_p - \sigma_b$$

$$\therefore \sigma_a = \frac{\sigma_{\max} - \sigma_{\min}}{2} = \sigma_b$$

$$\sigma_m = \frac{\sigma_{\max} + \sigma_{\min}}{2} = \sigma_p$$

$$\therefore \quad r = \sigma_a / \sigma_m = \sigma_b / \sigma_p = \frac{M(d/2)/\left(\frac{\pi d^4}{64}\right)}{P/\left(\frac{\pi}{4}d^2\right)}$$

Note that d is still unknown.

Using the Goodman equation,

$$\frac{S_a}{S_e} + \frac{S_m}{S_{ut}} = 1$$

We know that $S_a = \text{SF}(\sigma_a)$, $S_m = \text{SF}(\sigma_m)$, $S_e = 40$, and $S_{ut} = 110$. Thus, we now have only the diameter as an unknown that must be solved for. To do this, we will use the lead model, which will simplify the otherwise difficult task of solving nonlinear equations. All of the necessary equations are written in the Rules Sheet on page 183, and the diameter required using one theory at a time is obtained by solving the corresponding equations. The results are shown on page 184. The file name of the lead model is *exp9-2.tk*, or *fatig.tk*.

9.7 Haigh Diagram

In the previous section, we saw the effect of the mean (tensile) stress on the endurance limit. The mean stress reduces the amplitude load capacity of a machine element. Since the test required to generate the curve of amplitude vs. mean stress for 10^6 cycles can be expensive, we used empirical equations developed by Goodman, Gerber, and Soderberg, as well as the quadratic equation. Note that the Soderberg method is very conservative; actual test data fall between the Goodman and Gerber curves. Most machine components have a small mean stress compared to their amplitude stress, and in such cases there is very little difference between the various theories.

RULES SHEET

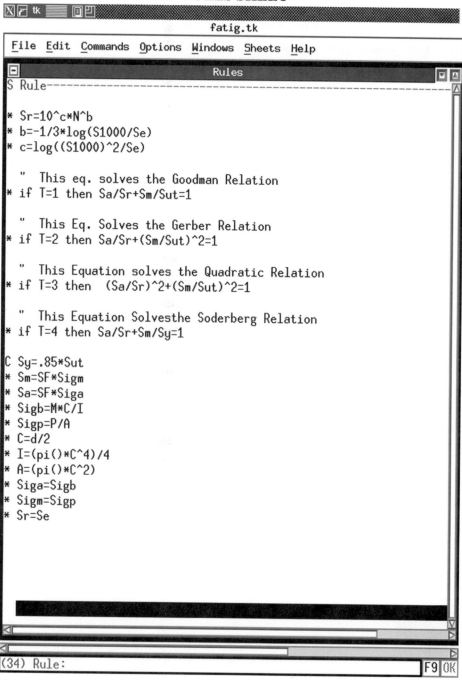

fatig.tk

File Edit Commands Options Windows Sheets Help

Rules

```
S Rule-------------------------------------------------------------

*  Sr=10^c*N^b
*  b=-1/3*log(S1000/Se)
*  c=log((S1000)^2/Se)

   "  This eq. solves the Goodman Relation
*  if T=1 then Sa/Sr+Sm/Sut=1

   "  This Eq. Solves the Gerber Relation
*  if T=2 then Sa/Sr+(Sm/Sut)^2=1

   "  This Equation solves the Quadratic Relation
*  if T=3 then   (Sa/Sr)^2+(Sm/Sut)^2=1

   "  This Equation Solvesthe Soderberg Relation
*  if T=4 then Sa/Sr+Sm/Sy=1

C  Sy=.85*Sut
*  Sm=SF*Sigm
*  Sa=SF*Siga
*  Sigb=M*C/I
*  Sigp=P/A
*  C=d/2
*  I=(pi()*C^4)/4
*  A=(pi()*C^2)
*  Siga=Sigb
*  Sigm=Sigp
*  Sr=Se
```

(34) Rule: F9 OK

VARIABLES SHEET

```
[X][tk]  [□][凹]
                              fatig.tk

 File  Edit  Commands  Options  Windows  Sheets  Help
[-]                              Variables                            [□][□]
St Input----  Name---  Output---  Unit-----  Comment----------------------------
    1          T                              Select Desired Failure Theory
                                              (1=Goodman,2=Gerber,
                                              3=Quadratic,4=Soderberg)

   40000       Se                  psi

  110000       Sut                 psi        Ultimate Tensile Stress
               c
               N                              Number of Life Cyles
               b
               S1000               psi        Yield Stress after 1000 cycles
   85000       Sy                  psi
               Sa      22887.602   psi        Safe Amplitude Stress
               Sr      40000       psi        Required Yield Stress
               Sm      47059.095   psi        Safe Mean Stress
    2          SF
               Sigm    23529.548   psi        Actual mean yield stress
               Siga    11443.801   psi        Actual amplitude stress

                                              Note:  All stresses may be expressed
                                                     as psi, ksi, Pa, or MPa.

               Sigb    11443.801   psi
    5000       M                   lb-in
               I       3.3506922   in^4
               Sigp    23529.548   psi
   50000       P                   lb
               A       2.1249877   in^2
               d       1.6448764   in
               C       .82243821   in




[█████████]
[◁|                                                                        [▷]
[◁|                                                                        [▷]
 (35) Input:                                                           [F9][OK]
```

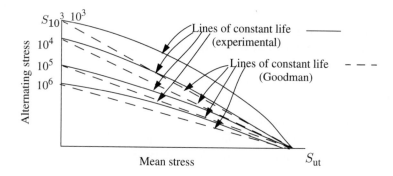

Figure 9–7 Haigh diagram showing Goodman equations

For the different theories using $S_e = 10^6$ cycle for the amplitude stress as the limiting point on the vertical (y) axis and S_{ut} as the limiting point on the horizontal (x) axis led to curves that were good for a life N of 10^6 cycles. Similar curves can be obtained for $N = 10^3$, $N = 10^4$, and $N = 10^5$ cycles by knowing S_{1000}, S_{10^4}, S_{10^5}, respectively—that is, the fatigue strengths that can be obtained from the S–N diagram. (See Fig. 9–2.) These values will represent points on the amplitude (y) axis while S_{ut} or S_{yp} will be the intercepts on the mean (x) axis. We can accomplish all of this with the aid of an empirical equation such as the Goodman equation.

The Goodman equation for $N = 10^3$ cycles will be

$$\frac{S_a}{S_{1000}} + \frac{S_m}{S_{ut}} = 1$$

where $S_m = SF\ \sigma_m$ is the mean stress in a machine element, $S_a = SF\ \sigma_a$ is the amplitude stress in a machine element, SF is the safety factor, S_{1000} is the fatigue strength at 1000 cycles of life, σ_a is the actual amplitude stress in the machine element, and σ_m is the actual mean stress in the machine element. Similarly, replacing S_{1000} by S_{10^4} will quickly yield a fatigue life curve of 10^4 cycles. Fig. 9–7 shows the experimental curves (solid) and Goodman curves (dotted) for $N = 1,000, 10,000, 100,000$, and $1,000,000$ cycles. Such a diagram is called a Haigh diagram.

The Goodman curves are straight lines joining, for example, S_{1000} and S_{ut}. The Goodman curves should be used in the absence of experimental evidence, as they will yield conservative results.

The following problem illustrates the use of the Haigh diagram for Gerber, quadratic, and Soderberg criteria for $N = 10^3$, 10^4, 10^5, and 10^6 cycles. Using such diagrams or equations, one can come up with a figure for the life of a machine component.

Example 9-3

Given a machine element subjected to $\sigma_m = 20.0$ ksi and $\sigma_a = 25$ ksi, if the ultimate stress $S_{ut} = 140$ ksi, $S_e = 65$ ksi, and $S_{1000} = 115$ ksi, determine the life in number of cycles of such a machine element, using SF = 2.0.

Solution

For a safety factor of 2,

$$S_m = 2\,(\sigma_m) = 40 \text{ ksi}$$

$$S_a = 2(\sigma_a) = 50 \text{ ksi}$$

Now we use the Goodman equation to obtain the required fatigue strength S_f; that is,

$$\frac{S_a}{S_f} + \frac{S_m}{S_{ut}} = 1$$

$$\frac{50}{S_f} + \frac{40}{140} = 1 \qquad \therefore S_f = 70 \text{ ksi}$$

However,

$$S_f = 10^c N^b \qquad (\text{for } 10^3 \le N \le 10^6)$$

where

$$c = \log_{10} \frac{(S_{1000})^2}{S_e} = 2.3084823$$

$$b = -(1/3) \log_{10} \left(\frac{S_{1000}}{S_e}\right) = -.0825948$$

(*Note:* One can use $S_{1000} = 0.9 S_{ut}$, and $S_e = 0.5 S_{ut}$ in the absence of Moore test results.) Thus,

$$S_f = 10^c N^b \text{ gives } N = 0.408 \times 10^6 \text{ cycles}$$

Similarly, N can be found using the Gerber, quadratic, and Soderberg equations. Some of the problems at the end of the chapter require you to determine the life of an element using the Haigh diagram for all other theories.

9.8 Multi-Axial Fatigue

So far, we have limited the study of fatigue behavior to materials and machine components subjected to a one-dimensional state of stress. However, real-world problems generally involve loads that create stresses in three directions (called 3-D stresses). To make matters worse, not only the magnitude, but also the direction of a 3-D stress changes as a function of time. Therefore, if we have a machine element in which 3-D stresses change their magnitudes as well as their directions in this manner, then we say that we have a complex state of stress. Engineering components such as cranks, axles, and propeller shafts are subjected to complex states of stress at notches or joint connections. Very little research has been carried out to study fatigue behavior under a complex state of stress. The approach we set forth next is adopted for the solution of a fatigue problem involving a 3-D state of stress that changes its magnitude, but not its direction as a function of time. Fatigue behavior under such stresses is also called *multiaxial fatigue*.

Among all the fatigue theories developed for multiaxial stress states, one that is very commonly used and that predicts fatigue behavior reasonably well is called the equivalent *Von Mises stress criterion*. Although the magnitudes of principal stresses change as a function of time, they are assumed not to do so when one applies this criterion.

The analysis which follows is based on the assumption that one knows the principal stresses σ_1, σ_2, and σ_3 at maximum and minimum loads; that is, $\sigma_{1,max}$, $\sigma_{2,max}$, $\sigma_{3,max}$, $\sigma_{1,min}$, $\sigma_{2,min}$, and $\sigma_{3,min}$ are known. The procedure to solve such problems involves four steps.

STEP 1

Determine σ_{1a}, σ_{2a}, and σ_{3a}, the principal amplitude stresses.

(Note that $\sigma_a = \dfrac{\sigma_{max} - \sigma_{min}}{2}$)

STEP 2

Determine σ_{1m}, σ_{2m}, and σ_{3m} the mean principal stresses.

(Note that $\sigma_m = \dfrac{\sigma_{max} + \sigma_{min}}{2}$)

STEP 3

Using values of amplitude and mean stresses, find the equivalent (Von Mises) amplitude and mean stresses, utilizing the equations

$$\sigma_{ea} = \frac{1}{\sqrt{2}}[(\sigma_{1a} - \sigma_{2a})^2 + (\sigma_{2a} - \sigma_{3a})^2 + (\sigma_{3a} - \sigma_{1a})^2]^{\frac{1}{2}}$$

and

$$\sigma_{em} = \frac{1}{\sqrt{2}}[(\sigma_{1m} - \sigma_{2m})^2 + (\sigma_{2m} - \sigma_{3m})^2 + (\sigma_{3m} - \sigma_{1m})^2]^{\frac{1}{2}}$$

STEP 4

Knowing σ_{ea}, and σ_{em}, go to, for example, the Goodman diagram, and find the safety factor using the following approach:

$$\frac{\sigma_{ea}}{\sigma_{em}} = \frac{S_a}{S_m}$$

(Slope remains the same)

$$\frac{S_a}{S_e} + \frac{S_m}{S_{ut}} = 1$$

(Equation of Goodman line)

Knowing σ_{ea}, σ_{em}, S_e, and S_{ut}, solve for S_a and S_m. The safety factor is given by

$$SF = \frac{S_a}{\sigma_{ea}} = \frac{S_m}{\sigma_{em}}$$

The next two examples will help you understand the foregoing procedure.

Example 9-4

A uniaxial bending fatigue test results in 10^6 cycles of life for a material subjected to 40-ksi sine wave loading. If the Von Mises criterion for multiaxial fatigue is applicable, find the equivalent completely reversed torsion fatigue strength for a material with the same life.

Solution

For the uniaxial bending fatigue test,

$\sigma_{1a} = 40$ ksi

$\sigma_{2a} = \sigma_{3a} = 0$

and since the test involves sine wave loading,

$\sigma_{1m} = \sigma_{2m} = \sigma_{3m} = 0$

$$\therefore \sigma_{ea} = \frac{1}{\sqrt{2}}[(\sigma_{1a} - \sigma_{2a})^2 + (\sigma_{2a} - \sigma_{3a})^2 + (\sigma_{3a} - \sigma_{1a})^2]^{\frac{1}{2}}$$

$$= \frac{1}{\sqrt{2}}(40^2 + 0^2 + (-40)^2)^{\frac{1}{2}}$$

$\sigma_{ea} = 40$ ksi

and

$\sigma_{em} = 0$

If the material is subjected to a cyclic pure fatigue shear stress τ, then, from Mohr's circle, it is clear that

$\sigma_{1a} = \tau$

$\sigma_{2a} = 0$

$\sigma_{3a} = \tau$

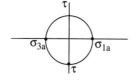

$$\sigma_{ea} = \frac{1}{\sqrt{2}}((\tau)^2 + (-\tau)^2 + (-2\tau)^2)^{\frac{1}{2}}$$

$$= \sqrt{3}\ \tau \quad \text{(for torsion load)}$$

and we know $\sigma_{em} = 0$ (for torsion load).

Using the Von Mises criterion, we have

$$\sigma_{ea})_{\text{Axial}} = \sigma_{ea})_{\text{Shear}}$$

$$40 = \sqrt{3}\ \tau$$

Hence,

$\tau = 23.095$ ksi

This test shows that fatigue endurance limit in shear is $\left(\frac{1}{\sqrt{3}}\right)$ times the endurance limit in bending fatigue, providing that the Von Mises criterion works for 3-D stresses.

Example 9-5

The pressure cylinder shown in the following diagram is subjected to a pressure p that varies from 0.1 to 1 ksi.

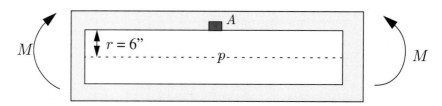

The cylinder rotates at 100 rps and is subjected to a bending stress of 50 ksi at r = 6". If the bending endurance limit is 80 ksi, find the safety factor. Assume that S_{ut} = 120 ksi.

Solution

An element at A is subjected to hoop, longitudinal, and radial stresses as shown in the following diagram

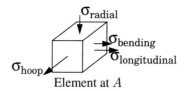

Element at A

Now, assuming that the cylinder is thin walled, it follows that

$$\sigma_{\substack{max \\ hoop}} = \frac{P_{max}r}{t} = \frac{1(6)}{0.25} = 24 \text{ ksi}$$

$$\sigma_{\substack{min \\ hoop}} = \frac{P_{min}r}{t}$$

$$= \frac{0.1(6)}{0.25}$$

$$\sigma_{\substack{min \\ hoop}} = 2.4 \text{ ksi}$$

$$\sigma_{\substack{longitidual \\ max}} = \frac{Pr}{2t}$$

$$\sigma_{\substack{max \\ hoop}} = 12 \text{ ksi}$$

Similarly,

$$\sigma_{\substack{longitidual \\ minimum}} = 1.2 \text{ ksi}$$

$$\sigma_{bending} = \pm 50 \text{ ksi}$$

Hoop, bending, and longitudinal stresses are cyclic and are shown on the following diagram of the element:

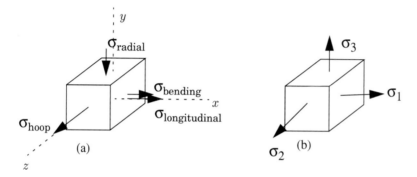

(a)   (b)

The diagram shows that

$$\sigma_{\substack{Total \ longitidual \\ maximum}} = 12 + 50 = 62 \text{ ksi}$$

$$\sigma_{\substack{Total \ longitidual \\ minimum}} = 1.2 - 50 = -48.8 \text{ ksi}$$

The cube in (b) shows the direction of the principal stresses σ_1, σ_2, and σ_3. Now let us find the maximum and minimum hoop stress:

$$\sigma_{\substack{Total \ max \\ hoop}} = 24 \text{ ksi}$$

$$\sigma_{\substack{Total \ min \\ hoop}} = 2.4 \text{ ksi}$$

Next, we find the amplitude and mean principal stress σ_1:

$$\sigma_{1a} = \frac{\sigma_{max} - \sigma_{min}}{2} = \frac{62 + 48.8}{2} = 55.4 \text{ ksi}$$

$$\sigma_{1m} = \frac{\sigma_{max} + \sigma_{min}}{2} = \frac{62 - 48.8}{2} = 6.6 \text{ ksi}$$

Similarly,

$$\sigma_{2a} = \frac{24 - 2.4}{2} = 10.8 \text{ ksi}$$

$$\sigma_{2m} = \frac{24 + 2.4}{2} = 13.2 \text{ ksi}$$

We see from the stressed element that $\sigma_{3a} = \sigma_{3m} = 0$ (The stress in the radial direction is assumed to be zero for a thin-walled cylinder.) Therefore,

$$\sigma_{ea} = \frac{1}{\sqrt{2}}[(\sigma_{1a} - \sigma_{2a})^2 + (\sigma_{2a} - \sigma_{3a})^2 + (\sigma_{3a} - \sigma_{1a})^2]^{\frac{1}{2}}$$

$$= \frac{1}{\sqrt{2}}[(55.4 - 10.8)^2 + (10.8 - 0)^2 + (0 - 55.4)^2]^{\frac{1}{2}}$$

$$= \frac{1}{\sqrt{2}}[(1989.16) + (116.64) + (3069.16)]^{\frac{1}{2}}$$

$$\sigma_{ea} = 51.0 \text{ ksi}$$

$$\sigma_m = \frac{1}{\sqrt{2}}[(\sigma_{1m} - \sigma_{2m})^2 + (\sigma_{2m} - \sigma_{3m})^2 + (\sigma_{3m} - \sigma_{1m})^2]^{\frac{1}{2}}$$

$$= \frac{1}{\sqrt{2}}\left[(6.6 - 13.2)^2 + (13.2 - 0)^2 + (0^2 - 6.6)^2\right]^{\frac{1}{2}}$$

$$= \frac{1}{\sqrt{2}}[(43.5) + (174.24) + (43.56)]^{\frac{1}{2}}$$

$$\sigma_{em} = 11.43 \text{ ksi}$$

Now, we know that

$$r = \frac{\sigma_{ea}}{\sigma_{em}} = \frac{S_a}{S_m} = \frac{50.87}{11.43} = 4.451$$

and the Goodman criterion is

$$\frac{S_a}{S_e} + \frac{S_m}{S_{ut}} = 1 \, ;$$

Dividing by S_m, we get

$$\frac{S_a/S_m}{S_e} + \frac{1}{S_{ut}} = 1/S_m \, .$$

Substituting known values, we obtain

$$\frac{4.451}{80} + \frac{1}{120} = 1/S_m$$

Solving for S_m yields

$$S_m = 15.78$$

$$\therefore \text{SF} = \frac{S_m}{\sigma_{em}} = \frac{15.63}{11.43}$$

$$= 1.38$$

Note that this solution is based on the assumption that the Von Mises equivalent stress criterion, along with Goodman criterion, works for the 3-D state of stress. It is recommended that one back up theoretical results with some experimental evidence.

9.9 Summary of Effects of Loading on Fatigue Life

A reduction in fatigue life is possible:

a. if the mean stress σ_m is tensile.

b. if the size of a machine element is larger than the Moore test specimen.

c. if the surface of a machine element is rougher than that of the Moore test specimen.

d. if a machine element is chrome or nickel plated.

e. if a machine element is decarburized due to forging, hot rolling, or any other process.

f. if a machine element was ground severely.

An improvement in fatigue life is possible:

a.   if a machine element is flame and induction hardened.

b.   if a machine element is carefully shot peened or cold rolled.

c.   if a machine element has surface compressive residual stresses due to overload, especially in the vicinity of notches.

Always be cautious, and assume that the use of an endurance limit as a safe stress below which fatigue will not occur is just an indication and not a guarantee. Such a limit may disappear due to a corrosive environment, a rise or fall in the temperature of a machine element, the development of surface cracks from poor treatment, or the cooling of a ductile material below its transition temperature.

We can assert the following about the four methods that purport to account for the effect of mean stress on fatigue life:

The Goodman equation is conservative.

The Soderberg equation is even more conservative.

The quadratic equation is not conservative enough.

Only the Gerber equation is close to the experimental data, and it is usually recommended for reasonably nonconservative results. For reasonably conservative results, however, we use the Goodman equation.

9.10 Important Equations

In this section, we summarize the important equations we have developed in this chapter.

Alternating Stress Relations

$$\Delta\sigma = (\sigma_{max} - \sigma_{min}) = \text{stress range} \qquad \Delta S = \text{SF}(\Delta\sigma) \qquad (1)$$

$$\sigma_a = \frac{\sigma_{max} - \sigma_{min}}{2} = \text{stress amplitude} \qquad S_a = \text{SF}(\sigma_a) \qquad (2)$$

$$\sigma_m = \frac{\sigma_{max} + \sigma_{min}}{2} = \text{mean stress} \qquad S_m = \text{SF}(\sigma_m) \qquad (3)$$

$$r = \frac{\sigma_a}{\sigma_m} = \text{amplitude ratio} \tag{4}$$

Mean Stress Correction Relationships

Soderberg (United States, 1930):

$$\frac{S_a}{S_e} + \frac{S_m}{S_y} = 1 \tag{5}$$

Quadratic:

$$\left(\frac{S_a}{S_e}\right)^2 + \left(\frac{S_m}{S_y}\right)^2 = 1 \tag{6}$$

Goodman (England, 1899):

$$\frac{S_a}{S_e} + \frac{S_m}{S_u} = 1 \tag{7}$$

Gerber (Germany, 1874):

$$\frac{S_a}{S_e} + \left(\frac{S_m}{S_u}\right)^2 = 1 \tag{8}$$

$$S_a = (\text{SF})\sigma_a$$

$$S_m = (\text{SF})\sigma_m$$

where SF is the safety factor, σ_a is the amplitude stress in a machine element, σ_m is the mean stress in the machine element, and S_e is the endurance limit of the machine element for which the preceding equations are being used.

Relationship Between S_e' the Bending Moore Endurance Limit and Various Other Loadings

$$S_{e'}(\text{axial}) \approx 0.70 \, S_{e'} \tag{9}$$

$$\tau'_e(\text{torsion}) \approx 0.577 \, S_{e'} \tag{10}$$

PROBLEMS

9–1)   An automobile crankshaft 2″ in diameter is subjected to a bending stress of ± 120 ksi. If S_{ut} = 100 ksi, determine the life of the crankshaft in cycles. Assume that the shaft has a very smooth surface and that if necessary, the Gerber criteria is applicable.

9–2)   A machine element is subjected to cyclic stress as shown in the following diagram:

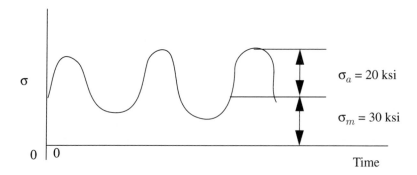

Use the Goodman criterion to determine the safety factor. Take S_e' = 100 ksi and S_{ut}= 220 ksi. Assume a round steel rod subjected to pure bending, diameter of shaft is given as 3″.

9–3)   A forged axle, subjected to pure axial stress as shown in the following diagram, is to be designed for 10^6 cycles of life, with a safety factor of 1.6.

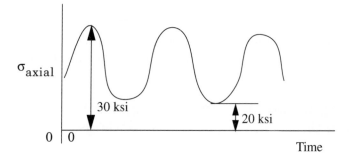

Determine the diameter of the axle's shaft if S_e' = 90 ksi and S_{ut} = 180 ksi.

9–4) A machined shaft is subjected to 10,500 in-lb of cyclic torque. Determine the diameter of shaft if a safety factor of 1.3 is required and if $S_e' = 100$ ksi, $S_{ut} = 220$ ksi, $S_{1000} = 190$ ksi, and a life of 10^5 cycles is required.

9–5) Consider a hot-rolled machine element subjected to a state of stress as given by the equation

$$\sigma_{ij})_{max} = \begin{bmatrix} 50 & 0 & 0 \\ 0 & 20 & 0 \\ 0 & 0 & 30 \end{bmatrix} \qquad \sigma_{ij})_{min} = \begin{bmatrix} 5 & 0 & 0 \\ 0 & 2 & 0 \\ 0 & 0 & 3 \end{bmatrix}$$

Find the safety factor based on the Von Mises criterion, if $S_e' = 90$ ksi and $S_{ut} = 170$ ksi. Use the Gerber criterion to account for the mean stress effect. Finally, solve the problem using the Goodman and Soderberg criteria.

9–6) A hollow shaft is rotating under loads, as in Problem 9–5. Determine the outside diameter D if $D_i = 0.8D$ with a safety factor of 1.67. The shaft material has the following mechanical and fatigue properties:

D_i = inside diameter

$S_{yp} = 80$ ksi

$S_{ut} = 120$ ksi

$S_e = 50$ ksi (corrected)

9–7) Use the data from Problem 9–6 and the Goodman, Gerber, Soderberg, and quadratic theories to find the outside and inside diameter of the shaft mentioned in that problem if

   a. $D_i = 0.95\ D$

   b. $D = 12"$.

   c.   Find D and D_i if $D–D_i = 2t$ and $t = 0.1$ (Draw a sketch of the hollow cross-section.)

9–8) For Problem 9–7,

   a.   List t from 0.1" to 1" and find the corresponding cross-sectional area A.

   b.   Plot A vs. t. Find the optimum A for this range of thickness.

9–9)   Given the data from Problem 9–5,

      a.   Determine, using various theories, the Haigh diagrams (Gerber, quadratic and Soderberg; take $S_{yp} = .85 S_{ut}$).

      b.   Use TK Solver software to determine the number of cycles for all four of the theories.

9–10)   A thick-walled pressure cylinder is to be used in a machine. The cylinder is subjected to a variable cyclic pressure and bending moment that are in phase with each other, as shown in the following graph.

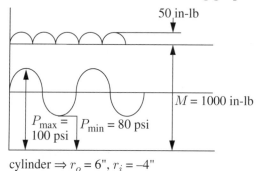

cylinder $\Rightarrow r_o = 6"$, $r_i = -4"$

Determine:

      a.   the amplitude and mean pressure

      b.   the amplitude and mean moment

      c.   the amplitude and mean principle stresses at A

      d.   the equivalent amplitude and mean (Von Mises) stresses.

9–11)   In a machine component, at the critical location the equivalent Von Mises amplitude and mean stresses σ_{ea} and σ_{em} are found to be 22,100 psi and 37,492 psi, respectively. Determine whether the element is safe for infinite life. If not, determine the possible number of cycles to failure if $S_{ut} = 80$ ksi and the fully corrected endurance limit is 40 ksi. (Use the Goodman criteria and the S–N relation, $S_f = 10^c N^b$.)

Curved Beams

10.1 Introduction

Curved beams in the form of hooks, brackets, and split rings are frequently used as machine elements. When such machine elements are subjected to an end couple M such that bending takes place in the plane of curvature, as shown in Fig. 10–1, the stress distribution σ_y is not linear on either side of the neutral axis, but increases more rapidly on the inner side. The following nomenclature is used while discussing stresses in curved beams.

NOMENCLATURE

R_n: radius to the neutral axis (NA)

$\sigma_{y\theta}$: stress a distance y in the direction θ

y: distance from NA to a fiber at a distance y

R_g: radius to the center-of-gravity (CG) axis of given cross-section

e: distance between NA and CG

ρ: radius to distance y

R_o: outside radius

M: external moment

10.2 Derivation of Stress in a Curved Beam

Since the bending stress formula for a curved beam is not the same as that for a straight prismatic beam, derived using straight-line geometry, the derivation to be presented shortly is for a curved beam subjected to an external moment M. It can be seen that many new terms apply, as shown in the preceding nomenclature in the derivation of the formula for bending stress. The derivation for the

199

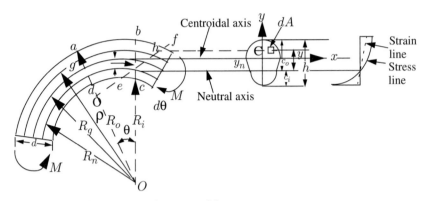

Figure 10–1   Geometry of a curved beam

bending stress at a distance y from the neutral axis (NA) for a curved beam subjected to a bending moment M is based on a stress geometry, as shown in Fig. 10–1. The figure clearly indicates that (1) the neutral axis does not coincide with the centroidal axis, (2) but the strain is proportional to the distance y from the neutral axis.

Let us assume that the beam's cross section at an angle θ with respect to an axis oa is stretched by an angle $d\theta$ due to an external bending moment M. Then, the stretching at a distance y is $yd\theta$.

From the figure, we see that the strain at distance y is given by

$$\varepsilon_y = \frac{yd\theta}{(R_n + y)\theta}$$

where y is taken to be positive if it is the distance toward the outer fiber measured from NA.

Now for the one-dimensional stress state, $\sigma = E\varepsilon$; therefore,

$$\sigma_{y\theta} = E\varepsilon_y = \frac{Eyd\theta}{(R_n + y)\theta} \tag{10-1}$$

For equilibrium, $\Sigma F = 0$ on cb and $\Sigma M = 0$.

For $\Sigma F = 0 = \int_A (\sigma_{y\theta} dA) = 0$

But $\sigma_{y\theta} = \dfrac{Eyd\theta}{(R_n + y)\theta}$

Hence,

$$\int \sigma_{y\theta}(dA) = \frac{Ed\theta}{\theta} \int_A \frac{ydA}{(R_n + y)} = 0$$

This equation shows that

$$\int_A \frac{ydA}{(R_n + y)} = 0 \text{ as } \frac{Ed\theta}{\theta} \neq 0$$

We will use

$$\int_A \frac{ydA}{(R_n + y)} = 0 \tag{10-2}$$

later on.

Now we will sum moments about NA due to the stress $\sigma_{y\theta}$. We get

$$\int_A \sigma_{y\theta} dAy = \frac{Ed\theta}{\theta} \int_A \frac{y^2 dA}{(R_n + y)} = M \qquad \text{as } \sigma_{y\theta} = \frac{Eyd\theta}{(R_n + 4)\theta}$$

Letting $\dfrac{y^2}{R_n + y} = y - \dfrac{R_n y}{(R_n + y)}$, we obtain

$$M = \frac{Ed\theta}{\theta}\left[\int_A ydA - R_n \int_A \frac{ydA}{R_n + y}\right]$$

Recall that the second term in the square brackets equals zero, as shown in Equation (10–2). Therefore,

$$M = \frac{Ed\theta}{\theta}\left[\int_A ydA\right]$$

But $\int_A ydA = Ae$. Consequently,

$$M = \frac{Ed\theta}{\theta}[Ae]$$

where e is the distance from the NA to the CG of the cross section.

Solving for E, the modulus of elasticity, we get

$$E = \frac{M\theta}{d\theta eA} \tag{10-3}$$

But

$$\sigma_{y\theta} = E\left[\frac{yd\theta}{(R_n + y)\theta}\right].$$

Substituting for E from Equation (10–3) results in

$$\sigma_{y\theta} = \frac{M\theta}{d\theta eA}\left[\frac{yd\theta}{(R_n + y)\theta}\right]$$

Simplifying, we obtain

$$\sigma_{y\theta} = \frac{My}{eA(R_n + y)} \tag{10–4}$$

This is a general formula for the bending stress at a distance y due to an external moment M.

Now we can find the stress in the extreme fibers at distances C_i and C_o:

$$\sigma_{i\theta} = \frac{MC_i}{eA(R_n - C_i)} \tag{10–5}$$

$$\sigma_{i\theta} = \frac{MC_i}{eAR_i} \tag{10–6}$$

Similarly,

$$\sigma_{o\theta} = \frac{MC_o}{eAR_o} \tag{10–7}$$

where C_i is the distance from NA to the innermost fiber, C_o is the distance from NA to the outermost fiber, $\sigma_{i\theta}$ is the stress at the innermost fiber, and $\sigma_{o\theta}$ is the stress at the outermost fiber.

However, we still need an equation to determine e, which is still an unknown. This can be done by finding the radius to NA. From Equation (10–2),

$$\int_A \frac{ydA}{R_n + y} = 0$$

But $y = \rho - R_n$, or $\rho = y + R_n$. Substituting for y and $R_n + y$ in Equation (10–2), we get

$$\int_A \frac{(\rho - R_n)dA}{\rho} = \int_A dA - \int_A \frac{R_n dA}{\rho} = 0$$

On integration and simplification, we rearrange to solve for R_n and obtain

$$R_n = \frac{A}{\int_A \frac{dA}{\rho}}. \tag{10–8}$$

R_n can be determined using Equation (10–8), as we can find A and $\int_A dA/\rho$ for a given cross section.

Now we have an equation relating e to R_n and R_g:

$$e = R_g - R_n \tag{10-9}$$

$$e = R_g - \frac{A}{\int_A \dfrac{dA}{\rho}}$$

Knowing e, y, A, M, and R_n we can now solve for the stress at a distance y for a curved beam:

$$\sigma_{y\theta} = \frac{My}{eA(R_n + y)}$$

Note that e depends upon R_g, A, and $\int_A dA/\rho$. The last quantity can be a difficult integration for certain cross sections. Therefore, a lead model was developed that can solve for stresses in 17 different cross sections.

First, let us consider an example and solve it conventionally. Then we will solve the same problem using the lead model.

Example 10-1

Let the rectangular cross-sectional curved beam shown in the following diagram have an inside fiber radius of 3" subjected to a pure bending moment $M = 40,000$ in-lb.

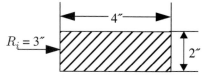

Determine R_n, e, $\sigma_{i\theta}$ and $\sigma_{o\theta}$.

Solution

We know

$$R_n = \frac{A}{\int_A dA/\rho}$$

We will determine $\int_A dA/\rho$ for a rectangular cross section.

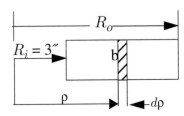

However, from the foregoing figure, we see that

$$\int_A dA/\rho = \int_{R_i}^{R_o} \frac{bd\rho}{\rho} = b|ln\rho|_{R_i}^{R_o} = b[ln\ R_o - ln\ R_i] = b[ln\ (R_o/R_i)]$$

Therefore,

$$R_n = \frac{A}{bln(R_o/R_i)}$$

But $A = 4 \times 2 = 8$ in^2, $b = 2''$, $R_i = 3''$, and $R_o = 7''$. Hence,

$$R_n = \frac{8}{2ln(7/3)}$$

Thus, $R_n = 4.72109$. (You'll need to calculate R_n to at least five decimal places.) Now,

$$e = R_g - R_n$$

$$= 5 - 4.72109 = 0.27891''$$

(Again, you need to carry the e value to at least five decimal places.) Next,

$$C_i = R_n - R_i$$

$$= 4.72109 - 3$$

$$= 1.721$$

and

$$\sigma_{i\theta} = \frac{MC_i}{eAR_i}$$

$$= \frac{40,000(-1.721)}{0.27911(8)(3)}$$

$$\sigma_{i\theta} = -10,276\ \text{psi}$$

Similarly,

$$\sigma_{o\theta} = \frac{MC_o}{eAR_o}$$

$$= \frac{40,000(2.279)}{0.27911(8)(7)}$$

$$= 5,832 \text{ psi}$$

As stated earlier, a curved-beam analysis, design, or optimization requires great accuracy in calculating e, R_n, etc. Again, if the cross section of a curved beam changes, for example, from a square beam to a T-beam or to an I-beam, all the calculations would have to be repeated. Thus, the work is cumbersome and susceptible to errors.

To avoid such problems, the following analysis of curved beams will be based on making use of well-devised equations to solve for 17 different cross sections. Explanations are given in detail on the next few pages, and it is recommended that the reader study them carefully. An understanding of them will save time in solving problems related to curved beams and will increase the likelihood to obtain error-free solutions.

Some of the previous figures, equations, and explanations pertaining to curved beams are repeated for review and convenience.

10.3 Analysis, Design, and Optimization for Curved Beams of Various Cross Sections Using a TK Lead Model

We begin this section with a short review of what we have studied so far about curved beams and computer complementary equations to determine bending stresses in different cross sections of such beams. In a curved beam subjected to an in-plane bending moment M, the stress at any distance y from the neutral axis is given by

$$\sigma_\theta = \frac{My}{Ae(R_n + y)} \tag{10–10}$$

where σ_θ is the stress in the θ-direction, M is the external moment, A is the cross-sectional area of the beam, R_n is the radius to the neutral axis (NA), e is the distance between NA and the centroidal axis, and y is the distance from the NA to the stress σ_θ. The distribution of stress in the beam is shown in Fig. 10–1.

Equation (10–10) is sensitive to even small variations in e; thus considerable care must be exercised in determining e accurately. The equation for e is

$$e = R_g - R_n \tag{10–11}$$

where R_g is the radius to the centroidal axis of a cross section of the curved beam and R_n is the radius to the neutral axis of the beam and is given by

$$R_n = \frac{A}{\int_A dA/\rho} \tag{10–12}$$

Equation (10–12) shows that one has to perform integration by parts if the area A is not a continuous function of ρ. Again, if the cross section changes, the value of e changes. Thus, the task of calculating e becomes cumbersome. Sometimes, not only the dimensions of a given cross section change, but also the type of cross section may change. For example, one may want to go from a T-section to a circular cross section. Accordingly, one may want to ask "What if?" and "How can I?" questions. To answer such questions would require extensive programming, and even then, not all the design questions might be answered. To minimize these difficulties, we present the necessary equations to *analyze, design,* and *optimize* curved beams. While these equations were developed for the geometry shown in Fig. 10–2(a), which consists of four trapezoids stacked on top of each other, they can be applied to many different cross sections, including the 17 forms shown in Fig. 10–2(b). This is possible because each of these sections can be made from the four trapezoids by varying their dimensions. For the equations developed, and also for the implementation of these equations and a description of the variables used in them, see Tables 1 and 2.

To answer "What if" and "How can I?" questions, TK Solver is used. Examples of the analysis, design, and optimization of curved beams using different cross sections are provided.

Sample Solutions Using a Lead Model

The next three examples present the three modes of operation of the lead-model implementation of the curved-beam equations. We use the lead model called *ex10-1.tk,* or *curvedb.tk*. The three modes are *analysis, design,* and *optimization*. Analysis consists of calculating forces or stresses for a known geometry. Design consists of determining one dimension of a geometry for known forces and stresses. Finally, optimization is used to determine the geometry that meets certain force and stress requirements with the smallest cross-sectional area and, therefore, using the least amount of material. A Variables Sheet corresponding to each of these cases is included.

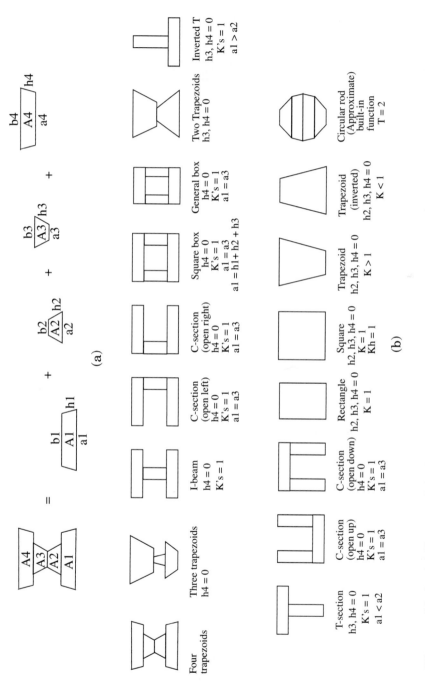

Figure 10–2 (a) Trapezoidal I-beam cross section used for developing equations required for solving problems involving curved beams. (b) Different cross sections that can be analyzed, designed, and optimized using the theory and lead model

TABLE 1

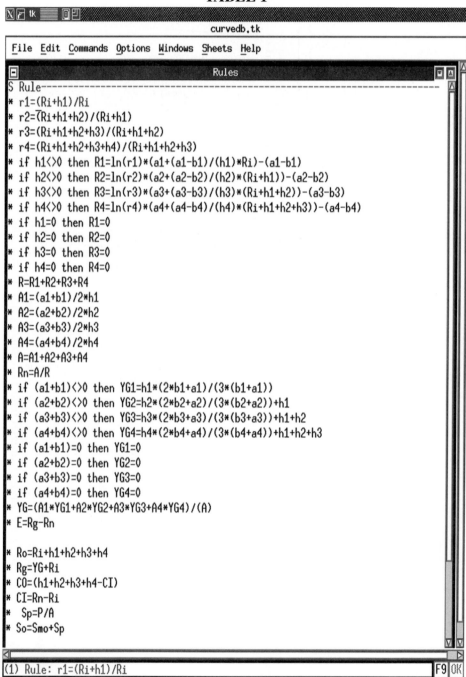

```
curvedb.tk

File  Edit  Commands  Options  Windows  Sheets  Help

                              Rules
S Rule--------------------------------------------------------------
* r1=(Ri+h1)/Ri
* r2=(Ri+h1+h2)/(Ri+h1)
* r3=(Ri+h1+h2+h3)/(Ri+h1+h2)
* r4=(Ri+h1+h2+h3+h4)/(Ri+h1+h2+h3)
* if h1<>0 then R1=ln(r1)*(a1+(a1-b1)/(h1)*Ri)-(a1-b1)
* if h2<>0 then R2=ln(r2)*(a2+(a2-b2)/(h2)*(Ri+h1))-(a2-b2)
* if h3<>0 then R3=ln(r3)*(a3+(a3-b3)/(h3)*(Ri+h1+h2))-(a3-b3)
* if h4<>0 then R4=ln(r4)*(a4+(a4-b4)/(h4)*(Ri+h1+h2+h3))-(a4-b4)
* if h1=0 then R1=0
* if h2=0 then R2=0
* if h3=0 then R3=0
* if h4=0 then R4=0
* R=R1+R2+R3+R4
* A1=(a1+b1)/2*h1
* A2=(a2+b2)/2*h2
* A3=(a3+b3)/2*h3
* A4=(a4+b4)/2*h4
* A=A1+A2+A3+A4
* Rn=A/R
* if (a1+b1)<>0 then YG1=h1*(2*b1+a1)/(3*(b1+a1))
* if (a2+b2)<>0 then YG2=h2*(2*b2+a2)/(3*(b2+a2))+h1
* if (a3+b3)<>0 then YG3=h3*(2*b3+a3)/(3*(b3+a3))+h1+h2
* if (a4+b4)<>0 then YG4=h4*(2*b4+a4)/(3*(b4+a4))+h1+h2+h3
* if (a1+b1)=0 then YG1=0
* if (a2+b2)=0 then YG2=0
* if (a3+b3)=0 then YG3=0
* if (a4+b4)=0 then YG4=0
* YG=(A1*YG1+A2*YG2+A3*YG3+A4*YG4)/(A)
* E=Rg-Rn

* Ro=Ri+h1+h2+h3+h4
* Rg=YG+Ri
* CO=(h1+h2+h3+h4-CI)
* CI=Rn-Ri
*  Sp=P/A
* So=Smo+Sp

(1) Rule: r1=(Ri+h1)/Ri                                        F9 OK
```

TABLE 1 (Continued)

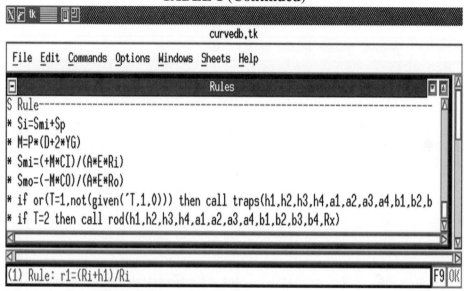

curvedb.tk

File Edit Commands Options Windows Sheets Help

| Rules |

```
S Rule-----------------------------------------------------------
* Si=Smi+Sp
* M=P*(D+2*YG)
* Smi=(+M*CI)/(A*E*Ri)
* Smo=(-M*CO)/(A*E*Ro)
* if or(T=1,not(given('T,1,0))) then call traps(h1,h2,h3,h4,a1,a2,a3,a4,b1,b2,b
* if T=2 then call rod(h1,h2,h3,h4,a1,a2,a3,a4,b1,b2,b3,b4,Rx)
```

(1) Rule: r1=(Ri+h1)/Ri   F9 OK

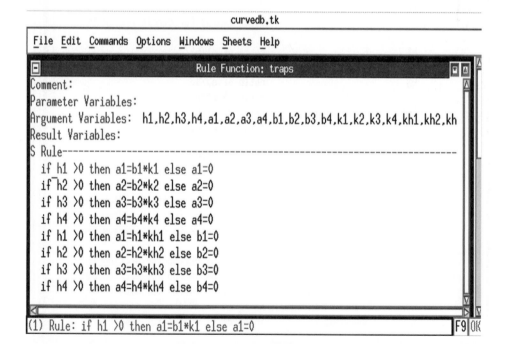

curvedb.tk

File Edit Commands Options Windows Sheets Help

| Rule Function: traps |

```
Comment:
Parameter Variables:
Argument Variables:  h1,h2,h3,h4,a1,a2,a3,a4,b1,b2,b3,b4,k1,k2,k3,k4,kh1,kh2,kh
Result Variables:
S Rule-----------------------------------------------------------
  if h1 >0 then a1=b1*k1 else a1=0
  if h2 >0 then a2=b2*k2 else a2=0
  if h3 >0 then a3=b3*k3 else a3=0
  if h4 >0 then a4=b4*k4 else a4=0
  if h1 >0 then a1=h1*kh1 else b1=0
  if h2 >0 then a2=h2*kh2 else b2=0
  if h3 >0 then a3=h3*kh3 else b3=0
  if h4 >0 then a4=h4*kh4 else b4=0
```

(1) Rule: if h1 >0 then a1=b1*k1 else a1=0   F9 OK

TABLE 2

```
X  tk       
                              curvedb.tk

 File  Edit  Commands  Options  Windows  Sheets  Help

                              Variables
St Input---- Name--- Output--- Unit----- Comment------------------------
             T                            geom. type:1=traps, 2=rod
  -
    .1       a1                 in         base width of sec. 1 (in,ft,m,mm)
    .2       b1                 in         top width of section 1 (in,ft,m,mm)
    .1       h1                 in         height of section 1 (in,ft,m,mm)
             a2       0         in         base width of section 2 (in,ft,m,mm)
             b2       0         in         top width of section 2 (in,ft,m,mm)
    0        h2                 in         height of section 2(in,ft,m,mm)
             a3       0         in         base width of section 3 (in,ft,m,mm)
             b3       0         in         top width of section 3 (in,ft,m,mm)
    0        h3                 in         height of section 3 (in,ft,m,mm)
             a4       0         in         base width of section 4 (in,ft,m,mm)
             b4       0         in         top width of section 4 (in,ft,m,mm)
    0        h4                 in         height of section 4 (in,ft,m,mm)

    3.1      Rx                 in         outer radius for circle rod or tube

             k1       .5                   ratio of a1/b1
             k2                            ratio of a2/b2
             k3                            ratio of a3/b3
             k4                            ratio of a4/b4
             kh1      1                    ratio of a1/h1
             kh2                           ratio of a2/h2
             kh3                           ratio of a3/h3
             kh4                           ratio of a4/h4

             r1       1.0333333            ratio of outer to inner radius sec. 1
             r2       1                    ratio of outer to inner radius sec. 2
             r3       1                    ratio of outer to inner radius sec. 3
             r4       1                    ratio of outer to inner radius sec. 4

             R1       .00490951 in         integral of dA/rho for section 1
             R2       0         in         integral of dA/rho for section 2
             R3       0         in         integral of dA/rho for section 3
             R4       0         in         integral of dA/rho for section 4
             R        .00490951 in         integral of dA/rho for entire area

(1) Input:                                                        F9 OK
```

TABLE 2 (Continued)

```
┌─────────────────────────────────────────────────────────────────────────┐
│ X  tk      ▦                                                              │
├─────────────────────────────────────────────────────────────────────────┤
│                              curvedb.tk                                   │
├─────────────────────────────────────────────────────────────────────────┤
│  File  Edit  Commands  Options  Windows  Sheets  Help                     │
├─────────────────────────────────────────────────────────────────────────┤
│ ▣                              Variables                            ▣ ▣   │
│ St Input---- Name--- Output--- Unit----- Comment-------------------       │
│              A1      .015      in^2     cross sectional area of section 1 │
│              A2      0         in^2     cross sectional area of section 2 │
│              A3      0         in^2     cross sectional area of section 3 │
│              A4      0         in^2     cross sectional area of section 4 │
│              A       .015      in^2     tot. cross sec.ar.(in^2,ft^2,m^2,mm^2) │
│                                                                           │
│              YG1     .05555556 in       dist. to CG of section 1 (in,ft,m,mm) │
│              YG2     0         in       dist. to CG of section 2 (in,ft,m,mm) │
│              YG3     0         in       dist. to CG of section 3 (in,ft,m,mm) │
│              YG4     0         in       dist. to CG of section 4 (in,ft,m,mm) │
│              YG      .05555556 in       dist. fr. inner fib. to CG (in,ft,m,mm │
│                                                                           │
│              Rn      3.0552924 in       radius to neutral axis (in,ft,m,mm) │
│              Rg      3.0555556 in       radius to CG (in, ft, m, mm)       │
│              E       .0002632  in       distance between neutral axis and CG │
│              D                 in       dist. from load to center of curvature │
│                                                                           │
│  3           Ri                in       radius to inner fiber (ft, m, mm) │
│              Ro      3.1       in       radius to outer fiber (ft, m, mm) │
│                                                                           │
│              CI      .05529235 in       distance from neutral axis to inside │
│              CO      .04470765 in       distance from neutral axis to outside │
│                                                                           │
│              M                 lb-in   applied moment (lb-in, lb-ft, Nm)  │
│              P                 lb       Axial Load (tension)(lb, N, kN)    │
│                                                                           │
│              Si                psi      Stress@Inner Fiber (psi, ksi, Pa, MPa) │
│              So                psi      Stress@ Outer Fiber(psi, ksi, Pa, MPa) │
│              Ks                         Negative Ratio of So/Si            │
│                                                                           │
│                                                                           │
│              Sp                psi      Principal Stress (psi, ksi, Pa, MPa) │
│                                                                           │
│              Smo               psi      Outside Ben. Mom. (psi, ksi, Pa, MPa) │
│              Smi               psi      Inside Ben. Mom. (psi, ksi, Pa, MPa) │
│ ◄                                                                       ►  │
├─────────────────────────────────────────────────────────────────────────┤
│ (1) Input:                                                       F9  OK   │
└─────────────────────────────────────────────────────────────────────────┘
```

TABLE 3

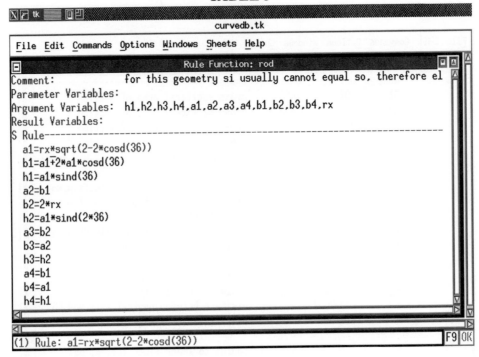

```
 X  tk        curvedb.tk

 File  Edit  Commands  Options  Windows  Sheets  Help

                    Rule Function: rod
 Comment:               for this geometry si usually cannot equal so, therefore el
 Parameter Variables:
 Argument Variables:  h1,h2,h3,h4,a1,a2,a3,a4,b1,b2,b3,b4,rx
 Result Variables:
 S Rule--------------------------------------------------------------
   a1=rx*sqrt(2-2*cosd(36))
   b1=a1+2*a1*cosd(36)
   h1=a1*sind(36)
   a2=b1
   b2=2*rx
   h2=a1*sind(2*36)
   a3=b2
   b3=a2
   h3=h2
   a4=b1
   b4=a1
   h4=h1

 (1) Rule: a1=rx*sqrt(2-2*cosd(36))                                F9 OK
```

Example 10-2

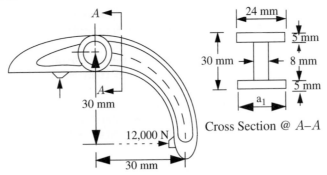

Cross Section @ *A–A*

For the rocker arm shown in the preceding diagram, determine the maximum tensile stress in section *A–A*.

Solution

As can be seen from the diagram, this problem involves the analysis of a curved I-beam with known dimensions and known loads. The I-beam can be simulated as three trapezoids. The following are the input conditions that must be set:

$a1 = .024$ (width of base of first section)

$b1 = .024$ (width of top of first section)

$h1 = .005$ (height of first section)

$a2 = .008$ (width of base of second section)

$b2 = .008$ (width of top of second section)

$a3 = .024$ (width of base of third section)

$b3 = .024$ (width of top of third section)

$h3 = .005$ (height of third section)

$h4 = 0$ (height of fourth section—set to 0 to eliminate)

$Rg = .03$ (radius to center of gravity)

$D = 0$ (distance to center of curvature)

$P = 12,000$ (applied axial load in N)

Be sure to delete all other input fields by entering a space.

Now press F9 to solve. This should produce the following output values:

Si = 205,866,505 (stress at inner fiber, 205.8 MPa tension; see output field in Table 2)

So = –68,622,168 (stress at outer fiber, 68.6 MPa compression)

Note that other values, such as R_g, e, etc., are produced in the output field in Table 2.

Example 10-3

The following figure shows a portion of a C-clamp.

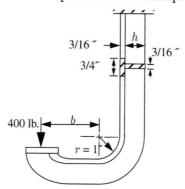

For the maximum force of 400 lb to be exerted by the screw, what is the minimum web height h if the maximum allowable tension stress is 30 ksi and $b = 2''$.

Solution

For this problem, all but one of the dimensions will be entered, along with the applied load and resulting stress. Accordingly, the following values must be entered into the Variables Sheet of the lead model called *curvedb.tk*, with necessary changes:

$a1$ = .75 (width of base of first section; see Table 4)

$b1$ = .75 (width of top of first section)

$h1$ = .2875 (height of first section)

$a2$ = .1875 (width of base of second section)

$b2$ = .1875 (width of top of second section)

$h3$ = 0 (height of third section—set to 0 to eliminate)

$h4$ = 0 (height of fourth section—set to 0 to eliminate)

Ri = 1 (radius of curvature for inner fiber)

D = 2 (distance from load to center of curvature)

P = 400 (applied axial load)

Si = 30,000 (allowable maximum tensile stress, psi)

The new model called *ex10-3.tk* is on your disk. Upon solving, we get

$h2$ = .65775

Thus, the web height must be approximately 0.66 inch. (See the output field in Table 4.)

Example 10-4

The circular clip in the following diagram can be made with a trapezoidal cross section of radial thickness 0.125:

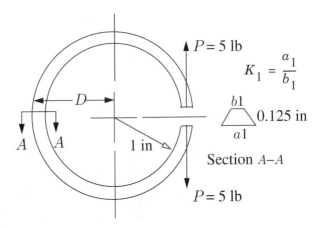

TABLE 4

X	루	tk		回 인					

ex10-3.tk

File Edit Commands Options Windows Sheets Help

			Variables			回 回	凸

St	Input----	Name---	Output---	Unit-----	Comment-------------------------------
		T			geom. type:1=traps, 2=rod
	.75	a1		in	base width of sec. 1 (in,ft,m,mm)
	.$\overline{75}$	b1		in	top width of section 1 (in,ft,m,mm)
	.1875	h1		in	height of section 1 (in,ft,m,mm)
	.1875	a2		in	base width of section 2 (in,ft,m,mm)
	.1875	b2		in	top width of section 2 (in,ft,m,mm)
		h2	.65775049	in	height of section 2(in,ft,m,mm)
		a3	0	in	base width of section 3 (in,ft,m,mm)
		b3	0	in	top width of section 3 (in,ft,m,mm)
	0	h3		in	height of section 3 (in,ft,m,mm)
		a4	0	in	base width of section 4 (in,ft,m,mm)
		b4	0	in	top width of section 4 (in,ft,m,mm)
	0	h4		in	height of section 4 (in,ft,m,mm)
		Rx		in	outer radius for circle rod or tube
		k1	1		ratio of a1/b1
		k2	1		ratio of a2/b2
		k3			ratio of a3/b3
		k4			ratio of a4/b4
		kh1	4		ratio of a1/h1
		kh2	.2850625		ratio of a2/h2
		kh3			ratio of a3/h3
		kh4			ratio of a4/h4
		r1	1.1875		ratio of outer to inner radius sec. 1
		r2	1.5538951		ratio of outer to inner radius sec. 2
		r3	1		ratio of outer to inner radius sec. 3
		r4	1		ratio of outer to inner radius sec. 4
		R1	.12888769	in	integral of dA/rho for section 1
		R2	.0826434	in	integral of dA/rho for section 2
		R3	0	in	integral of dA/rho for section 3
		R4	0	in	integral of dA/rho for section 4
		R	.21153109	in	integral of dA/rho for entire area

(3) Input: .75

F9 OK

TABLE 4 (Continued)

X ʎ tk	⊡⊡		ex10-3.tk

File Edit Commands Options Windows Sheets Help

```
                                   Variables                                    ⊡⊡⊿
St Input----  Name---  Output---  Unit-----  Comment----------------------------
             A1       .140625    in^2       cross sectional area of section 1
             A2       .12332822  in^2       cross sectional area of section 2
             A3       0          in^2       cross sectional area of section 3
             A4       0          in^2       cross sectional area of section 4
             A        .26395322  in^2       tot. cross sec.ar.(in^2,ft^2,m^2,mm^2)

             YG1      .09375     in         dist. to CG of section 1 (in,ft,m,mm)
             YG2      .51637524  in         dist. to CG of section 2 (in,ft,m,mm)
             YG3      0          in         dist. to CG of section 3 (in,ft,m,mm)
             YG4      0          in         dist. to CG of section 4 (in,ft,m,mm)
             YG       .29121536  in         dist. fr. inner fib. to CG (in,ft,m,mm

             Rn       1.2478223  in         radius to neutral axis (in,ft,m,mm)
             Rg       1.2912154  in         radius to CG (in, ft, m, mm)
             E        .04339303  in         distance between neutral axis and CG
   2         D                   in         dist. from load to center of curvature

   1         Ri                  in         radius to inner fiber (ft, m, mm)
             Ro       1.8452505  in         radius to outer fiber (ft, m, mm)

             CI       .24782233  in         distance from neutral axis to inside
             CO       .59742815  in         distance from neutral axis to outside

             M        1316.4861  lb-in      applied moment (lb-in, lb-ft, Nm)
   400       P                   lb         Axial Load (tension)(lb, N, kN)

   30000     Si                  psi        Stress@Inner Fiber (psi, ksi, Pa, MPa)
             So       -35698.01  psi        Stress@ Outer Fiber(psi, ksi, Pa, MPa)
             Ks                             Negative Ratio of So/Si


             Sp       1515.4201  psi        Principal Stress (psi, ksi, Pa, MPa)

             Smo      -37213.43  psi        Outside Ben. Mom. (psi, ksi, Pa, MPa)
             Smi      28484.58   psi        Inside Ben. Mom. (psi, ksi, Pa, MPa)
```

(3) Input: .75   F9 OK

The inner radius of the clip should be 1 inch, and when the clip is loaded with a maximum load of 5 lb, the maximum tensile stress should not exceed 40,000 psi. Determine the optimum dimensions of the clip.

Solution

This problem is somewhat more complex than the previous two. To solve it, we must calculate the dimensions that satisfy the stress requirements, while varying one of the parameters. Then, that parameter can be plotted against the cross-sectional area. The value of the parameter that corresponds to the minimum cross-sectional area is obviously the optimum solution for the given conditions.

The first values to be set are the geometrical constraints, which should be entered into the Variables Sheet as follows:

$h1$ = .125 (height of first section; see Table 5)

$h2$ = 0 (height of second section—set to 0 to eliminate)

$h3$ = 0 (height of third section—set to 0 to eliminate)

$h4$ = 0 (height of fourth section—set to 0 to eliminate)

Ri = 1 (radius of curvature for inner fiber)

Next, the stress and force restraints must be established. To do this, enter the following given values into the Variables Sheet:

P =5 (applied axial load; see Table 5)

Si = 40,000 (allowable maximum tensile stress—output field)

It can be seen from the figure of the clip that the highest moment will occur opposite the gap at section A–A. The 5-lb applied load will act at this point, combined with a moment equal to the load times twice the radius to the CG. The bending moment is calculated according to the equation

$$M = P \times (D + R_g) \quad \text{(see Table 1)}$$

where D is the eccentric distance. The following restraint equation must be entered into the Rules Sheet to satisfy this condition:

$$D = R_g$$

Next, it must be decided what parameter to vary. By varying the value of k1, the ratio of the widths of the bottom and top of the trapezoid, a wide range of possible sections can be examined. This requires the creation of a list for k1. To do this, we open the List Sheet. Then we type in the name of the list to be created—k1 in this case—and press Enter. With that line highlighted, we select Commands from the menu bar and then select the Fill List command. Now we must input the range of values to be considered. We enter a starting value of 1

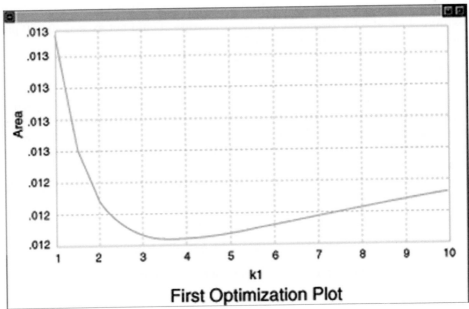

Figure 10–3 First optimization plot

and an ending value of 10, with a step size of 0.5. Then we press Fill List. In the List Sheet, there should now be a 19 next to k1. This means that there are 19 values in that list. Next, return to the Variables Sheet and put an L into the status field next to k1, thereby identifying k1 as a list variable. Also, we put a I into the input field of k1 (you will not see I). Then put an L and an O in the Status Field for A.

Finally, it will be necessary to input a guessed value for b1. Simply putting a G into the status field as mentioned before will not work for list solving. To enter the guessed value, the Variables Subsheet corresponding to the variable b1 must be opened. In this subsheet, the First Guess field must be set. We enter a value of 0.01 and then, when the computer is ready, press <SHIFT>F9 or click on list solve in the option menu to list solve.

The computer will now solve for values corresponding to each of the 19 values in the k1 list. To interpret the data, it is now necessary to produce a plot of A vs. k1. To do this, we open the Plot Sheet, enter the word "Area" or any other appropriate plot title on the first line, and press Enter, whereupon we enter the corresponding subsheet. In this subsheet, we type k1 in the X-Axis List field and A under Y-Axis. Then we press F7 to display the plot. From this first optimization plot shown in Fig. 10–3, it can be seen that the area is smallest when k1 is between 3.5 and 4. To get a better approximation of this optimal value, the input list for k1 can be changed to show only values from 3.5 to 4. Plotting the results

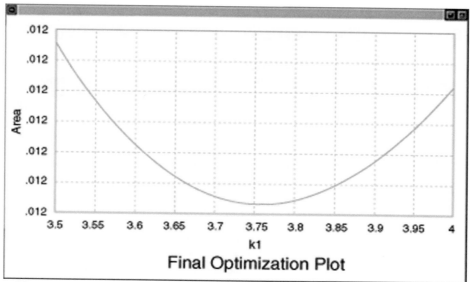

Figure 10–4   Final optimization plot

the optimal value of k1 is nearly 3.76, as shown in the second optimization plot in Fig. 10–4.

It is then necessary to determine the values of a1 and b1 that correspond to this optimal value of k1. To do this, we enter the value of 3.76 into the input field for k1 in the Variables Sheet, and then press F9 to solve. This should produce the following results:

a1 = .152077 in (see Table 5)

b1 = .04044601 in (Output Field)

The cross-sectional area is then about 0.01203 sq in. Now we can conclude that

1. The single set of equations that was developed can be used to analyze, design, and optimize curved beams of 17 different cross sections subjected to external loads and/or moments.

2. By using a software package called TK Solver, one can *analyze*, *design*, and *optimize* machine elements in a very short period of time.

3. TK Solver allows designers to get answers to "What if?" and "How can I?" questions without writing elaborate programs.

4. The equations employed in analysis, design, and optimization must be well understood for the engineer to use them

TABLE 5

ex10-4.tk

File Edit Commands Options Windows Sheets Help

Variables

St	Input----	Name---	Output---	Unit-----	Comment----------------------------------
		T			geom. type:1=traps, 2=rod
		a1	.152077	in	base width of sec. 1 (in,ft,m,mm)
		b1	.04044601	in	top width of section 1 (in,ft,m,mm)
	.125	h1		in	height of section 1 (in,ft,m,mm)
		a2	0	in	base width of section 2 (in,ft,m,mm)
		b2	0	in	top width of section 2 (in,ft,m,mm)
	0	h2		in	height of section 2(in,ft,m,mm)
		a3	0	in	base width of section 3 (in,ft,m,mm)
		b3	0	in	top width of section 3 (in,ft,m,mm)
	0	h3		in	height of section 3 (in,ft,m,mm)
		a4	0	in	base width of section 4 (in,ft,m,mm)
		b4	0	in	top width of section 4 (in,ft,m,mm)
	0	h4		in	height of section 4 (in,ft,m,mm)
		Rx		in	outer radius for circle rod or tube
L	3.76	k1			ratio of a1/b1
		k2			ratio of a2/b2
		k3			ratio of a3/b3
		k4			ratio of a4/b4
		kh1	1.216616		ratio of a1/h1
		kh2			ratio of a2/h2
		kh3			ratio of a3/h3
		kh4			ratio of a4/h4
		r1	1.125		ratio of outer to inner radius sec. 1
		r2	1		ratio of outer to inner radius sec. 2
		r3	1		ratio of outer to inner radius sec. 3
		r4	1		ratio of outer to inner radius sec. 4
		R1	.011467	in	integral of dA/rho for section 1
		R2	0	in	integral of dA/rho for section 2
		R3	0	in	integral of dA/rho for section 3
		R4	0	in	integral of dA/rho for section 4
		R	.011467	in	integral of dA/rho for entire area

(5) Input: .125 F9 OK

TABLE 5 (Continued)

```
X  tk        

                              ex10-4.tk

 File  Edit  Commands  Options  Windows  Sheets  Help

                            Variables
St Input----  Name---  Output---  Unit-----  Comment---------------------------
              A1       .01203269  in^2       cross sectional area of section 1
              A2       0          in^2       cross sectional area of section 2
              A3       0          in^2       cross sectional area of section 3
              A4       0          in^2       cross sectional area of section 4
L             A        .01203269  in^2       tot. cross sec.ar.(in^2,ft^2,m^2,mm^2)

              YG1      .05042017  in         dist. to CG of section 1 (in,ft,m,mm)
              YG2      0          in         dist. to CG of section 2 (in,ft,m,mm)
              YG3      0          in         dist. to CG of section 3 (in,ft,m,mm)
              YG4      0          in         dist. to CG of section 4 (in,ft,m,mm)
              YG       .05042017  in         dist. fr. inner fib. to CG (in,ft,m,mm

              Rn       1.0493322  in         radius to neutral axis (in,ft,m,mm)
              Rg       1.0504202  in         radius to CG (in, ft, m, mm)
              E        .00108794  in         distance between neutral axis and CG
              D        1.0504202  in         dist. from load to center of curvature

    1         Ri                  in         radius to inner fiber (ft, m, mm)
              Ro       1.125      in         radius to outer fiber (ft, m, mm)

              CI       .04933222  in         distance from neutral axis to inside
              CO       .07566778  in         distance from neutral axis to outside

              M        10.504202  lb-in      applied moment (lb-in, lb-ft, Nm)
    5         P                   lb         Axial Load (tension)(lb, N, kN)

    40000     Si                  psi        Stress@Inner Fiber (psi, ksi, Pa, MPa)
              So       -53554.48  psi        Stress@ Outer Fiber(psi, ksi, Pa, MPa)
              Ks                             Negative Ratio of So/Si



              Sp       415.53476  psi        Principal Stress (psi, ksi, Pa, MPa)

              Smo      -53970.01  psi        Outside Ben. Mom. (psi, ksi, Pa, MPa)
              Smi      39584.465  psi        Inside Ben. Mom. (psi, ksi, Pa, MPa)

(5) Input: .125                                                          F9 OK
```

effectively. TK does not allow the program to be used as a "black box."

5.   The ideas developed in this chapter can be extended to analyze, design, or optimize other machine elements as effectively as shown here.

6.   The savings in time and, possibly, a resulting error-free solution will allow engineers to do more design work in less time.

PROBLEMS

10–1)   Work parts (a) and (b) by hand, then do them using the TK model. The TK file name is *curvedb.tk*.

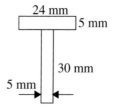

24 mm

5 mm

30 mm

5 mm

a.   Same as Example 10-1, but with preceding cross section.

b.   Same as Example 10-2, except that $P = 5,000$ lb and $M = 40,000$ in-lb, applied at the 5000-lb location.

10–2)   A piston ring with a 6″ inside diameter and a cross section as shown in the following diagram is subjected to a load $P = 3$ lb. (For detailed figure see Example 10-4.)

b

0.5″   0.25″

$R_i = 3″$

a.   Determine R_g the radius to CG, R_n, and the maximum bending stress if $b = 1$.

b.   If the radius is changed to 4″ and P to 5 lb, determine the width b, if the allowable stress is 50,000 psi.

10–3) Given the following tappet, determine the height h if the allowable stress is 200 MPa. (Note load is given in lbs. Do not change it to Newtons, let TK do this.)

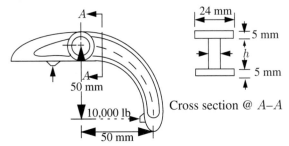

Cross section @ A–A

10–4) Given the following C clamp, determine h if the maximum allowable stress is 30 ksi.

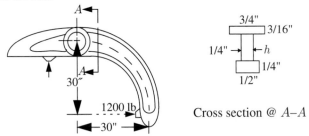

Cross section @ A–A

10–5) For a safety factor of 2 and $b = 2.5''$, find the minimum value of h for the T-beam shown in the following diagram if the allowable $S_{yp} = 40,000$ psi.

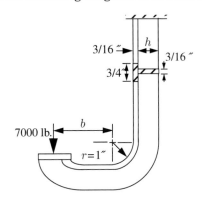

10–6)   Determine the dimension b, if C clamp to carry a load $P = 8000$ lb. The
dimension b should be such that the yield point stress, which is 40,000
psi, is not exceeded either in tension or in compression. Use other
dimensions as shown in the following figure.

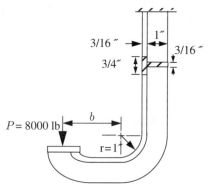

Thin- and Thick-Walled Cylinders

*T*hin- and thick-walled cylinders are used in machine elements such as pressure tanks, hydraulic cylinders, gun barrels, rocket housings, etc. Such cylinders create 3-D stresses when they are subjected to internal and external pressure.

In the next section, we derive the equations for stresses in thin-walled cylinders subjected to internal and external pressure.

11.1 Thin-Walled Cylinder

A cylindrical rubber balloon can be considered an ideal thin-walled cylinder subjected to an external pressure, P_o, and an internal pressure, P_i. (The external pressure can be atmospheric pressure.) In Fig. 11–1, for the element A, if one neglects the radial stress σ_r (which may be done because the cylinder is very thin) and if one also assumes that the external pressure P_o is much less than the internal pressure P_i (nearly zero), then we have only two types of internal stresses to deal with:

1. Tangential stress σ_t

2. Longitudinal stress σ_l

If we cut the cylinder along B–B, then we see a section, as shown in Fig. 11–2(a), in which an internal stress σ_t must balance the internal pressure P_i for static equilibrium. Therefore,

$$\sigma_t(2t)l = P_i(2r)(l)$$

It follows that

$$\sigma_t = \frac{2P_i r l}{2tl} = \frac{P_i r}{t} \tag{11–1}$$

225

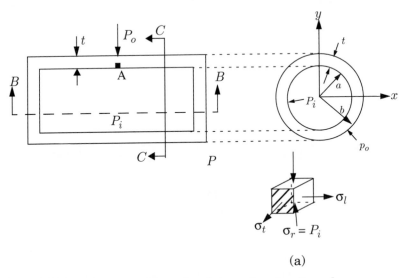

Figure 11-1   Cylinder subjected to internal and external pressures, (a) stressed element at A.

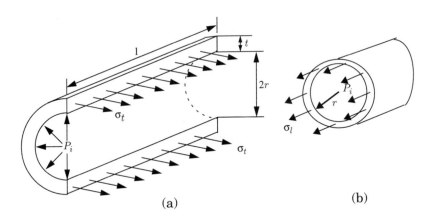

Figure 11-2   (a) Longitudinal cross section of cylinder. (b) Transverse.

Similarly, if we cut the cylinder along C–C as shown in Fig. 11–2(b), then, for transverse equilibrium, one needs σ_l, the longitudinal stress, which can be related to an internal pressure P_i by the equation

$$\sigma_l(2\pi r t) = P_i(\pi r^2)$$

or

$$\sigma_l = \frac{P_i r}{2t} \tag{11-2}$$

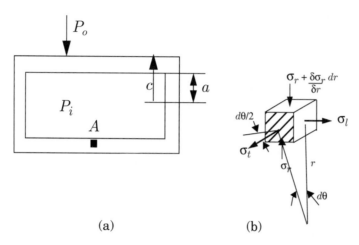

Figure 11–3   Cross-section of thick walled cylinder and stress element A

Note that the transverse stress σ_t is also called the *hoop stress*. The longitudinal stress σ_l is half the hoop stress. The hoop stress is a critical stress, as it causes failure in thin-walled cylinders. Also, using the standard sign convention for a thin-walled cylinder subjected only to an internal pressure, we can easily see that the principal stress $\sigma_1 = \sigma_t$, the other principal stress $\sigma_2 = \sigma_l$, and $\sigma_3 = 0$. Therefore, the maximum shear stress developed internally is one-half of the hoop stress σ_t, that is,

$$\tau_{max} = P_i r / 2t = (\sigma_1 - \sigma_3)/2.$$

11.2 Thick-Walled Cylinders

When a thick-walled cylinder (see Fig. 11–3(a)) is subjected to an internal and an external pressure, stresses in the tangential, longitudinal, and radial directions are created. If the cylinder is homogeneous and isotropic, and if we select an element A from it, the stresses acting on such an element can be shown to be σ_l, σ_t, and σ_r—the longitudinal, hoop, and radial stress, respectively.

The element shown in Fig. 11–3(b) is in equilibrium. Therefore, by summing the forces in the r-direction and realizing that $\sin d\theta/2 = d\theta/2$ for small angles, we find that

$$\sigma_r r d\theta - (\sigma_r + d\sigma_r)(r + dr)d\theta - 2\sigma_t \frac{d\theta}{2} = 0$$

Neglecting second-order terms results in

$$\sigma_r d\theta - (\sigma_r r - \sigma_r dr - r d\sigma_r)d\theta - \sigma_t d\theta = 0$$

Rearranging and canceling terms gives

$$\sigma_t - \sigma_r - \frac{r d\sigma_r}{dr} = 0 \tag{11-3}$$

However, another equation is needed, since there are two unknowns: σ_t and σ_r. Such an equation can be obtained by realizing that, in the absence of longitudinal stresses, the strain in the longitudinal direction is constant; that is,

$$\varepsilon_z = -v\frac{\sigma_t}{E} - v\frac{\sigma_r}{E}$$

$$\varepsilon_z = -\frac{v}{E}(\sigma_t + \sigma_r)$$

where v is Poisson's ratio and E is the modulus of elasticity of the cylinder. (One can assume a longitudinal stress σ_l that is still a constant; therefore, it does not change the derivation.)

On rearranging the foregoing equation, we can show that

$$-\sigma_t - \sigma_r = -2c_1 \tag{11-4}$$

where

$$-2c_1 = \varepsilon_z\frac{E}{v}$$

Adding Equation (11–3) to (11–4) produces

$$-2\sigma_r - r\frac{d\sigma_r}{dr} = -2c_1 \tag{11-5}$$

and multiplying Equation (11–5) by r gives

$$-2r\sigma_r - r^2\frac{d\sigma_r}{dr} = -2c_1 r$$

But

$$(d/dr)(r^2\sigma_r) = 2r\sigma_r + r^2\frac{d\sigma_r}{dr}$$

Therefore,

$$\frac{d}{dr}(r^2\sigma_r) = 2c_1 r \tag{11-6}$$

Integrating gives

$$r^2\sigma_r = c_1 r^2 + c_2$$

$$\sigma_r = c_1 + c_2/r^2$$

We can evaluate c_1 and c_2 by applying the boundary conditions $\sigma_r = P_i$ at $r = a$ and $\sigma_r = P_o$ at $r = c$, which yields

$$c_1 = \frac{P_i a^2 - P_o c^2}{c^2 - a^2} \qquad c_2 = \frac{a^2 c^2 (P_o - P_i)}{c^2 - a^2}$$

By substituting for c_1 and c_2, we then get

$$\sigma_r = \frac{P_i a^2 - P_o c^2 + (a^2 c^2 (P_o - P_i)/r^2)}{c^2 - a^2} \qquad (11-7)$$

Similarly, by realizing that $\sigma_t = 2c_1 - \sigma_r$, we can show that

$$\sigma_t = \frac{P_i a^2 - P_o c^2 - (a^2 c^2 (P_o - P_i)/r^2)}{c^2 - a^2} \qquad (11-8)$$

Notice that simplified expressions can be obtained for σ_r and σ_t if either P_i or P_o is zero. Simplified equations are unnecessary, however, since TK Solver needs equations in the forms shown in Equations (11–7) and (11–8), and all one has to do is set P_i or P_o equal to zero if it is zero.

We have earlier shown that, for an element on a thick-walled cylinder, the equation for longitudinal stress can be shown to be

$$\sigma_l = \frac{P_i a^2}{c^2 - a^2} \qquad (11-9)$$

We can now conclude that the required derivation for the radial stress (σ_r), hoop stress (σ_t), and longitudinal stress (σ_l) in a thick-walled cylinder with end caps and subjected to an internal pressure (P_i) and external pressure (P_o) has been completed.

The following example applies the preceding equations to problems involving thick-walled cylinders.

Example 11-1

A thick-walled cylinder with flat end caps is subjected to an internal $P_i = 500$ psi and an external pressure $P_o = 15$ psi. If the internal radius $a = 6''$ and the external radius $c = 8''$, then, for radial distances $r = a$, $(a + c)/2$ and c:

    a.   Determine σ_r, σ_t, and σ_l at the locations mentioned.

    b.   Determine the distribution of σ_r , σ_t, and σ_l across the thickness of the cylinder.

c.   Determine the maximum shear stress, principal stress, von Mises stress, and principal strain at $r = a$.

d.   Solve parts (a) and (b) using the lead model by incrementing r by $(c - a)/20$. Obtain a plot of σ_r vs. r, σ_t vs. r, and σ_l vs. r.

Solution

We know that

$$\sigma_r = \frac{p_i a^2 - p_o c^2 + a^2 c^2 (p_o - p_i)/r^2}{c^2 - a^2}$$

Now, for $r = 6"$, $P_i = 500$ psi, $P_o = 15$ psi, $a = 6"$, and $c = 8"$,

$$\sigma_r = \frac{500(6^2) - 15(8^2) + 6^2 8^2 (15 - 500)/(6)^2}{8^2 - 6^2}$$

$\sigma_r = -500$ psi

Similarly, σ_t and σ_l can be found by using Equations (11–8) and (11–9). We get

$\sigma_t = 1717.14$ psi

$\sigma_l = 360.00$ psi

For $r = 7"$, we get

$\sigma_t = -205$ psi

$\sigma_t = 1423.03$ psi

$\sigma_l = 360.00$ psi

Similarly, for $r = 8"$,

$\sigma_r = -15$ psi

$\sigma_t = 1232.14$ psi

$\sigma_l = 360.00$ psi

This problem has been partly solved using the lead model called *ex11-1.tk*, or *twc.tk*. See the Rules and Variables Sheets for the solution. Try to solve the rest of the example by modifying the original lead model.

RULES

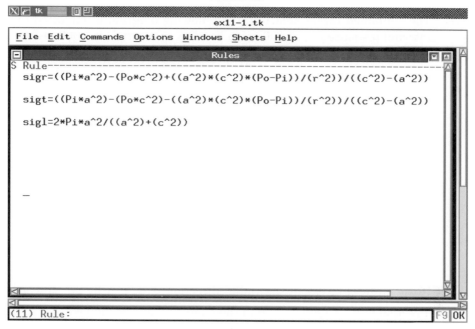

```
X⌐ tk      回凹
                        ex11-1.tk
 File  Edit  Commands  Options  Windows  Sheets  Help
 ⊟                         Rules                        回△
S Rule─────────────────────────────────────────────────
  sigr=((Pi*a^2)-(Po*c^2)+((a^2)*(c^2)*(Po-Pi))/(r^2))/((c^2)-(a^2))

  sigt=((Pi*a^2)-(Po*c^2)-((a^2)*(c^2)*(Po-Pi))/(r^2))/((c^2)-(a^2))

  sigl=2*Pi*a^2/((a^2)+(c^2))



  ─



(11) Rule:                                          F9 OK
```

VARIABLES

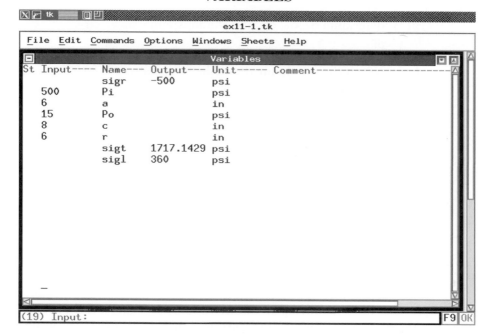

```
X⌐ tk      回凹
                        ex11-1.tk
 File  Edit  Commands  Options  Windows  Sheets  Help
 ⊟                      Variables                       回△
St Input────  Name───  Output───  Unit─────  Comment──────────
             sigr     -500       psi
   500       Pi                  psi
   6         a                   in
   15        Po                  psi
   8         c                   in
   6         r                   in
             sigt     1717.1429  psi
             sigl     360        psi











  ─


(19) Input:                                         F9 OK
```

PROBLEMS

11–1) The simplified rocket shown in the following diagram is subjected to an internal pressure p = 10,000 psi, a bending moment M = 5,000 in-lb, and a torque of 10,000 in-lb.

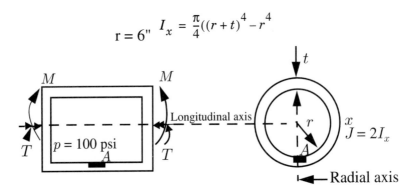

$$I_x = \frac{\pi}{4}\left((r+t)^4 - r^4\right)$$
$$r = 6"$$
$$J = 2I_x$$

If the state of stress in the radial direction is $\sigma_r = p$, in the tangential direction is $\sigma_t = (pr)/t$, and in the longitudinal direction is $\sigma_l = pr/2t$ on an element at A, then show all the possible stresses acting on an element at A, and

    a.   determine σ_1, σ_2, σ_3, and the maximum shear stress in terms of t.

    b.   Now if the maximum allowable normal stress is 80,000 psi and the maximum allowable shear stress is 47,000 psi, determine the thickness t required to support all the preceding loads with a safety factor equal to 3.6, based on the applied pressure, moment, and torque.

11–2) Do Problem 11–1, assuming that the thick-walled cylinder stress equations apply, and internal pressure is increased to 800 psi.

11–3) Solve Problem 11–2 using the lead model *twc.tk*, and check your answers.

For the following two problems, you need to have an understanding of the theory of failure.

11–4) A pressure cylinder having an internal radius of 6" and a thickness of 1" is subjected to an internal pressure of 5,000 psi, a twisting torque of

50,000 in-lb, and a bending moment of 50,000 psi. The cylinder is shown in the following diagram:

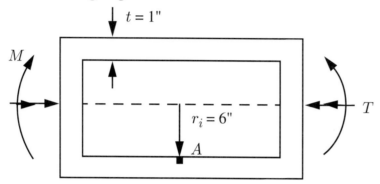

If the transverse stress due to internal pressure, σ_t is 32,692.31 psi, the longitudinal stress due to internal pressure, σ_l is 13,846.15 and the radial stress due to internal pressure, σ_r is 5000 psi at $r = r_i = 6"$,

    a.   Determine the bending and torsional stress at $r = 6"$.

    b.   Superimpose the preceding stresses on the pressure stresses, and determine the principal stresses σ_1, σ_2, and σ_3 and the maximum shear stress.

        (Answer: $\sigma_1 = 32693.92$, $\sigma_2 = 14190.21$, $\sigma_3 = -5000$).

    c.   Determine the safety factor using the normal stress theory, maximum shear stress theory, and distortion theory if $S_{yp} = 40,000$ psi. (Answer: $SF_n = 1.223$, $SF_s = 1.06$, $SF_e = 1.225$)

11–5)   Assume that the cylinder of Problem 11–4 has a yield-point stress in tension of 40,000 psi and a yield-point stress in compression of 80,000 psi. Using the modified Mohr theory, determine the safety factor if the cylinder now has $\sigma_1 = 33,000$ psi, $\sigma_2 = 15,000$, and $\sigma_3 = 50,000$ psi as resultant stresses.

11–6)   Given a thick-walled cylinder and $\sigma_{yp} = 40,000$ psi, $P_i = 774$ psi, $r_i = 6"$, $SF_{energy} = 4.0$, $M = 10,000$ in-lb, and $T = 10,000$ in-lb, determine r_o and t. Do this first by hand, assuming that $t = 0.53''$, and solve for SF_{energy}. If your answer is near 4, then go to TK, type in the Rules Sheet necessary equations, and find the exact value of t. Check your results for the modified Mohr theory if $\sigma_{ypC} = 1.2\sigma_{ypT}$. Make sure that the Rules Sheet is correctly modified. Plot σ_l, σ_t, and σ_r as a function of r ($r \rightarrow r_i$ to r_o).

11–7)   Prove Equation (11–8).

Design of Machine Components

12.1 Introduction

*D*esigning machines involves designing machine components or machine elements and then assembling them to create a machine. A thorough background in mechanics, including the mechanics of materials, is needed for these designs. The student should therefore review the theory of statics, dynamics, and mechanics of materials, plus the first 11 chapters of this textbook.

A machine is a device for doing useful work. Early machines were relatively crude. Modern machines, such as internal combustion engines, turbines, jet engines, rockets, space shuttles, robots, etc., are far more sophisticated than machines of earlier generations and can perform much more challenging tasks. Ironically, today's machines become tomorrow's crude machines, despite the fact that the reliability of today's machines is very high.

There are many kinds of machines in existence, and each one fulfills a specific task. At times, machine components are classified on the basis of the task they perform. Table 12–1 is just one of the many ways to classify machine components.

Each machine element must be designed for its service life. General considerations in the design of machine elements are as follows:

1. Kinematic analysis: Motion of the parts or kinematics of the machine, especially if the parts or the machine is subjected to linear or angular acceleration.

2. Kinetics: Force and strength analysis of the type of load and the stresses caused by the load; can be carried out experimentally or analytically. (See Chapters 3 and 4.)

3. Strength and rigidity analysis: Selection of materials based on strength and rigidity.

4. Form and size of parts.

Table 12–1

FUNCTION OF COMPONENT	NAME OF COMPONENT
Fastening/joining	Rivets, screws, nuts, bolts, pins, snap rings, keys, washers, couplings, universal joints, etc.
Energy storing	Springs (coil, torsion bar, beam, leaf, disk), flywheels, etc.
Power transmission and speed control	Belts, pulleys, chains, sprockets, ropes, gears, power screws, linkages, etc.
Carrying and transmission of rotary motion	Axial shafts, bearings, linkages, etc.
Transmission and control of rotary power	Pistons, connecting rods, flywheels, cams, brakes, and clutches
Sealing-enclosing and controlling fluid-serving elements	Seals, gaskets, valves, bushings, fittings, piston rings, hand wheels, levers, steering wheels, cranks, linkages, etc.

5. Frictional resistance, lubrication, wear, corrosion, and environmental deterioration.

6. Convenience and economic factors.

7. Use of standard parts.

8. Workshop facilities and manufacturability.

9. Cost of planning, materials, and construction.

10. Safety.

11. Ecology: avoiding toxic waste, air and water pollution, and solid waste.

12.2 Concept of Work, Power, and Energy

Machine design requires understanding concepts involving work and power. The work performed by an object moving under the action of a force is given by

$$W = \int_{r_1}^{r_2} \overline{F} \cdot d\overline{r} \text{ (a scalar product)}$$

$= \bar{F} \cdot \bar{q}$ (force multiplied by displacement in the direction of the force)

If F and q are in the same direction, then $W = Fq$. Note also that $W = T\theta$ where T is the torque, θ is the angular displacement in radians, and the work input is equal to the work output plus work lost due to friction, etc. Therefore, all machines have an efficiency of less than 100%. The efficiency of a machine is defined as

$$\eta = \frac{\text{work output}}{\text{work input}}$$

and the percent efficiency as

$$\eta\% = \frac{\text{work output}}{\text{work input}} \times 100$$

Power is defined as the capacity of a machine to do work in a unit of time. One horsepower is defined as the capacity of a machine to do 550 ft-lb of work per second, or 33,000 ft-lb of work per minute; that is,

$$\text{HP} = \frac{2\pi n T}{550}$$

If n is in revolutions per second (rps) and T is in ft-lb and

$$\text{HP} = \frac{2\pi n T}{33,000}$$

If n is in revolutions per minute (rpm) and T is in ft-lb and

$$\text{HP} = \frac{Fv}{33000}$$

or

$$\text{HP} = \frac{T\omega}{550}$$

or

$$\text{HP} = \frac{Tn}{5252}$$

where F is in lb, v is in ft/min., ω is in radians per second, T is in ft-lb, and n is in rpm.

Similarly, work per unit time, or power, in International Standard (SI) units is defined as *watts*. One watt is equal to one newton-meter of work per second; that is,

Watt = 1 Nm/sec

Therefore, 1,000 watts, commonly called 1 kilowatt (kW), is 1,000 Nm/sec. Also, 1kWh = 3.6×10^6 joules, 1 BTU = 1,055.87 joules

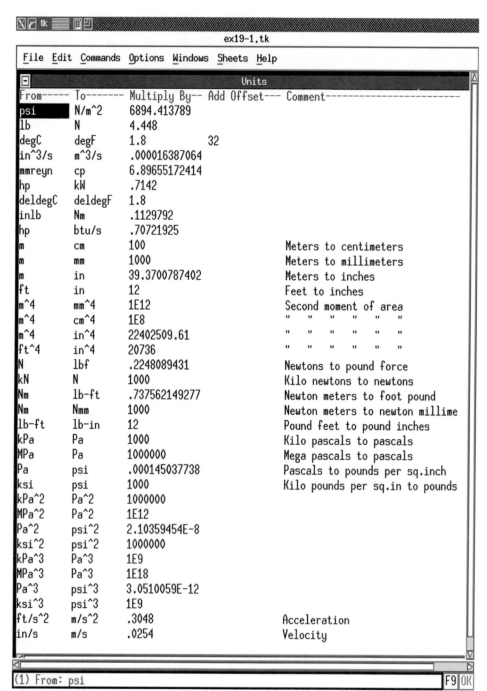

Figure 12–1

```
X  tk        
                         ex19-1.tk
 File  Edit  Commands  Options  Windows  Sheets  Help
 [-]                        Units
From-----  To-------  Multiply By--  Add Offset---  Comment-----------------
in/s       m/s        .0254                         Velocity
ft/s       in/s       12                             "       "
           mm/s       1000                           "       "
km/hr      m/s        .2777777778                    "       "
mph        ft/s       1.466666667                    "       "
km/hr      mph        .621371192146                  "       "
mile       ft         5280                          Distance
mile       m          1609.344                       "       "
g/cm^3     kg/m^3     1000                          Density
lbm/in^3   kg/m^3     27679.905                      "       "
lbm/ft^3   kg/m^3     16.018463                      "       "
slug/ft^3  kg/m^3     515.379                        "       "
Btu        J          1055.87                       Energy
ft-lb      J          1.3558179                      "       "
Kwh        J          3600000                        "       "
klb        lb         1000                          Force
klb        N          9.80665                        "       "
kip        N          4448.22161526                  "       "
lbm        kg         .45359237                     Mass
slug       kg         14.5939029                     "       "
slug       lbm        32.1740484744                  "       "
ton        kg         907.18474                      "       "
ton        lbm        2000                           "       "
Btu/s      W          1054.35026449                 Power
ft-lb/min  W          .022596966                     "       "
ft-lb/s    W          1.3558179                      "       "
Hp         ft-lb/s    550                            "       "
Hp         W          745.699845                     "       "
Hp         kW         .745699845                     "       "
W          kW         .001                           "       "
atm        N/m^2      101325                        Pressure
bar        N/m^2      100000                         "       "
R          K          1.8                           Temperature
day        s          86400                         Time
hr         s          3600                           "       "
min        s          60                             "       "

(1) From: psi                                              F9 OK
```

Figure 12-2

```
┌────────────────────────────────────────────────────────────────┐
│ X 🗗 tk ▓▓▓ 🔲🖺                                                    │
│                            ex19-1.tk                            │
├────────────────────────────────────────────────────────────────┤
│  File  Edit  Commands  Options  Windows  Sheets  Help          │
├────────────────────────────────────────────────────────────────┤
│ □                          Units                             △  │
│ From----- To------- Multiply By-- Add Offset--- Comment--------- │
│ day       s         86400                       Time            │
│ hr        s         3600                        "    "          │
│ min       s         60                          "    "          │
│ ft^2/s    m^2/s     .09290304                   Viscosity       │
│ ft^3      m^3       .028316846592               Volume          │
│ gal       m^3       .003785411748               Liquid gallon   │
│ ft^3      in^3      1728                         Volumm          │
│ m^3       mm^3      1E9                          "    "          │
│ m^3       cm^3      1000000                      "    "          │
│ yard^3    m^3       .764554857984               "    "          │
│ Hz        rad/s     .159154943092               Frequency to radians per second │
│ rad/s     rpm       9.54929658551               Radians per second to revolutio │
│ GPa       Pa        1E9                          Giga pascals to pascals │
│ deg       rad       .0174                        Degrees to radians │
│ lbm-ft^2  kg-m^2    .0421                        Moment of inertia │
│ lbm-in^2  kg-mm^2   293                          "     "     "     " │
│ ◁│                                                            │▽ │
├────────────────────────────────────────────────────────────────┤
│ ◁│                                                            │▷ │
├────────────────────────────────────────────────────────────────┤
│ (1) From: psi                                         │F9│OK│   │
└────────────────────────────────────────────────────────────────┘
```

Figure 12–3

$$\dot{w} \text{ (Power in kW)} = \frac{Fv}{1000}$$

or

$$\dot{w} = \frac{T\omega}{1000}$$

or

$$\dot{w} = \frac{Tn}{9549}$$

where ω is in rad/sec, F is in N, v is in m/s, T is in Nm, and n is in rpm.

Other conversions from FPS to SI units are given in the previous pages. This Unit Sheet is used in many lead models you have utilized so far. In case you need to develop your own models, you can copy the Unit Sheet (see Figures 12–1 through 12–3) from the *stress3.tk* lead model to your model. You must make sure that the symbols you will use in the Variables Sheet match the symbols used in the Unit Sheet. For example, pounds per square inch must be typed as psi in the units column of the Variables Sheet, and not as lb/sq", PSI, etc.

Power Screws and Bolted Connections

13.1 Introduction

Bolts, nuts, cap screws, set screws, rivets, spring retainers, locking devices, and keys fall under the category of fasteners. In this chapter, we consider power screws and then bolts. Typical types of threads used in both of these are shown in Fig. 13–1. The definitions we will use in our study of the analysis or design of power screws and bolts, are as follows:

Pitch p: the distance between two adjacent thread forms (inches or millimeters).

Major diameter d: the largest diameter of a screw or a bolt (peak of threads).

Minor diameter d_r: the smallest diameter of a screw or a bolt (root of threads).

Mean diameter d_m: the average of d and d_r: $d_m = (d + d_r)/2$.

Most threads are made according to the right-hand rule (and hence are called right-hand threads) with clockwise tightening and counterclockwise loosening, unless otherwise specified.

The *pitch* p relates various diameters of a screw or a bolt. (However, one is cautioned to use actual values if given.) The chief relationships are:

$$d_r = d - 1.299038p \tag{13–1}$$

and

$$d_m = d - 0.659519p = \frac{d + d_r}{2} \tag{13–2}$$

241

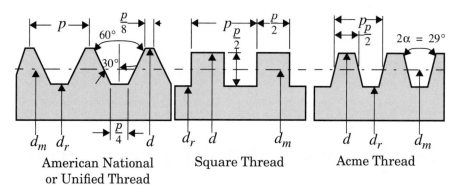

American National   Square Thread   Acme Thread
or Unified Thread

Figure 13–1   Types of thread used in power screws and bolts.

The tension (throat) diameter d_t is the average of d_r and d_m; that is, $d_t = (d_r + d_m)/$ 2. It is this diameter that is related to the tension (throat) area, by the equation

$$A_t = \pi d_t^2 / 4$$

An unthreaded rod having a diameter of d_t has the same strength as a threaded rod having minimum diameter d_r and maximum diameter d. The *proof load* $P_f = A_t S_p$, where S_p is called the proof stress in a bolt at failure by yielding.

13.2 Screw Elements

Power Screws

A power screw, shown in Fig. 13–2, is a device that converts rotational motion into linear motion. The linear motion is relatively slower than the rotational motion, but it creates a much larger output force compared to the input force, thus imparting mechanical advantage. Therefore, power screws are used to get mechanical advantage in lifting weights. Examples of power screws are the following:

Screw Jack: Lifts weights (e.g., a car jack)

Press: Presses parts (used, for example, in pressing of auto fenders)

Testing Machines: Exerts a large force (e.g., a universal tension machine)

Garbage Compactor: Compacts waste (e.g., to minimize the volume of waste put into the environment)

C-clamps: Holds a large force (used, for example, to bond two layers of sheets together)

Gear Screw: Used in aircraft (e.g., landing gears)

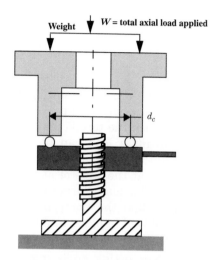

Figure 13–2 Power screw

Adjuster Screw: Used in aircraft (e.g., flap screws)

Micrometer Screw: Precise positioning of axial movement

Lathe Screw: Creates horizontal motion

The nomenclature (see Fig. 13–3) used in power screw analysis or design is as follows:

$$\tan\lambda = \frac{L}{\pi d_m}$$

λ = Lead angle

L = Lead

$L = n_t\, P$

$n_t = 1$ for a single-threaded screw

$n_t = 2$ for a double-threaded screw

$n_t = 3$ for a triple-threaded screw

P = Pitch

d_m = Mean diameter of contact

W = Total axial load applied = Σw

w = Portion of total axial load transmitted

n = Normal force due to weight w

nf = Frictional force

f = Coefficient of friction

q = Tangential force

If a power screw rotates at a constant velocity, it can be considered in static equilibrium, and therefore, static analysis is possible. Static analysis is required to determine the relation between the torque applied and the load lifted or lowered. The torque required to lift a load can be used to find the horsepower required if one knows the rotational speed either in radians per unit time, rpm, or rps. Most power is required at the time of starting a power screw to rotate because the static coefficient of friction is larger than the kinematic coefficient of friction. Therefore, we always find the horsepower required to start a power screw moving with an external load in place, and then, as the screw rotates with a constant angular velocity, the horsepower requirements will drop.

Fig. 13–3 shows the forces acting on a screw thread when one applies torque to lift a load W. Clearly, for every rotation πd_m, the nut rises or drops by an amount equal to the lead L. Thus, we can define the angle λ by $\tan\lambda = L/(\pi d_m)$.

If the forces in the x-direction and y-direction are summed and set equal to zero (because the system is in static equilibrium), then taking $\Sigma F_x = 0$ yields

$$q - nf\cos\lambda - n\cos\alpha_n\sin\lambda = 0$$

Rearranging, we get

$$q - n(f\cos\lambda + \cos\alpha_n\sin\lambda) = 0 \qquad (13\text{–}3)$$

Similarly, taking $\Sigma F_y = 0$ yields

$$w + nf\sin\lambda - n\cos\alpha_n\cos\lambda = 0$$

Rearranging again, we obtain

$$w + n(f\sin\lambda - \cos\alpha_n\cos\lambda) = 0 \qquad (13\text{–}4)$$

Solving Equation (13–4) for n results in

$$n = \frac{w}{\cos\alpha_n\cos\lambda - f\sin\lambda}$$

and now solving for q using Equation (13–3) and substituting for n from the preceding equation, we get

$$q = w\frac{f\cos\lambda + \cos\alpha_n\sin\lambda}{\cos\alpha_n\cos\lambda - f\sin\lambda}$$

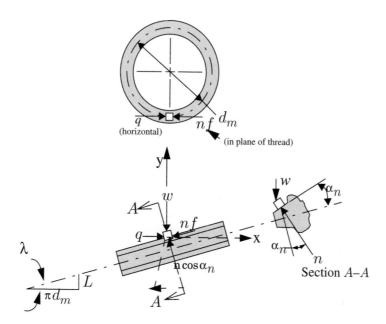

Figure 13–3 Power screw with loads and reactions.

Now we can find the torque that is required to lift the load w.

$$\Delta T = q \frac{d_m}{2}$$

$$= \frac{wd_m}{2} \frac{f \cos\lambda + \cos\alpha_n \sin\lambda}{\cos\alpha_n \cos\lambda - f \sin\lambda}$$

Dividing the numerator and the denominator by $\cos\lambda$, substituting $\tan\lambda = L/\pi d_m$, and rearranging yields

$$\Delta T = \frac{wd_m}{2}\left(\frac{f\pi d_m + L\cos\alpha_n}{\pi d_m \cos\alpha_n - fL}\right) \tag{13–5}$$

Here, w is a small part of the total W. Therefore, T is part of the total torque. However, $\Sigma w = W$, so replacing w with W results in the total torque required.

However, if one has a washer or bearing area of diameter d_c, as shown in Fig. 13–2, and if f_c is the coefficient of friction of the collar washer or bearing, then the torque required to overcome the collar friction is $(Wf_c d_c)/2$. Therefore, the total torque is

$$T_{\text{total}} = \frac{Wd_m}{2}\left(\frac{f\pi d_m + L\cos\alpha_n}{\pi d_m \cos\alpha_n - fL}\right) + \frac{Wf_c d_c}{2} \tag{13–6}$$

Now, for square threads, $\alpha_n = 0$. If we now replace $T_{\text{total}} = T_R$, where T_R is the torque required to raise a load W, we have

$$T_R = \frac{Wd_m}{2}\left(\frac{f\pi d_m + L}{\pi d_m - fL}\right) + \frac{Wf_c d_c}{2} \tag{13–7}$$

Thus, Equation (13–6) can be used to determine the torque T required to raise a weight W for any type of thread. By contrast, Equation (13–7) should be used for square threads only. Similarly, if we seek the torque required to lower the load, we change the direction of q and the friction force and take $\cos\alpha_n = 1$ for square threads. Then the torque required to lower the weight W for a square-threaded screw is given by

$$T_L = \frac{Wd_m}{2}\left(\frac{f\pi d_m - L}{\pi d_m + fL}\right) + \frac{Wf_c d_c}{2} \tag{13–8}$$

Ball bearings or thrust bearings can be used to reduce the coefficient of friction f_c. Similarly, one can use lubricant between the screw and nut to reduce the coefficient of friction f. However, too much reduction of friction may cause the load to start lowering on its own. For example, if we define the screw to be self-locking—that is, if T must be greater than zero in order to lower a weight W—then, assuming that $f_c = 0$ it follows that $f\pi d_m - L\cos\alpha_n > 0$. In other words, *self-locking* requires nonzero torque to lower a weight W. This shows that

$$f \geq \frac{L\cos\alpha_n}{\pi d_m} \text{ (for the screw to be self-locking)} \tag{13–9}$$

Therefore, for a square thread ($\cos\alpha_n = 0$),

$$f \geq \frac{L}{\pi d_m} \qquad \text{or} \qquad f \geq \tan\lambda$$

Overhauling Screw:　A power screw is said to be *overhauling* if friction is sufficiently low to enable the load to lower itself. This means that a positive torque is required for an overhauling screw to keep the load from lowering itself on its own.

A screw that is self-locking under static conditions may overhaul when exposed to vibration. Accordingly, special locking devices are used to prevent screws from overhauling.

In sum, if $f = \tan\lambda$, then free lowering is expected; if $f > \tan\lambda$, then self-locking is expected; and if $f < \tan\lambda$, then the load will lower itself (i.e., we have an overhauling screw).

The efficiency η of a power screw is defined as the ratio of the work output to the work input. Since the angle of rotation $\theta = 2\pi$ for one rotation, then

$$\eta = \frac{\text{work output}}{\text{work input}} = \left(\frac{WL}{2\pi T}\right)100\ \%$$

Neglecting collar friction and substituting for T yields

$$\eta = \frac{L}{\pi d_m}\left(\frac{\pi d_m \cos\alpha_n - fL}{\pi f d_m + L\cos\alpha_n}\right) \tag{13-10}$$

For square threads;

$$\eta = \frac{L}{\pi d_m}\left(\frac{\pi d_m - fL}{\pi f d_m + L}\right)$$

$$\eta = \frac{1 - f\tan\lambda}{1 + f\cot\lambda} \tag{13-11}$$

We can now plot η versus λ for a given f to study the effect of L on efficiency. (See Problem 13–2 at the end of the chapter.)

Static Screw Stresses

Subjected to an external torque, power screws are able to move a load W up or down. Such a load creates stresses in the threads of a power screw and in the threads of the nut. One of these stresses may damage the screw threads or nut threads. To avoid power screw failure, the nut and its threads are often made from a softer material, so that if excessive force is applied, the nut threads will get damaged and not the power screw threads. The damaged nut is inexpensive to replace, compared to replacing a damaged power screw.

Assuming that the nut will fail (before the power screw does) due to an accidental excessive load, it is important to know what type of stresses exist in a nut. It is also advisable to check the stresses in a power screw thread. Two types of stress are possible: bearing stress and shear stress. (Bending stress is negligible.)

Stresses in the Nut

One type of stress in a nut is bearing stress. We have

bearing stress $\sigma = W/A_b$ \hfill (13–12)

but bearing area of one thread = $A_b = (\pi/4)(d^2 - d_r^2)$ (see Fig. 13–4)

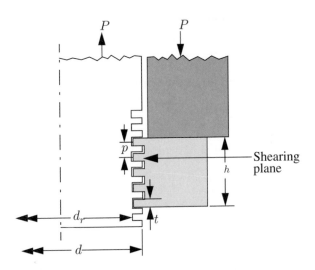

Figure 13–4 Cross section of a square thread

and

Number of threads in a nut = h/p

Therefore,

$$\text{total bearing area} = A_b = \frac{\pi}{4}(d^2 - d_r^2)h/p$$

where d is the outside diameter of the nut or screw thread, d_r is the inside diameter of the nut or screw thread, h is the thickness of a nut, and p is the pitch (distance between two consecutive threads) in the nut or screw.

Now taking

$$\sigma = \frac{W}{A_b}$$

we obtain

$$\sigma = \frac{4Wp}{\pi h(d^2 - d_r^2)} \qquad (13\text{–}13)$$

Similarly, the shear stress is given by

$$\tau = \frac{W}{\text{area of shear}}$$

but

$$\text{area of shear} = \pi d \frac{h}{p} t$$

Therefore,

$$\tau = \frac{Wp}{\pi d h t} \tag{13-14}$$

where t is the thickness (at the root) of the nut thread, h is the height of the nut, and p is the pitch in the nut.

One can use a similar approach to derive the equation for stresses in power screw threads.

Power screws are also subject to shear stresses due to the torque T produced in raising a load W. This torque stress is given by

$$\tau_T = \frac{k_c T C}{J} \tag{13-15}$$

where T is the external net torque applied (excluding collar friction torque), $C = d_r/2$, $J = (\pi/32)(d_r)^4$, d_r is the root diameter of a power screw, and k_c is the stress concentration factor (an average value of 1.3 to 1.6 is acceptable unless otherwise specified).

One should also realize that the axial load W may create buckling in a power screw, and therefore the length of the screw must be checked to see whether buckling will occur. To prevent buckling, the power screw should be made to act like a strut. (See Chapter 6). However, if the column length is such that buckling is possible, one can use a hollow or solid power screw and design it using either Johnson's formula or Euler's formula.

The formulas for stresses that we have just derived are also applicable to nut stresses in a bolted connection. (See Section 13.3 on page 251.)

The following example applies the preceding formulas to the design of power screws.

Example 13-1

A power screw having a *square double thread* and a major diameter $d = 40$ mm is to carry a 100-kN load. The friction coefficient for the screw is 0.16 and has lead $L = 10$ mm. The collar diameter $d_c = 100$ mm and has a friction coefficient of 0.12, and the nut is rotating at 5 rpm. Determine:

a. the pitch and the minimum friction required for nonoverhauling

b. the power required to raise the load

c. the efficiency of the screw

d. the lead angle λ

e. the bearing stress in the thread if the height of the nut is 40 mm

f. the torque shear stress.

Solution

Since the screw is double threaded,

$$\text{pitch} = \frac{L}{2} = \frac{10}{2} = 5\,\text{mm}$$

$$d_m = d - 0.50p = 40 - 0.5(5) = 37.50\,\text{mm}$$

$$d_r = d - p = 40 - 5 = 35\,\text{mm}$$

The minimum friction is

$$f_r = \tan(L/((\pi)d_m))$$

$$= \tan\left(\frac{10}{(3.14)(37.50)}\right)$$

$$= 0.084 < 0.16 \text{ (nonoverhauling)}$$

Let us evaluate T_R in parts, that is,

$$A = \frac{Fd_m}{2} = \frac{100000(38.5)}{2(1000)} = 1925\,\text{Nm}$$

$$B = f\pi d_m + L\cos(1) = (0.12)(3.14)(36.70) + 10(1) = 24$$

$$C = \pi d_m - fL = (3.14)(36.70) - 0.16(10) = 114.00$$

(*Note*: The values of B and C were not divided by 1000.)

Now, the torque required to overcome friction in the screw is

$$T_s = A \times B/C = \frac{1925(29.00)}{114.00}$$

$$T_s = 473.42\,\text{Nm}$$

The torque required to overcome friction in the collar is

$$T_c = Ff_c d_c/2$$

$$= \frac{100000}{1000}(0.12)(100)/2$$

$$T_c = 600\,\text{Nm}$$

    a.   The total torque required to raise the load W is
$$T_R = T_s + T_c = 473.2 + 600$$
$$= 1073.2\,\text{Nm}$$

    b.   Power input $P_I = \dfrac{T_R 2\pi n}{60(1000)} = \dfrac{1073.2(2\pi)5}{60000}$
$$= 0.5260\,\text{kW}$$

Power output is

$$P_o = \frac{FL_v}{6000}$$

where

velocity $L_v = ((nL/60)/1000)$ m/sec

$$P_o = \frac{100000(5)(10)}{60000}$$

$P_o = 0.0834$ kW

c. $\eta = \frac{0.0834}{0.5620}(100)$

= 15.02%

(Note: $\eta = \frac{WL}{2\pi T}(100)$ could be used)

d. $\lambda = \text{atan}(L/\pi d_m) = \text{atan}(10/\pi(36.70))$

= 4.9°

e.   Number of threads in nut = 40/5 = 8 = N_t

$$\tau_{shear} = \frac{Pp}{\pi d h t} = \frac{100000(5)}{\pi(40)(40)(2)}$$

$\tau_{shear} = 50.00$ MPa

The preceding problem is also solved using the TK lead model called *ex13-1.tk*, or *pscrew.tk*. See the Variables Sheet on page 253 for the answers.

13.3 Bolted Members

Bolted Connections

Bolts are used to connect two members temporarily. If a nut is tightened just enough to prevent separation of the two members with very little pretension in the bolt, and if an external load is applied to a bolted connection, all the load is transferred from one member to another by way of the bolt. Mechanical engineers, however, always tighten (pretension) a bolt to between 75% to 95% of its yield strength. When an external load is applied, some of the load in a bolted connection is taken up by members, and the rest of the load is taken up by the bolt, up to a certain limit. Such pretension in a bolt is created to prevent separation of the connected members and to prevent gases enclosed, say, in a cylinder, from

RULES SHEET

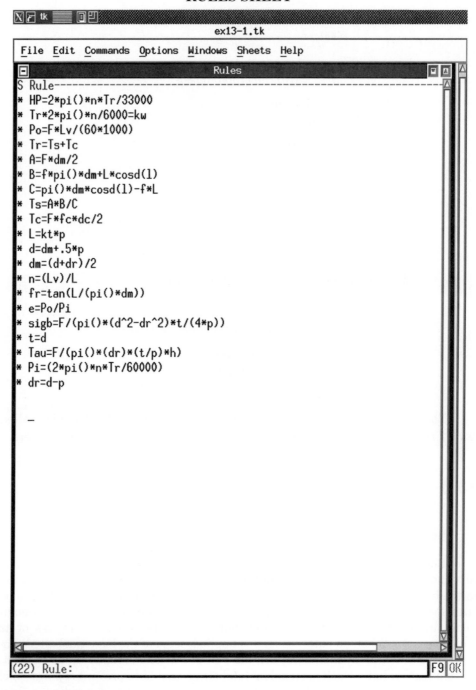

```
S Rule---------------------------------------------
* HP=2*pi()*n*Tr/33000
* Tr*2*pi()*n/6000=kw
* Po=F*Lv/(60*1000)
* Tr=Ts+Tc
* A=F*dm/2
* B=f*pi()*dm+L*cosd(l)
* C=pi()*dm*cosd(l)-f*L
* Ts=A*B/C
* Tc=F*fc*dc/2
* L=kt*p
* d=dm+.5*p
* dm=(d+dr)/2
* n=(Lv)/L
* fr=tan(L/(pi()*dm))
* e=Po/Pi
* sigb=F/(pi()*(d^2-dr^2)*t/(4*p))
* t=d
* Tau=F/(pi()*(dr)*(t/p)*h)
* Pi=(2*pi()*n*Tr/60000)
* dr=d-p
```

```
(22) Rule:                                    F9 OK
```

VARIABLES SHEET

```
X f tk       □ 凹
```

ex13-1.tk

File Edit Commands Options Windows Sheets Help

St	Input---	Name---	Output---	Unit---	Comment---
		HP	1.0143314	HP	Power Required
	5	n		rpm	Rate of Rotation
		T			
		kw	5.5788229		
		Ne			
		Po	.08333333		Power output
	100000	F			Applied Force
		v			
		Tr	1065.4767		Torque to raise load
		Ts	465.47669		Torque component from screw.
		Tc	600		Torque component from collar
		A	1875		Ts calculation variable
		dm	.0375	mm	Mean Screw Diameter
		B	.02884956		Ts calculation variable
	.16	f			Friction coefficient of scre
	10	L		mm	Lead of screw
	0	l			
		C	.11620972		Ts calculation variable
	.12	fc			Friction coefficient of coll
	100	dc		mm	Diameter of collar
		p	.005		Pitch of screw
		dr	35	mm	Inside diameter of screw
	40	d		mm	Outside diameter
		w			
		Lv	.05	m/sec	Velocity of screw
		fr	.08508709		Min. frict. coeff. for non-o
	2	kt			
		e	.1493744		Efficiency
		sigb	42441318		
		t	40	mm	Thickness at root of nut
		Tau	2842052.6		Shear stress in thread
		Pi	.55788229		Power input
	40	h		mm	Height of nut

(36) Input: F9 OK

leaking. A good example of where such pretension is required is when one connects a head to a block of an engine. The pretension in the bolts creates enough sealing pressure to prevent any internal pressure from leaking out.

Let us analyze bolted connections in which there is extensive pretension in bolts connecting two members. The derivation that follows is for a bolted connection in which the external load applied to the system is shared by the bolts and the members connected by the bolts. The aim is to enable the reader to determine what portion of the total external load P is carried by the bolt and what portion is carried by the connected members. The derivation is used in the lead model *bolt.tk*, developed and saved on the accompanying disk.

Consider, then, the bolted connection shown in Fig. 13–5. The connection is subjected to an external load P. The bolt has initial tension obtained by applying a torque and tightening the nut. Naturally, the initial tension creates compression in the members. When the external load P is applied to the joint, the load is divided between the bolt and the connected members. To know exactly what portion of P is carried by the bolt and what portion by the members, it is necessary to define certain terms used in the derivation. The following is the nomenclature used in deriving the required equations.

F_i: preload (initial tension) on bolt (see Fig. 13–5)

P: external load on bolted assembly

P_b: portion of P taken by bolt

P_m: portion of P taken by member

F_b: resultant bolt load = $F_i + P_b$

F_m: resultant member load = $P_m - F_i$

The application of the external load P, which is divided between the bolt and the members, creates additional elongation in the bolt and the members, that is,

$$\Delta\delta_b = P_b/K_b \tag{13–16}$$

where $\Delta\delta_b$ is the additional (tensile) elongation (the bolt is already elongated due to the initial tension F_i), P_b is the portion of P carried by the bolt, and K_b is the spring constant for the bolt. Similarly, the additional elongation in the members is

$$\Delta\delta_m = P_m/K_m \tag{13–17}$$

where P_m is the portion of P carried by the members and K_m is the spring constant for the members. (The members were already compressed due to initial tension.)

Now, the additional elongation in the bolt is equal to the additional elongation in the members; that is,

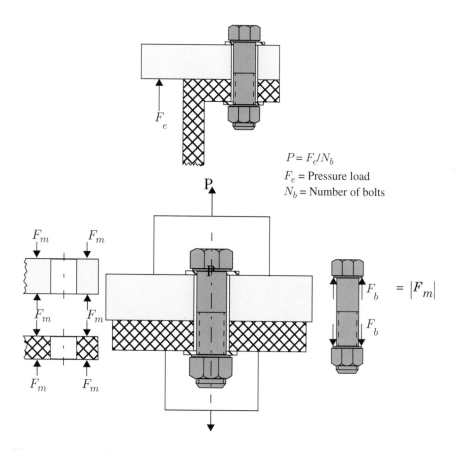

Figure 13–5 Bolted connection showing external load P, resultant load F_m on members, and resultant load F_b on bolt

$$\Delta\delta_b = \Delta\delta_m \tag{13–18}$$

Substituting Equations (13–16) and (13–17) into Equation (13–18) results in

$$P_b / K_b = P_m / K_m \tag{13–19}$$

But for static equilibrium, the total external load is equal to the sum of the loads taken by the members and the bolt; therefore,

$$P = P_m + P_b \tag{13–20}$$

Solving Equations (13–19) and (13–20) for P_b and P_m simultaneously reveals that

$$P_b = \frac{K_b}{K_b + K_m} P = K_1 P \tag{13–21}$$

where

$$K_1 = \frac{K_b}{K_b + K_m}$$

Similarly, it can be shown that

$$P_m = P - P_b$$

$$= P - \frac{PK_b}{K_b + K_m}$$

$$= P(1 - K_1)$$

so that

$$P_m = K_2 P \qquad\qquad (13\text{--}22)$$

where

$$K_2 = 1 - K_1 = \frac{K_m}{K_m + K_b} \qquad\qquad (13\text{--}23)$$

Summarizing, we see that the resultant load in the bolt is

$$F_b = P_b + F_i$$

$$F_b = K_1 P + F_i \qquad\qquad (13\text{--}24)$$

and the resultant load in the members is

$$F_m = P_m - F_i$$

$$F_m = K_2 P - F_i \qquad\qquad (13\text{--}25)$$

The distribution of F_b and F_m is shown in Fig. 13–6.

As the load P is increased, compression in the members approaches zero, and at A the members are on the verge of separation. At this point, any further increase in P has to be carried by the bolt alone. However, to prevent separation of the members or to make pressure-proof members, the resultant load in the members is always maintained in compression (i.e., it is not allowed to go to zero). This is necessary to prevent any leakage of fluids or gases.

The foregoing equations require that one know the values of K_b and K_m. Accordingly, we next consider how to determine these values. From an introductory course in the mechanics of materials, we learn that, for the bolt,

$$K_b = \frac{A_b E_b}{l_b} \qquad\qquad (13\text{--}26)$$

where K_b is the spring constant, A_b is the cross-sectional area of the bolt (based on the major diameter), l_b is the length of the bolt between the base of its head

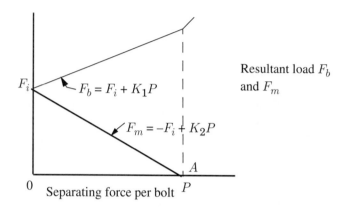

Figure 13–6   Distribution of resultant load in bolt and members

and the nut (which is equal to the thickness of the members), and E_b is the modulus of elasticity of the bolt. Similarly,

$$K_m = \frac{A_m E_m}{l_m}$$ (13–27)

where A_m is the cross-sectional area of each member subjected to compression, l_m is the length of each member, and E_m is the modulus of elasticity of the members.

It is obvious that the value of A_m is not as easy to obtain as that of A_b, because A_m changes as a function of the thickness of the members. The simpler, but still reasonable, method presented next allows one to determine K_m, the spring stiffness of the members, by accounting for the change in A_m.

Spring Stiffness of Bolted Members

Consider two members of length l_1 and l_2 connected by a bolt of diameter d. We assume that washers are introduced between the bolt head and the members, as well as between the nut and the members. We also assume that the washers have a diameter equal to d_w. Let the load distribution in the member due to the load in the bolt be of a conical type making an angle α, as shown in Fig. 13–7. Now, the area at a distance x for member 1 is based on geometry and is

$$A_{m1x} = \pi(r_x^2 - (d/2)^2)$$

But

$$r_x = \frac{d_w}{2} + x \tan \alpha$$

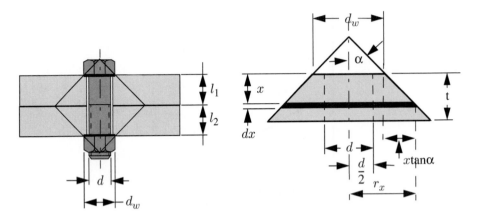

Figure 13-7 Pressure distribution in members.

Therefore,

$$A_{m1x} = \pi\left[\left(\frac{d_w}{2} + x\tan\alpha\right)^2 - (d/2)^2\right]$$

Remembering that $a^2 - b^2 = (a + b)(a - b)$, we rearrange the preceding equation and get

$$A_{m1x} = \pi\left[\left(\frac{d_w + d}{2} + x\tan\alpha\right)\left(\frac{d_w - d}{2} + x\tan\alpha\right)\right] \qquad (13\text{-}28)$$

Earlier, we showed that the elongation in a member is given by

$$\Delta\delta_{m1} = \frac{P_{m1}dx}{E_{m1}A_{m1x}}$$

where "1" stands for member 1. It follows that

$$\int\Delta\delta_{m1} = \int\frac{P_{m1}dx}{E_{m1}A_{m1x}}$$

But $P_{m1} = P_m$ and $\delta_{m1} = \int\Delta\delta_{m1}$. So, substituting the value of A_{m1x} from Equation (13-28) and rearranging, we get

$$\delta_{m1} = \frac{P_m}{\pi E_{m1}}\int_0^{l_1}\frac{dx}{\left(\left(x\tan\alpha + \frac{d_w + d}{2}\right)\left(x\tan\alpha + \frac{d_w - d}{2}\right)\right)} \qquad (13\text{-}29)$$

Using a table of integrations, we can show that this expression reduces to

$$\delta_{m1} = \left(\frac{P_m}{\pi E_{m1} d \tan\alpha}\right) ln\left(\frac{(2l_1 \tan\alpha + d_w - d)(d_w + d)}{(2l_1 \tan\alpha + d_w + d)(d_w - d)}\right)$$

Now, we know that

$$K_{m1} = \frac{P_m}{\delta_{m1}} \tag{13-30}$$

$$K_{m1} = \frac{\pi E_{m1} d \tan\alpha}{\ln\left(\frac{(2l_1 \tan\alpha + d_w - d)(d_w + d)}{(2l_1 \tan\alpha + d_w + d)(d_w - d)}\right)}$$

Similarly, it can be shown that

$$K_{m2} = \frac{P_m}{\delta_{m2}}$$

so that

$$K_{m2} = \frac{\pi E_{m2} d \tan\alpha}{\ln\left(\frac{(2l_2 \tan\alpha + d_w - d)(d_w + d)}{(2l_2 \tan\alpha + d_w + d)(d_w - d)}\right)} \tag{13-31}$$

Now, if we assume that $l_1 = l_2$ and since the two members are in series, the resultant K_m is given by

$$1/K_m = 1/K_{m1} + 1/K_{m2}$$

or

$$K_m = \frac{K_{m1}K_{m2}}{K_{m1} + K_{m2}} \tag{13-32}$$

It is now possible to determine the values of K_m, K_b, etc. and to find the load shared by a bolt or members when the connection is subjected to an external load P. Note, however, that the above derivation was based on the assumption that $l_1 \approx l_2$. If l_1 and l_2 are different, the analysis must be modified.

The next example is presented to further illustrate the application of the equations we have derived.

Example 13-2

Given the following pressure cylinder and accompanying data, determine the number of bolts (N_b) required if the total load carried by N_b bolts is $F = 36$ kips ($F = PN_b$):

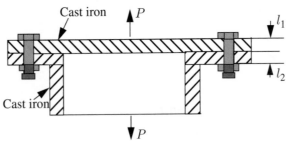

Data		
$E_b = 30 \times 10^6$ psi	$S_p = 85$ ksi	$l_b = 1.5$ in
$d = 5/8"$ in.	$F_i = 0.8\, S_p A_t$	$A_t = 0.226$ in^2
$l_1 = l_2 = 0.75$ in	$E_m = 12 \times 10^6$ psi	$d_w = 1.5d$
SF = safety factor = 2.4	$\alpha = 30^\circ$	

Solution

Since the modulus of elasticity and the lengths of the two connecting members are the same, it is obvious that $K_{m1} = K_{m2}$. Now,

$$K_{m1} = \frac{\pi(12 x 10^6)(0.625)(\tan 30^\circ)}{\ln\left[\dfrac{(2(0.75)\tan 30^\circ + 1.5(0.625) - 0.625)(1.5(0.625) + 0.625)}{(2(0.75)\tan 30^\circ + 1.5(0.625) + 0.625)(1.5(0.625) - 0.625)}\right]}$$

$$= 15.35 \times 10^6 \text{ lb/in}$$

But

$$1/K_m = 1/K_{m1} + 1/K_{m2}$$

$$K_m = \frac{K_{m1} K_{m2}}{K_{m1} + K_{m2}} = \frac{(15.35(10^6))^2}{30.70(10^6)}$$

$$= 7.68(10^6) \text{ lb/in}$$

Similarly,

$$K_b = \frac{A_b E_b}{l_b} = \frac{0.30730(10^6)}{1.5} = 6.1359(10^6) \ \text{lb/in}$$

$$K_1 = \frac{K_b}{K_b + K_m} = \frac{6.1359}{6.1359 + 7.673} = 0.4443$$

$$K_2 = \frac{K_m}{K_m + K_b} = \frac{7.673}{13.8089} = 0.5557$$

(Note that $K_1 + K_2 = 1$.)

It is now known that the initial tension in the bolt is equal to 80% of its proof load, that is,

$$F_i = 0.8 A_t Sp = 0.8(0.226)(85) = 15.368 \ \text{kip}$$

Accounting for the safety factor (SF), we find that the stress in the bolt is

$$\sigma_{bolt} = \frac{(\text{SF})K_1 P}{A_t} + \frac{F_i}{A_t}$$

(Note that F_i is not multiplied by safety factor SF, as F_i is the initial tension.)

The stress in the bolt must be equal to S_p, the proof stress; that is,

$$\sigma_{bolt} = \frac{(\text{SF})K_1(F/N_b)}{A_t} + \frac{F_i}{A_t} = S_p$$

$$\frac{(2.4)0.4443(36/N_b)}{0.226} + \frac{15.368 \times 10^3}{0.226} = 85000$$

Solving for N_b, we get $N_b = 9.99$, so we use 10 bolts.

The TK Solver solution of Example 13-2 uses the TK lead model called *ex13-2.tk*, or *bolt.tk*. Study carefully the output in the Variables Sheet; you will note that a large number of equations is handled well by TK Solver, and the model can be used to analyze, design, or optimize a similar bolted-connection problem.

Bolted Joints Subjected to Fatigue Loading

Anytime the load P is cyclic, a bolted connection must be designed as a fatigue member. Even though most fatigue loads are not cyclic in nature, in this text we assume them to be cyclic. In most cases bolted joints require that the initial tension F_i be non-zero. This initial tension is superimposed on it by the load contribution from P to the bolt as $P_b = K_1 P$. (See Fig. 13–8.) The combination of

RULES SHEET

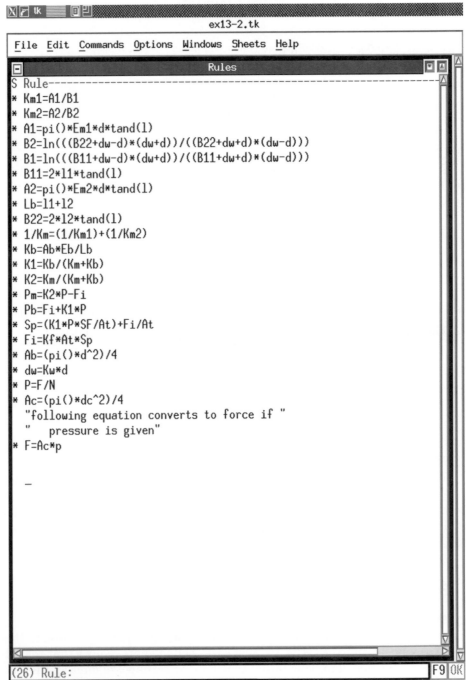

VARIABLES SHEET

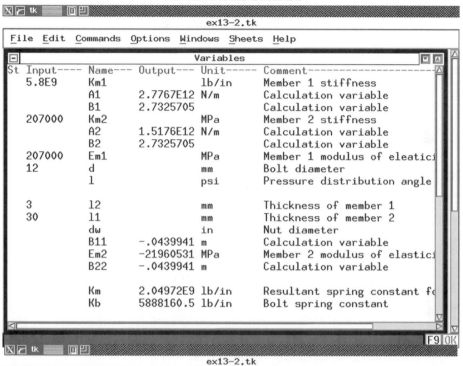

ex13-2.tk

| File | Edit | Commands | Options | Windows | Sheets | Help |

			Variables		
St	Input----	Name---	Output---	Unit-----	Comment--------------------
	5.8E9	Km1		lb/in	Member 1 stiffness
		A1	2.7767E12	N/m	Calculation variable
		B1	2.7325705		Calculation variable
	207000	Km2		MPa	Member 2 stiffness
		A2	1.5176E12	N/m	Calculation variable
		B2	2.7325705		Calculation variable
	207000	Em1		MPa	Member 1 modulus of eleatici
	12	d		mm	Bolt diameter
		l		psi	Pressure distribution angle
	3	l2		mm	Thickness of member 1
	30	l1		mm	Thickness of member 2
		dw		in	Nut diameter
		B11	-.0439941	m	Calculation variable
		Em2	-21960531	MPa	Member 2 modulus of elastici
		B22	-.0439941	m	Calculation variable
		Km	2.04972E9	lb/in	Resultant spring constant fd
		Kb	5888160.5	lb/in	Bolt spring constant

F9 OK

ex13-2.tk

| File | Edit | Commands | Options | Windows | Sheets | Help |

			Variables		
St	Input----	Name---	Output---	Unit-----	Comment--------------------
		Ab	.19634954	in^2	Bolt cross-sectional area
	3E7	Eb		psi	Bolt modulus of elasticity
		Lb	25.4	mm	Length of the bolt
		K1	.00286444		Resultant bolt stiffness rat
		K2	.99713556		Resultant member stiffness r
L		Pm	5168.248	lb	Portion of P taken by member
	8482	P		lb	Total load carried by member
		Fi	14632.229	N	Preload on bolt
		Pb	3313.752	lb	Portion of P taken by bolt
	140	Sp		MPa	Proof stress
	.18	At		in^2	Area
L		N	2.8102112		Number of bolts
L	.9	Kf			Initial tension factor
	1.5	Kw			Ratio of nut diameter to bol
		F	106028.75	N	Total load carried by N bolt
		SF	15.04334		Safety factor
		Ac	.01767146	m^2	Area
	8	p		MPa	Pressure
	150	dc		mm	

F9 OK

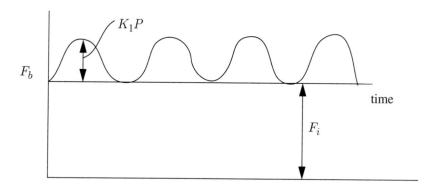

Figure 13–8 The resultant load on a bolt as a function of time

the initial tension and the load causes tension mean stress σ_m and amplitude stress σ_a in the bolt.

Earlier, we showed that the resultant load in a bolt is given by the equation

$$F_b = F_i + K_1P \tag{13–33}$$

where F_b is the resultant tensile load in the bolt, F_i is the initial tension in the bolt, $K_1 = K_b/(K_m + K_b)$, K_b is the stiffness of the bolt, K_m is the stiffness of the member, and P is the external load. In this case, it is known that the bolt is subjected to minimum and maximum loads; that is,

$$F_{\min} = F_i \tag{13–34}$$

and

$$F_{\max} = F_b = F_i + K_1P \tag{13–35}$$

Therefore

$$F_a = (F_{\max} - F_{\min})/2 = (K_1P)/2$$

and

$$F_m = (F_{\max} + F_{\min})/2 = F_i + F_a$$

We can now determine the amplitude and mean stresses in a bolt. Let

$$\sigma_a = \frac{F_a}{(A_t)} \qquad \text{(Since } \sigma = P/A\text{)}$$

Substituting Equations (13–34) and (13–35) into the preceding equation and rearranging, we get

$$\sigma_a = \frac{K_1P}{2A_t} \tag{13–36}$$

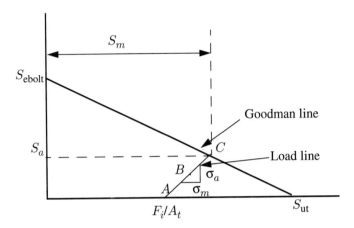

Figure 13–9 A Goodman diagram for a bolt

Also,

$$\sigma_m = \frac{F_m}{A_t}$$

$$\sigma_m = \frac{F_i}{A_t} + \frac{K_1 P}{2A_t}$$

So

$$\sigma_m = \frac{F_i}{A_t} + \sigma_a \qquad (13\text{–}37)$$

Now, to predict fatigue behavior, the endurance limit of the bolt is required. The endurance limit will account for K_f, the fatigue notch factor, as well as standard reduction factors, i.e., $S_{ebolt} = S_e/K_f$, where $S_e = k_a\, k_b\, k_c\, k_d\, S_e'$, in which the K's and S_e' are the reduction factors and the Moore test endurance limit, respectively. Knowing S_{ebolt} and S_{ut}, the ultimate tensile strength of a bolt, one can use the Goodman diagram, which can be drawn as shown in Fig. 13–9.

From Equations (13–36) and (13–37), it can be seen that Equation (13–37) is made up of a constant term F_i/A_t and a variable term $(K_1 P)/2A_t$, and Equation (13–36) is made up of a variable term $(K_1 P)/2A_t$. Thus, if we draw a line from F_i/A_t with $\sigma_a = \sigma_m = (K_1 P)/2A_t$, then we have what is called the *load line*, as shown in Fig. 13–9. From the figure, it can be seen that the safety factor is given by the equation

$$SF = \overline{AC}/\overline{AB} = \frac{S_a}{\sigma_a}$$

If we multiply Equation (13–37) by the safety factor SF, then, knowing that $SF\sigma_m = S_m$ and $SF\sigma_a = S_a$, and rearranging, we get

$$S_a = S_m - \frac{F_i}{A_t} \qquad\qquad (13\text{--}38)$$

and the Goodman equation is

$$\frac{S_a}{S_{eb}} + \frac{S_m}{S_{ut}} = 1 \qquad\qquad (13\text{--}39)$$

Equations (13–38) and (13–39) can be solved simultaneously to obtain S_m and S_a. When we solve, we get

$$S_a = \frac{S_{ut} - F_i/A_t}{(1 + S_{ut}/S_{eb})} \qquad\qquad (13\text{--}40)$$

Now, the safety factor n is given by

$$SF = S_a/\sigma_a \qquad\qquad (13\text{--}41)$$

It is important, however, to check that the resultant load in a bolt, F_b, does not exceed the yield strength S_y To check the safety factor against yielding, let us develop the required equation. We know that

$$SF = \frac{S_y}{F_b/A_t} \qquad\qquad (13\text{--}42)$$

where S_y is the yield point stress in psi, F_b is the resultant load per bolt in lb, and A_t is the area of the throat in, for example, square inches. Therefore,

$$SF = \frac{S_y}{\dfrac{F_i}{A_t} + \dfrac{K_1 P}{A_t}}$$

or

$$SF = \frac{S_y}{(\sigma_a + \sigma_m)} \qquad\qquad (13\text{--}43)$$

Make sure you check Equations (13–41) and (13–43) and use the SF that is minimum.

The next example shows how the theory is applied in designing a bolt for fatigue loading.

Example 13-3

Given a bolted joint and the following data, determine the safety factor and check for failure against yield stress (use Goodman's criterion):

DATA		
K_b = 6.78 Mlb/in	A_t =.226 in^2	P = 0 to 5 kip (cyclic)
K_m = 17.4 Mlb/in		F_i =.75 F_p
S_p = 85 ksi	S_{ut} = 120 ksi	S_{eb} = 18.6 ksi

Solution

We have

$$F_i = 0.75 F_p = 0.75 A_t S_p = 0.75(0.226)(85) = 14.4 \text{ klb}$$

$$\sigma_a = \frac{K_1 P_{max}}{2 A_t}$$

But

$$K_1 = \frac{K_b}{K_b + K_m} = \frac{6.78}{6.78 + 17.4} = 0.280$$

$$\sigma_a = \frac{0.280(5)}{2(0.226)} = 3.10 \text{ ksi}$$

$$\sigma_m = \frac{K_1 P}{2 A_t} + \frac{F_i}{A_t} = \sigma_a + \frac{F_i}{A_t} = 3.10 + \frac{14.4}{0.226}$$

$$= 3.10 + 63.72 = 66.82 \text{ ksi}$$

$$S_a = \frac{S_{ut} - F_i / A_t}{1 + (S_{ut}/S_{eb})} = \frac{120 - 63.72}{1 + (120/18.6)} = 7.55 \text{ ksi}$$

Therefore,

$$SF = S_a / \sigma_a = \frac{7.55}{3.10} = 2.44$$

Let us check the safety factor against yielding:

$$\sigma_{bolt} = \frac{F_b}{A_t} = \frac{F_i + K_1 P}{A_t} = \frac{14.4 + 0.280(5)}{0.226} = \frac{15.80}{0.226} = 69.91 \text{ksi}$$

$$\sigma_{yp} = 0.85(S_{ut}) = 0.85(120) = 102.00$$

$$SF = \frac{102}{69.91} = 1.46$$

Thus, the real safety factor SF = 1.46, which is based upon yielding, is lower than the safety factor based on the Goodman criterion. This is not always the case.

The preceding example is also solved using TK Solver and the lead model called *ex13-3.tk*, or *boltf.tk*. The Rules and Variables Sheets are shown on pages 269 and 270.

Example 13-4

For the pressure cylinder shown in the following figure, let $A = 0.9$ m, $B = 1$ m, $C = 1.10$ m, $D = 20$ mm, and $E = 25$ mm.

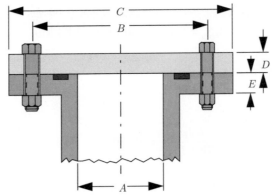

The cylinder is made of cast iron ($E = 96$ GPa), and the head is made of low-carbon steel. There are 36 bolts having $S_{eb} = 162$ MPa, $S_{ut} = 1040$ MPa, and $S_p = 830$ MPa tightened to 75% of proof load. The bolts are M10 × 1.5 = 10 mm diameter and $p = 1.5$ mm. During use, the pressure in the cylinder fluctuates between 0 and 550 kPa. Using the Gerber relation, find the factor of safety necessary to guard against a fatigue failure of the bolt. (Take $A_t = 58$ mm².)

Solution

$$F_i = K_f A \ _t S \ _p$$

$$F_i = (0.75) \ (58 \ \text{mm}^2) \ (830 \ \text{MPa}) = 36{,}105 \ \text{N}$$

$$K_{m1} = 3.50293 \times 10^9 \ \text{N/m}$$

$$K_{m2} = 1.42524 \times 10^9 \ \text{N/m}$$

(K_{m1} and K_{m2} were solved using the lead model that was previously presented.) Now,

RULES SHEET

```
ex13-3.tk

 File  Edit  Commands  Options  Windows  Sheets  Help

S Rule----------
* Km1=A1/B1
* Km2=A2/B2
* A1=pi()*Em1*d*tand(l)
* B1=ln(((B11+dw-d)*(dw+d))/((B11+dw+d)*(dw-d)))
* B11=2*l1*tand(l)
* A2=pi()*Em2*d*tand(l)
* B2=ln(((B22+dw-d)*(dw+d))/((B22+dw+d)*(dw-d)))
* B22=2*l2*tand(l)
* 1/Km=(1/Km1)+(1/Km2)
* Kb=A*Eb/Lb
* K1=Kb/(Km+Kb)
* K2=Km/(Km+Kb)
* Pm=K2*P-Fi
* Pb=Fi+K1*P
*  (K1*P*n/At)+Fi/At=Sp
* Fi=Kf*At*Sp
* A=(pi()*d^2)/4
* dw=Kw*d
* P=F/N
* F=(pi()/4)*(dcy^2)*p
* SigA=(K1*P)/(2*At)
* SigM=(Fi/At)+K1*P/(2*At)
C Sa/(Sm-Fi/At)=1
* Sa=(Sut-Fi/At)/(1+Sut/Seb)
C if T=1 then (Sa/Seb)+(Sm/Sut)=1
* if T=2 then  (Sa/Seb)+(Sm/Sut)^2=1
* Nf=Sa/SigA
* Seb=(ka*kb*kc*kd*1/kf)*.5*Sut
* Nyp=Sigyp/(SigB)
* Sigyp=.85*Sut
* SigB=(Fi+K1*P)/At
* Lb=l1+l2

   _

(34) Rule:                                                        F9 OK
```

VARIABLES SHEET

```
X ⌐ tk      ▣ ▤
```

ex13-3.tk

```
File  Edit  Commands  Options  Windows  Sheets  Help
```

```
▣                              Variables                              ▣ ▣
St Input----  Name---  Output---  Unit-----  Comment------------------------
              Km1                  lb/in      Member 1 stiffness
              A1                   lb/in      Calculation variable
              B1                              Calculation variable
              Km2                  lb/in      Member 2 stiffness
              A2                   lb/in      Calculation variable
              B2                              Calculation variable

              Em1                  psi        Member 1 modulus of elasticity
              d                    in         Bolt diameter
              l                               Pressure dist. angle (alpha)

              l1                   in         Member 1 thickness
              dw                   in         Nut diameter
              B11                             Calculation variable
              Em2                  psi        Member 2 modulus of elasticity
              B22                             Calculation variable
              l2                   in         Member 2 thickness
   17400000   Km                   lb/in      Resultant member stiffness
    6780000   Kb                   lb/in      Resultant bolt stiffness
              A                    in^2       Bolt shaft area
   3E7        Eb                   psi        Bolt modulus of elasticity
              Lb                   in         Bolt length
              K1        .28039702             Resultant bolt stiffness ratio
              K2        .71960298             Resultant member stiffness ratio
L             Pm        -10809.48  lb         Portion of P seen by members
   5000       P                    lb         External load
              Fi        14407.499  lb         Bolt preload (initial tension)
```

```
(43)  Input: 18600                                                   F9 OK
```

```
X ⌐ tk      ▣ ▤
```

ex13-3.tk

```
File  Edit  Commands  Options  Windows  Sheets  Help
```

```
▣                              Variables                              ▣ ▣
St Input----  Name---  Output---  Unit-----  Comment------------------------
              Fi        14407.499  lb         Bolt preload (initial tension)
              Pb        15809.484  lb         Portion of P seen by bolts
   85000      Sp                   psi        Proof stress
   .226       At                   in^2       Throat area
              N                               Number of bolts
   .75        Kf                              Initial tension factor
              Kw                              Ratio of nut dia. to bolt dia.
              F                    lb         Force
   3.5        dcy                  in         Diameter
              p                    psi        Pressure
              SigA      3101.7372  psi        Amplitude stress
              SigM      66851.737  psi        Mean stress
              Sa        7548.7013  psi        Amplitude stress
              Sm                   psi        Mean stress
   1          T                               T=1 for Goodman; T=2 for Gerber
   18600      Seb                  psi        Endurance limit
   120000     Sut                  psi        Ultimate tensile strength
              Nf        2.4337011             Safety factor based on Goodman criteri
              ka                              Endurance limit red. factor
              kb                              Endurance limit red. factor
              kc                              Endurance limit red. factor
              kd                              Endurance limit red. factor
              kf                              Fatigue notch factor

              Sigyp     102000     psi        Yield point Stress
              SigB      69953.474  psi        Max. load in bolt
              Nyp       1.458112              Safety factor based on yielding
                        96081.581
```

```
(51)  Name:                                                          F9 OK
```

$$K_b = \frac{A_b E_b}{L_b}$$

where A_b is the cross-sectional area of the bolt, L_b is the length of the bolt, E_b is the modulus of elasticity of the bolt. Thus,

$$K_b = \frac{(\pi/4)(0.01)^2(207 \times 10^9)}{0.045} = 3.61 \times 10^8 \, N/m$$

$$\frac{1}{K_m} = \frac{1}{K_{m1}} + \frac{1}{K_{m2}}$$

$$K_m = \frac{K_{m1} K_{m2}}{K_{m1} + K_{m2}} = 1.01 \times 10^9 \, N/m$$

$$K_1 = \frac{K_b}{K_m + K_b} = \frac{3.61 \times 10^8}{1.0130568 \times 10^9 + 3.61 \times 10^8}$$

$$K_1 = 0.263$$

$$K_2 = \frac{K_m}{K_m + K_b} = \frac{1.0130568 \times 10^9}{1.013 \times 10^9 + 3.61 \times 10^8} = 0.737$$

$$\sigma_a = \frac{K_1 P}{2A_t}$$

where P is the load on the cylinder. Also,

$$P = \frac{F}{N} = \frac{\pi/4(d_{cylinder})^2 P_{cylinder(max)}}{36}$$

$$= \frac{(\pi/4)(0.9)^2 550000}{36} = 9719.302 \, N$$

$$\sigma_a = \frac{K_1 P}{2A_t} = \frac{(0.263)(9719.302)}{2(58 \times 10^{-6} m^2)} = 22.025 \, MPa$$

$$\sigma_m = \frac{F_i}{A_t} + K_1 \frac{P}{2A_t}$$

$$\sigma_m = \frac{36105}{58} + 0.263\left(\frac{9719.302}{2(58)}\right) = 64.45 \text{ MP}_a$$

Finally,

$$\frac{S_a}{S_m - F_i/A_t} = 1$$

$$\frac{S_a}{S_e} + \left(\frac{S_m}{S_{ut}}\right)^2 = 1$$

Rearranging the preceding two equations and solving for S_a and S_m results in

$$S_a = 86.67 \text{ MPa}$$

So

$$SF = \frac{S_a}{\sigma_a}$$

and we have

$$SF = \frac{86.67}{22.025} = 3.93 \quad \text{(based on the Gerber relation)}$$

Now we check the safety factor based on the proof stress $S_p = 830$ MPa. The maximum stress in the bolts is $\sigma_{max} = \sigma_a + \sigma_m \approx 22 + 644 = 666$ MPa. Therefore, $SF = 830/666 \approx 1.2 < 3.93$. Thus the real safety factor is 1.20 and not 3.93. This happens often, and one must check the safety factor based on the proof stress.

PROBLEMS

Solve Problems 13–1 through 13–4 using your calculator. Then solve the same problems using the lead models employed in the sample problems, in order to check your hand-calculated results.

13–1) Find the power required to drive a 40-mm power screw having double-lead threads with a pitch of 6 mm. The nut moves at a velocity of 48 mm/s and is to move a load of $F = 10$ kN. The static frictional coefficients are 0.10 for the threads and 0.15 for the collar. The frictional diameter of the collar is 60 mm.

13–2) A single square-threaded power screw has an input power of 3 kW at a speed of 1 rev/s. The screw has a diameter of 36 mm and a pitch of 6 mm. The frictional coefficients are 0.14 for the threads and 0.09 for the collar, with a collar radius of 45 mm. Find the axial resisting load F and the combined efficiency of the screw and collar.

13–3) A square-threaded power screw with a double thread is used to raise a load of 3,000 lb. The screw has a mean diameter of 1 in and four threads per inch. The mean diameter of the collar is 1.5 in. The static coefficient of friction is estimated as 0.12 for both the thread and the collar. Determine the horsepower required to make the screw rotate at 10 rpm.

13–4) A square-threaded power screw with a double thread is used to raise a load of 5,500 lb. The screw has a mean diameter of 1.5 in and four threads per inch. The mean diameter of the collar is 2 inches. The static coefficient of friction is estimated to be 0.10 for both the thread and the collar.

    a.   What is the major diameter of the screw?

    b.   Estimate the screw torque required to raise the load.

    c.   If collar friction is eliminated, what minimum value of thread coefficient of friction is needed to prevent the screw from over-hauling?

    d.   Estimate shear and bearing stress in the screw thread if the thickness of the nut is equal to the major diameter of the screw.

    e.   If the screw is rotating at 15 rpm, determine the horse power required to raise the load.

13–5) Plot a graph of efficiency η vs. λ for $f = 0.1$. Let λ go from $10°$ to $80°$. Find the value of λ that will yield maximum efficiency.

13–6) The following figure illustrates the connection of a cylinder head to a pressure vessel using N bolts having a diameter of 150 mm and a confined-gasket seal:

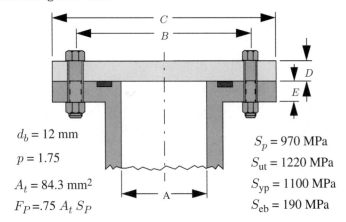

$d_b = 12$ mm

$p = 1.75$

$A_t = 84.3$ mm^2

$F_P = .75\, A_t\, S_P$

$S_p = 970$ MPa

$S_{ut} = 1220$ MPa

$S_{yp} = 1100$ MPa

$S_{eb} = 190$ MPa

The effective sealing diameter is 150 mm. Other dimensions are $A = 100$, $B = 200$, $C = 300$, $D = 20$, and $E = 25$, all in millimeters. The cylinder is used to store gas at a static pressure of 6.0 MPa. Determine the number of bolts required. Use a safety factor of 2, and assume that the initial tension in the bolt is equal to $0.9S_pA_t$. You are advised to use the lead model *ex13-4.tk*.

13–7) In the figure for Problem 13–6, let $A = 0.9$ m, $B = 1.2$ m, $C = 1.20$ m, $D = 25$ mm, and $E = 25$ mm. The cylinder is made of cast iron ($E = 96$ GPa) and the head is made of low-carbon steel. There are 40 bolts having 10 mm diameter and are tightened to 75 percent of proof load. During use, the pressure in the cylinder fluctuates between 10 and 550 kPa. Using the Gerber relation, find the factor of safety necessary to guard against a fatigue failure of the bolt. Check the safety factor based on proof stress ($S_p = 830$ MPa, $S_{ut} = 1040$ MPa, $S_{yp} = 940$ MPa, $S_{eb} = 162$ MPa).

Solve Problems 13–8 and 13–9 using your calculator. Then solve them using the lead models and check your hand-calculated results.

13–8) The figure illustrated in Problem 13–6 shows the connection of a cylinder head to a pressure vessel using 10 bolts and a confined-gasket seal. The effective sealing diameter is 150 mm. Other dimensions are $A = 100$, $B = 200$, $C = 300$, $D = 20$, and $E = 25$, all in millimeters. The cylinder is used to store gas with a variable pressure of 0 to 8 MPa. Bolts with a diameter of 12 mm have been selected. This provides an acceptable bolt spacing. What safety factor SF results from the selection of bolts? Use data from Problem 13–6.

13–9) In the following figure, let $A = 0.9$ m, $B = 1$ m, $C = 1.10$ m, $D = 20$ mm, and $E = 25$ mm.

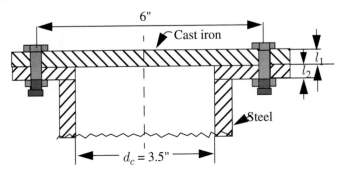

The cylinder is made of cast iron (E = 96 GPa), and the head is made of low-carbon steel. There are 36 bolts tightened to 85 percent of proof load. During use, the pressure in the cylinder fluctuates between 20 and 650 kPa. Using the Goodman relation, find the factor of safety necessary to guard against a fatigue failure of the bolt. Also solve the problem using the Gerber criterion. The pressure cylinder has a cap made of cast iron having E = 18 × 10⁶ psi. The cylinder and bolts are made of steel with E = 30 × 10⁶ psi. If the length of the bolt is $l_1 = l_2 = 0.5''$, the diameter = 0.5″, A_t = 0.18 in², the proof stress = 40,000 psi, the number of bolts = 10, and F_i = 0.75$A_t S_p$, determine the safety factor if the cylinder is subjected to cyclic pressure changing from 0 to 2000 psi. Assume that S_{eb}= 18,600 psi (corrected) and S_{ut} = 120,000.

13–10) Determine number of bolts for Problem 13–9 if F_i =.75F_p, $l_1 = l_2 = 150$ mm, and n = 2.4. Then determine the safety factors for α = 30° to 45° (for N as obtained in 1), in increments of 2.5°.

13–11) For Problem 13–10, if it is now desired that the thicknesses l_1 = 1" and l_2 =.5", what will happen to the safety factor (keeping other input values the same)?

13–12) For Problem 13–10, if l_1 =.5", l_2 =.1" and n = 2.4, how many bolts (N) are required, keeping other input values the same. Add an equation to obtain the safety factor based on the proof stress.

13–13) The pressure cylinder shown in the following diagram has a cap made of cast iron with E = 18 × 10⁶ psi. The cylinder and bolts are made of steel with E = 30 × 10⁶ psi. If the length of the bolt is $l_1 = l_2 = 0.6''$, the diameter = 0.5", A_t = 0.18 in², the proof stress = 40,000 psi, the number of bolts = 10, and F_i = 0.75$A_t S_p$, determine the safety factor if the cylinder is subjected to cyclic pressure changing from 100 to 2000 psi. Assume that S_{eb} = 18,600 psi (corrected) and S_{ut} = 120,000.

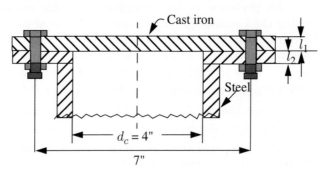

Mechanical Springs

14.1 Introduction

Mechanical springs are elastic members whose primary functions are to exert force or torque and to store or absorb elastic energy. Use of various types of springs in engineering structures and machines is extensive. In an automobile, for example, there are more than 100 machine components that have one form of spring or another. Major classes of springs are as follows:

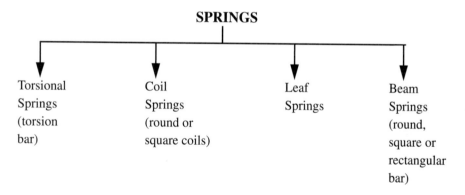

SPRINGS			
Torsional Springs (torsion bar)	Coil Springs (round or square coils)	Leaf Springs	Beam Springs (round, square or rectangular bar)

14.2 Torsional Springs and Stresses

Torsional springs are used as suspension springs in automobiles. They are also used as energy-storing devices in many mechanical components and in various other machines. The torsional springs are made from solid or hollow, straight, cylindrical bars as shown in Fig. 14–1.

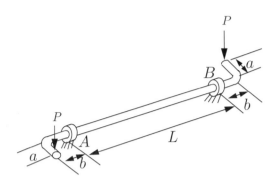

Figure 14–1   Torsional spring subjected to load P

Such springs are popular in applications requiring the absorption of heavy loads. Fig. 14–1 shows that the torque T applied to the bar is equal to the load P multiplied by the length of the lever arm a. This torque produces a maximum shear stress in the central portion of the spring (AB) and is given by the expression

$$\tau = \frac{Tc}{J} \tag{14–1}$$

where τ is the maximum shear stress at extreme fiber, J is the polar moment of inertia $= (\pi/32)[D^4 - d^4]$, D is the external diameter, d is the internal diameter (not shown in Fig. 14–1), $c = D/2$ (the distance to an extreme fiber), and $T = Pa$.

Bending is present in the shaft because the moment M is equal to Pb but this bending is usually neglected if b is small compared to a. It is also recommended that hollow tubes not be used as torsion bar springs as they may buckle if the wall is too thin.

The spring constant for the torsional spring is defined as

$$k = T/\theta$$

But we learned earlier that a straight solid or hollow circular shaft has an angle of twist θ related to an external torque and mechanical and geometrical properties, that is,

$$\theta = TL/JG$$

L is the length between bearings (AB, assuming b is very small), G is the shear modulus, and J is the polar moment of inertia. Therefore,

$$k = \frac{T}{TL/JG}$$

$$k = JG/L \tag{14–2}$$

Design of torsional springs subjected to static torque requires that the maximum shear stress, given by Equation (14–1), not exceed the shear stress in a simple tension test, which is $\sigma_{yp}/2$. If it is based on the energy approach, then the allowable shear stress is equal to 0.57 σ_{yp}. Similarly, shafts subjected to fatigue loads can be designed appropriately using Equation (14–1) and one of the fatigue criteria such as the Goodman diagram. Equation (14–2) allows one to have a desired spring constant by varying J, G, or L.

14.3 Helical Compression Springs

In earlier times, automobile manufacturers used leaf springs in cars and trucks. Since the 1960s, helical compression springs have been used instead. The main reason for this change is that helical compression springs do not transfer friction force to a supporting mass as leaf springs do. This is a great advantage in an automobile suspension as it keeps sudden jolts from being transferred to the passengers. It has also been proven that an independent suspension system performs better than a solid-axle suspension system. Independent suspension design requires the use of helical compression springs. Therefore the use of coil springs in place of leaf springs has become common in modern automobiles. Helical compression springs are also softer in nature and give passengers a better ride. Extensive use of helical compression or tension springs can be seen in aircraft, ships, and many types of machines.

Fig. 14–2(a) shows a helical compression spring with a special nib end, loaded in compression with an external load P. The spring has a mean coil diameter D and a wire diameter d. It has a free length L and active and dead coils as shown. The dead coils are, by definition, those which are not accounted for in an analysis or design of spring stresses the way the active coils are.

Let us now derive the equations necessary to analyze or design a helical compression spring subjected to an external load P.

If the cut portion, as shown in Fig. 14–2(b), is assumed to be in static equilibrium, then the resisting torque is equal to $P\ (D/2)$ and the resisting shear ($V = P$) is present. Using the equations studied in an elementary mechanics course, we see that additive torsional and shear stresses are present on an element at location A, shown in an expanded view in Fig. 14–2(c). (At location B they counteract each other.)

$$\tau_{max})_A = \tau_T + \tau_V \quad \text{(they are additive as they are colinear)}$$

$$\tau_{max})_A = \frac{Tc}{J} + \frac{VQ}{It} \tag{14–3}$$

where $T = P\ (D/2)$, $J = \pi d^4/32$ (note that solid wire is used in helical compression springs), d is the wire diameter, $A = \pi d^2/4$, $V = P$, and

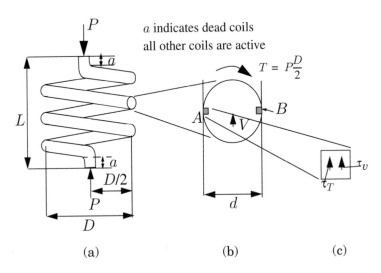

Figure 14–2   Helical compression spring with nib end

$$Q = \left(\frac{\pi d^2}{4}\right)\left(\frac{1}{2}\right)\left(\frac{4d/2}{3\pi}\right) \quad \text{(for solid wire)}$$

$I = \pi d^4/64$

$t = d$

Instead of using such a cumbersome formula, one can derive an easier formula by expanding the above formula (which includes the stress concentration effect) to find the maximum shear stress in a coil spring. The equation, called the *Lewis equation*, is given below.

$$\tau_{max} = K_s 8PD/\pi d^3$$

where K_s is called the shear stress multiplication factor (or Lewis factor). K_s is related to the spring index C, defined as the ratio of the coil diameter to the wire diameter ($C = D/d$). The equation for K_s is

$$K_s = 1 + 0.615/C \tag{14–4}$$

The factor K_s is used in springs subjected to static loads. Therefore, Equation (14–4) is used if a coil spring is subjected to a static load.

It is important to note that the curvature effect in Equation (14–4) has not been accounted for. The curvature effect increases the resultant shear stress on the inside of the spring. Therefore, there was a need to develop an equation to account for this curvature as well as the effect of dynamic (fatigue) loads on these springs. Wahl proposed the factor K_w, which is given by

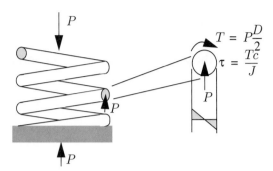

Figure 14–3   Coil spring subjected to a concentric load P

$$K_w = \frac{4C-1}{4C-4} + \frac{0.615}{C} \tag{14–5}$$

Thus K_w now includes the effect of the curvature as well as stress concentration. In sum, the maximum shear stress in a spring wire due to a dynamic load P is given by

$$\tau = K_w \frac{8PD}{\pi d^3} \tag{14–6}$$

Use Equation (14–6) in the design of coil springs subjected to cyclic (fatigue) loads.

14.4 Deflection of Coil Springs

To obtain the deflection due to a load P in a helical compression spring with N active coils, wire diameter d, coil diameter D, and shear modulus G, it is always best to use Castigliano's theorem, which states that the deflection due to a load P in the direction of P is given by the partial derivative of the total strain energy in the spring with respect to the load P, i.e.,

$$\delta = \frac{\partial U}{\partial P}$$

where δ is the deflection due to external load(s) at P in the direction of P and U is the total energy stored in the spring due to external loads.

In the case of helical compression springs, as shown in Fig. 14–3, the energy is stored in the spring due to torsional shear stress and shear stress due to vertical shear. If we neglect the energy due to vertical shear stress ($\tau = VQ/It$) and consider the energy due to torque (shear stress due to torque) then

$$\delta = \frac{\partial U}{\partial P} = \int_0^L \frac{T}{GJ}\left(\frac{\partial T}{\partial P}\right)dL \tag{14-7}$$

Let us find the values of the variables in the above integration.

T = torque = P $(D/2)$

Therefore,

$$\frac{\partial T}{\partial P} = D/2$$

$$L = \pi DN$$

where N is the number of active coils.

Now taking the polar moment of inertia for a solid circular cross section as

$$J = \pi d^4/32$$

Equation (14-7) can be rewritten as

$$\delta = \int_0^{\pi DN} \frac{P(D/2)(D/2)dL}{G(\pi d^4/32)}$$

Integrating and substituting the limits produces

$$\delta = 8PD^3N/(d^4 G) \tag{14-8}$$

Now, finding the spring constant k_{cs} for the coil spring is simple, since

$$k_{cs} = P/\delta$$

$$k_{cs} = \frac{d^4 G}{8D^3 N}$$

but

$$C = D/d,$$

therefore

$$k_{cs} = dG/8NC^3 \tag{14-9}$$

To increase k, increase d and/or G. You could also decrease N and/or C.

Let us now study the free length of a spring required for easy operation. The solid length of the coil spring is

$$l_{solid} = (N_d + N_a)d = N_T d \tag{14-10}$$

where N_d is the number of dead coils, N_a is the number of active coils, N_T is the total number of coils ($N_T = N_d + N_a$), and d is the wire diameter.

The deflected length of the coil is

$$\delta = P/k_{cs} \tag{14-11}$$

To prevent clashing of the coils, an extra length equal to, say, 10% of the total length is needed. Therefore, the total length of the coil spring is

$$l_T = N_T d + \frac{P}{k_{cs}} + 0.1 l_T$$

or

$$l_T = 1.11[N_T d + P/k_{cs}] \tag{14–12}$$

Now the designer can determine the required wire diameter d, coil diameter D, and the total length l_T, as design parameters.

Let us also find the natural frequency of the coil spring. The critical frequency of the helical compression spring is given by the equation

$$f = (1/4)\sqrt{k_{cs} g / w} \tag{14–13}$$

where g is the acceleration due to gravity, w is the weight of the spring $(\pi^2 d^2 D N_a \rho/4)$, N_a is the number of active coils, and ρ is the weight (not mass) density.

It is always advised that the applied load frequency should be far away in either direction from the natural frequency of the spring.

14.5 Helical Compression Springs Subjected to Fatigue Loads

Thus far we have assumed that the external load applied on a helical compression spring was a static load. In cases of dynamic load applied to such springs, the following method must be used in the spring design. Most springs are subjected to a dynamic load whose frequency as well as amplitude changes as a function of time. However, in this study we will assume that the frequency as well as amplitude are not functions of time. As shown in Fig. 14–4, one can assume the load as a function of time in such cases. If one knows the maximum and minimum load, i.e., P_{max} and P_{min}, then the amplitude load P_a is given by

$$P_a = (P_{max} - P_{min})/2 \tag{14–14}$$

and the mean load P_m is given by

$$P_m = (P_{max} + P_{min})/2 \tag{14–15}$$

Therefore, the amplitude shear stress and the mean shear stress can be obtained using the formulas already derived but using P_a and P_m, as shown in the following equations:

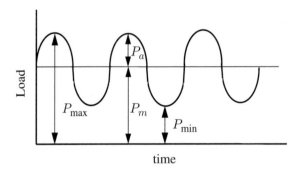

Figure 14–4 Load vs. time curve for cyclic loading

$$\tau_a = K_w \frac{8 P_a D}{\pi d^3} \tag{14–16}$$

and

$$\tau_m = K_w \frac{8 P_m D}{\pi d^3} \tag{14–17}$$

Using the Goodman criterion (see Fig. 14–5), it can be shown that

$$\frac{S_a}{S_{se}} + \frac{S_m}{S_{sut}} = 1 \tag{14–18}$$

but $S_a = \text{SF}\tau_m$ and $S_m = \text{SF}\tau_m$, then

$$\frac{\text{SF}\tau_a}{S_{se}} + \frac{\text{SF}\tau_m}{S_{sut}} = 1$$

where SF is the safety factor, S_{se} is the endurance limit for a coil spring in shear (which can safely be assumed to be $\sigma_{yp}/2$ in the absence of actual test data), and S_{sut} is the ultimate shear strength of the wire.

If the Gerber approach (see Fig. 14–5) is used, then Equation (14–18) is modified to

$$\frac{\text{SF}\tau_a}{S_{se}} + \left(\frac{\text{SF}\tau_m}{S_{sut}}\right)^2 = 1 \tag{14–19}$$

Knowing τ_a, τ_m, S_{se}, S_{sut}, one can determine the safety factor SF; or given SF, one can determine the maximum or minimum load that can be applied on such a spring.

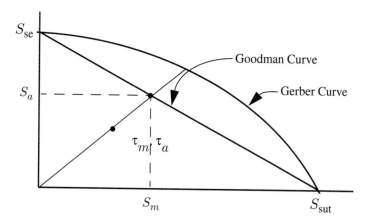

Figure 14–5 Typical Goodman and Gerber curves for coil springs

14.6 Effect of Initial Compression

There are many designs in which the spring has an initial compression P_i (i.e., a constant compression load). Such a spring has the load diagram shown in Fig. 14–6. In such cases, the following equations are obtained for amplitude and mean loads

$$P_a = \frac{P_{max} - P_{min}}{2}$$

but

$$P_{max} = P_i + P$$

$$P_{min} = P_i$$

Therefore,

$$P_a = \frac{P_i + P - P_i}{2}$$

$$P_a = P/2 \tag{14-20}$$

$$P_m = \frac{(P_i + P) + P_i}{2} \tag{14-21}$$

$$P_m = P_i + \frac{P}{2}$$

$$P_m = P_i + P_a \tag{14-22}$$

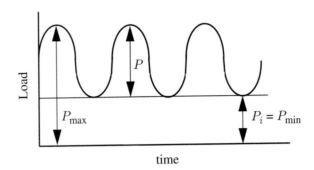

Figure 14–6 Load vs. time curve for a spring subjected to initial compression and external cyclic load

Knowing that $P_a = P/2$, now amplitude shear stress is

$$\tau_a = K_w \frac{4PD}{\pi d^3}.$$

The initial shear stress is

$$\tau_i = K_w 8P_i D/\pi d^3.$$

Similarly, we can show, referring to Equations () and above two equations, that the mean stress is

$$\tau_m = \tau_i + \tau_a \tag{14–23}$$

$$\tau_m = K_w \frac{8P_i D}{\pi d^3} + K_w \frac{4PD}{\pi d^3}$$

$$\tau_m = K_w \left[\frac{8P_i D}{\pi d^3} + \frac{4PD}{\pi d^3} \right]. \tag{14–24}$$

Using Equation (14–23) and rearranging, we get

$$\tau_a = \tau_m - \tau_i.$$

Multiply the above equation by a safety factor SF (except τ_i—why?) and we obtain

$$\text{SF}\tau_a = \text{SF}\tau_m - \tau_i$$

but

$$\text{SF}\tau_a = S_{sa}$$

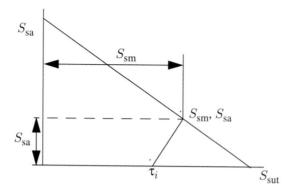

Figure 14–7 Use of Goodman diagram if initial compression stress τ_i exists

and

$$\text{SF}\tau_m = S_{sm}$$

Therefore,

$$S_{sa} = S_{sm} - \tau_i \tag{14–25}$$

If we use the Goodman criterion, then we have a second equation of the type

$$\frac{S_{sa}}{S_{se}} + \frac{S_{sm}}{S_{sut}} = 1 \tag{14–26}$$

or, if we use the Gerber criterion, then

$$\frac{S_{sa}}{S_{se}} + \left(\frac{S_{sm}}{S_{sut}}\right)^2 = 1 \,. \tag{14–27}$$

Now, solve for S_{sa} and S_{sm} from two simultaneous equations, (14–25) and either (14–26) or (14–27). The safety factor SF is then given by

$$\text{SF} = \frac{S_{sa}}{\tau_a} \tag{14–28}$$

All of the above appropriate equations can be used in an analysis or design of helical compression springs. However to protect against yielding, the following must also be true:

$$\tau_a + \tau_m \le S_{sy} \tag{14–29}$$

where S_{sy} is the yield point shear strength of the spring ($\approx 0.5\, S_{yp}$). Therefore the safety factor based on yielding is

$$SF = \frac{S_{sy}}{\tau_a + \tau_m} \tag{14–30}$$

Use a minimum of two safety factors—one obtained, for example, using the Goodman criterion equation (14–27); and one obtained using the yield criterion equation (14–29).

Thus we have sufficient equations to develop a mathematical model to analyze or design helical compression springs subjected to static or dynamic loads. Let us take some example problems to understand the application of these equations and then solve these examples using lead models.

First we will consider an example problem based on dynamic load but no initial compression. The second example is based on some initial compression and a superimposed dynamic load.

Example 14-1

A helical compression spring is to be designed for a fatigue load which varies between P_{min} = 500 lb and P_{max} = 5,000 lb. If the wire from which the spring is made has a shear ultimate strength of 50,000 psi, a corrected endurance limit of 30,000 psi, and a modulus of elasticity value of 30×10^6 psi, determine the required wire diameter and spring stiffness. The maximum deflection in the spring is not to exceed three inches and the coil diameter is assumed to be six inches. The safety factor is 1.6.

Solution

First let us determine amplitude and mean loads

$$P_a = \frac{P_{max} - P_{min}}{2}$$

$$P_a = \frac{5000 - 500}{2}$$

= 2250 lbs

$$P_m = \frac{P_{max} + P_{min}}{2}$$

= (5000 + 500)/2

P_m = 2750 lbs

Now we can determine amplitude and mean shear stresses using

$$\tau_a = K_w \frac{8(2250)6}{\pi d^3}$$

$$= 34,377 \, K_w/d^3$$

$$\tau_m = K_w \frac{8(2750)6}{\pi d^3}$$

$$= 42{,}017 \; K_w/d^3$$

(Note that we do not yet know K_w and d.)

If we decide to use the Goodman criterion, then we know:

$$\frac{SF\tau_a}{S_{se}} + \frac{SF\tau_m}{S_{sut}} = 1$$

$$\frac{1.6\left(34377\dfrac{K_w}{d^3}\right)}{30{,}000} + \frac{1.6\left(42017\dfrac{K_w}{d^3}\right)}{60{,}000} = 1 \qquad \text{(assuming } S_{se} = 0.5 S_{sut}\text{)}$$

Solving for K_w in terms of d, we get

$$K_w = 0.3385 \; d^3 \tag{A}$$

but we know

$$K_w = \left(\frac{4C-1}{4C-4}\right) + \frac{0.615}{C}$$

where $c = D/d$. Therefore,

$$K_w = \frac{4(6/d)-1}{4(6/d)-4} + \frac{0.615}{(6/d)}$$

or

$$K_w = \left(\frac{24/d-1}{24/d-4}\right) + 0.1025d \tag{B}$$

but

$$K_w = 0.3385 \; d^3 \tag{A}$$

Now set K_w of Equation (A) equal to K_w of Equation (B) as follows:

$$0.3385d^3 = \left(\frac{24/d-1}{24/d-4}\right) + 0.1025d$$

Solve the above non-linear equations using any of the methods available to us, we can find the roots, one of which is

$$d = 1.622 \text{ inches}$$

Now

$$C = D/d = 6/1.622$$

so

$$C = 3.7$$

Also, to determine the required active number of coils we know

$$N_a = \frac{dG}{8C^3 k_{cs}}$$

However, we need to find k_{cs}, C, and G, where

$$k_{cs} = \frac{P_{max}}{\delta_{max}}$$

$$= \frac{5000 \text{ lb}}{3 \text{ in}} \qquad \text{(given a 3" deflection corresponding to the maximum load)}$$

$$k_{cs} = 1666.67 \text{ lb/in}$$

Now to determine shear modulus, we know

$$G = \frac{E}{2(1+v)} \qquad \text{(assuming Poisson's Ration } v = 0.3)$$

$$= 30 \times 10^6/(2(1 + 0.3))$$

$$G = 11.538 \times 10^6 \text{ psi}$$

Therefore,

$$N_a = \frac{(1.622 \text{in})(11.538 \times 10^6 \text{psi})}{8(3.7)^3 (1666.67 \text{lb/in})}$$

$$N_a = 27.7$$

$$= 28 \text{ (the number of active coils)}$$

Assume 2 dead coils—one on each end ($N_d = 2$). Then $N_T = N_a + N_d$ or 30 total coils. Now

$$L_{solid} = N_T(d)$$

$$L_{solid} = 48.66 \text{ in}$$

$$\text{Total spring length} = L_T = L_{solid} + \frac{P_{max}}{k_{cs}} + 0.1 L_{solid}$$

$$L_T \approx 57.4 \text{ in}$$

Example 14-1 is solved using the lead model called *ex14-1.tk* or *spring.tk*. The Variables Sheet is shown next with the above answers matching the answers in the Variables Sheet.

VARIABLES SHEET

```
┌─────────────────────────────────────────────────────────────────────────┐
│▽                                    tk                                     │
├─────────────────────────────────────────────────────────────────────────┤
│              /tmp_mnt/home2/florida.me/srbhonsl/tk/ex14-1.tk              │
├─────────────────────────────────────────────────────────────────────────┤
│  File  Edit  Commands  Options  Windows  Sheets  Help                     │
├─────────────────────────────────────────────────────────────────────────┤
│ ▣                              Variables                            ▣ ▣   │
├─────────────────────────────────────────────────────────────────────────┤
```

St	Input	Name	Output	Unit	Comment
		G	11538462	psi	Shear modulus
		rho		lb/in^3	Density of spring material
		C	3.6996265		Spring index
	3E7	E			
		Do	7.6217853	in	Outside diameter of coil
	6	D		in	Mean diameter of coil
		d	1.6217853	in	Diameter of coil
	.3	v			
		k	1666.6667	lb/in	Spring constant of coil
		kw	1.4440492		Wahl spring constant
	1	ka			Modification factor to Sn
	1	kb			Modification factor to Sn
	1	kc			Modification factor to Sn
	1	kd			Modification factor to Sn
	1	ke			Modification factor to Sn
	.58	Cl			Load factor
	.9	Cg			Gradient factor
	1	Cs			Surface factor
		Ft	5000	lb	Total applied force
		Fa	2250	lb	Amplitude force
	5000	Fmax		lb	Maximum force
	500	Fmin		lb	Minimum force
		Fm	2750	lb	Mean force
		Fs		lb	Static load on spring
		Tmax	25862.069	psi	Maximum stress
		Taua	11637.931	psi	Amplitude stress
		Taum	14224.138	psi	Mean stress
	30000	Sse		psi	Endurance limit of spring
		Se`	30000	psi	Fat. strength for rotating bending
	55000	Ssyp		psi	Yield strength of spring
		Sus	16.671004	psi	Ultimate shear strength of spring
		Ssut	60000	psi	Ultimate tensile strength of spring

```
├─────────────────────────────────────────────────────────────────────────┤
│ ◁                                                                    ▷    │
├─────────────────────────────────────────────────────────────────────────┤
│ (1) Status:                                                      F9 OK    │
└─────────────────────────────────────────────────────────────────────────┘
```

VARIABLES (Cont.)

```
┌─────────────────────────────────────────────────────────────────────────┐
│▽                                    tk                                    │
├─────────────────────────────────────────────────────────────────────────┤
│              /tmp_mnt/home2/florida.me/srbhonsl/tk/ex14-1.tk             │
├─────────────────────────────────────────────────────────────────────────┤
│  File  Edit  Commands  Options  Windows  Sheets  Help                    │
├─────────────────────────────────────────────────────────────────────────┤
│□                                Variables                           ▣ ▲  │
├─────────────────────────────────────────────────────────────────────────┤
│St Input---- Name--- Output--- Unit----- Comment----------------------△   │
│             Ssa     18620.69  psi       Ampitude strength of spring      │
│             Ssm     22758.621 psi       Mean strength of spring          │
│                                                                          │
│  1.6        nl                          Safety factor based upon load    │
│  1.6        nf                          Safety factor based upon fatigue │
│             ny      2.1266667           Safety factor based upon yield   │
│             nb      9.3853465           Safety factor based upon buckling│
│                                                                          │
│  0          theta                       Angle of spring                  │
│                                                                          │
│             Lf      56.312079 in        Free length of spring            │
│             Ls      48.192799 in        Solid length                     │
│             Ld      3         in        Deflection length                │
│             Lc      5.1192799 in        Clash length                     │
│                                                                          │
│             Na      27.715893           Number of active coils           │
│  2          Nd                          Number of dead coils             │
│             Nt      29.715893           Total number of coils            │
│                                                                          │
│  .1         kcl                         Clash allowance                  │
│             f                 Hz        Frequency of spring loading      │
│  32.2       g                 ft/s^2    Gravity constant                 │
│             W                 lb        Weight of load                   │
│                                                                          │
│  1          T                           Method of fatigue (ex. T=1 Goodman)│
│                                                                          │
│             delta1            in        Change in height1                │
│             delta2            in        Change in height2                │
│             deltin  .3        in        Compression dist. to install spring│
│                                                                          │
│                                                                          │
│                                                                          │
│             N       27.715893                                            │
│                                                                          │
│  3          dlmax                                                    ▽   │
├─────────────────────────────────────────────────────────────────────────┤
│◁                                                                    ▷    │
├─────────────────────────────────────────────────────────────────────────┤
│(37) Status:                                                    F9 │OK│   │
└─────────────────────────────────────────────────────────────────────────┘
```

Now let us take another example in which the spring is subjected to an initial compression that is independent of time, and therefore need not be multiplied by the safety factor SF. Let us see how to solve such problems.

Example 14-2

An automobile suspension coil spring has an initial compression of 10 mm when it is placed in the rear suspension. Estimate the load-carrying capacity and the free length of the spring if the coil diameter is 150 mm and the wire diameter is 20 mm. The active number of turns is 10. Assume that a safety factor of 2.0 is required. The high strength steel has an ultimate stress (in tension) of 1200 MPa and the endurance (corrected) limit of 500 MPa. Take $E = 207 \times 10^9$ Pa.

Solution

$$G = \frac{E}{2(1+u)} = \frac{207 \times 10^9}{2(1+0.3)}$$

$$G = 79.601 \times 10^9 \text{ Pa}$$

$$C = D/d = 150/20 = 7.5$$

$$k_{cs} = \frac{dG}{8C^3 N}$$

$$= \frac{(20/1000)79.61 \times 10^9}{8(7.5)^3(10)}$$

$$K_{cs} = 47{,}200 \ N/m$$

Therefore,

$$F_i = k_{cs}\delta_i$$

$$= 47200\left(\frac{10}{1000}\right)$$

$$F_i = 472 \ N$$

Therefore,

$$\tau_i = \frac{K_w 8PD}{\pi d^3}$$

but

$$K_w = \frac{4C-1}{4C-4} + \frac{0.615}{C}$$

$$K_w = \frac{4(7.5)-1}{4(7.5)-4} + \frac{0.615}{7.5}$$

$$K_w = 1.1154 + 0.082$$

$$K_w = 1.1974$$

Therefore,

$$\tau_i = \frac{1.1974 \times 8(472)}{\pi\left(\dfrac{20}{1000}\right)^3}$$

$$\tau_i = 180 \text{ MPa}$$

Assume

$$S_{sut} = 0.57\,\sigma_{ult}$$

(This is the ultimate strength in shear.)

$$= 0.57(1200)$$

$$S_{sut} = 684 \text{ MPa}$$

Assume

$$S_{se} = 0.8\,S_e$$

(This is the shear endurance limit for the machine element.)

$$= 0.8(500)$$

$$= 400 \text{ MPa}$$

Now we know

$$S_{sa} = S_{sm} - \tau_i$$

and therefore

$$S_{sm} = S_{sa} + \tau_i$$

and

$$\frac{S_{sa}}{S_{se}} + \frac{S_{sm}}{S_{sut}} = 1 \qquad \text{(Goodman criterion)}$$

Therefore, substitute for S_{sm} in Goodman criterion and we get

$$\frac{S_{sa}}{S_{se}} + \frac{S_{sa} + \tau_i}{S_{sut}} = 1$$

$$\frac{S_{sa}}{400} + \frac{S_{sa} + 180}{684} = 1$$

$$S_{sa} = 186 \text{ MPa}$$

but

$$S_{sa} = \text{SF}\tau_a$$

and

$$\text{SF}\tau_a = \text{SF}K_w \frac{4PD}{\pi d^3}$$

Therefore,

$$186 = 1.6(K_w 4PD/\pi d^3)$$

$$P = \frac{186(\pi)\left(\dfrac{20}{1000}\right)^3}{(1.6)(4)(1.1974)(4)\left(\dfrac{150}{1000}\right)}$$

$$= 4065 \text{ N.}$$

Now

$$\delta = P/k_{cs} = \frac{4065}{47200} = 0.0861 \text{ m}$$

$$\delta = 86 \text{ mm}$$

To determine total length, let us find

$$N_T = 10 + 2 \qquad\qquad (2 \text{ dead coils})$$

$$= 12$$

Therefore,

$$l_T = [(N_T d) + \delta]1/0.9$$

$$= ([(10(20))] + 86)1/0.9 = 286/0.9 = 317.78$$

$$\approx 318 \text{ mm (about 12 inches)}$$

14.7 Design of Leaf Springs

Leaf spring design is based upon the design of a constant-stress triangular beam. This becomes obvious with the step-by-step analysis of such a beam, as shown in Fig. 14–8. First, find the bending stress in a triangular beam of width b, thickness t, and length L subjected to a load P at its tip end. The b end is fixed (shown in Fig. 14–8).

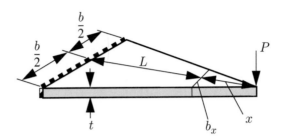

Figure 14-8   Constant stress beam

The moment at distance x is $M_x = P(x)$. The moment of inertia at distance x is

$$\frac{b_x t^3}{12}$$

but $b_x/x = b/L$, therefore $b_x = bx/L$.

The maximum bending stress at distance x along the beam is given by

$$\sigma_x = \frac{M_x t/2}{I_x}$$

but

$$I_x = \frac{b_x t^3}{12} = \frac{b}{L}x(t^3/12)$$

Therefore,

$$\sigma_x = \frac{P(x)(t/2)}{(x)\frac{b}{L}(t^3/12)}$$

$$\sigma_x = \frac{6PL}{bt^2} \tag{14-31}$$

Since P, L, b, and t are constant for a triangular beam, as shown in Fig. 14-8, then

$$\sigma_x = \text{constant}$$

Therefore, such a beam is called a *constant stress beam*.

Now, using Castigliano's theorem, the deflection under load P can be found. (Neglect shear deflection.)

$$\delta_p = \frac{\partial U}{\partial P}$$

We know

$$\frac{\partial U}{\partial P} = \int_o^L \frac{\partial M_x}{\partial P} \frac{M_x}{EI} dx \qquad (14\text{--}32)$$

but

$$M_x = P(x)$$

Therefore,

$$\frac{\partial M_x}{\partial P} = x$$

$$\therefore \delta_p = \int_o^L (x)\frac{P(x)}{E_x I_x} dx$$

but

$$I_x = \frac{(b_x)t^3}{12} = \frac{bx}{L}\frac{t^3}{12}$$

since $b_x = \dfrac{bx}{L}$. Therefore,

$$\delta_p = \int_o^l \frac{(x)P(x)dx}{E\left(\dfrac{bx}{L}\dfrac{t^3}{12}\right)} \qquad (14\text{--}33)$$

On integrating, we get

$$\delta_p = \frac{6PL^3}{Ebt^3}$$

The spring constant k for a leaf spring is defined as

$$k_{\text{leaf}} = P/\delta_p = \frac{Ebt^3}{6L^3} \qquad (14\text{--}34)$$

With this background, Example 14-3 shows how to design a leaf spring.

Example 14-3

Design a leaf spring with a safety factor of not less than 1 for an experimental cargo truck to carry a superimposed one-ton load. Assume the dead weight of the truck to be 2.5 tons. The maximum width of the leaf spring is limited to 3 inches, and the yield point strength of leaf spring material is $S_{yp} = 100,000$ psi.

Solution

Our solution is based on the following assumptions:

1. The leaf spring is to be made of steel having a modulus of elasticity $E = 30 \times 10^6$ psi.

2. The thickness of the spring is 1/4".

3. Seventy percent of the live load and thirty percent of the dead load is acting upon the back two springs and will create a 2" deflection under such a load.

4. An additional shock load equal to 1" deflection is assumed to exist.

5. The stress concentration factor $= 1.8$.

6. Shackles are assumed to create no bending moment at the supports.

7. The length of the leaf spring is 30".

$$\text{Load on each leaf spring} = \left[\left(\frac{2.5}{2}\right)(0.30) + (1)((0.70)/2)\right]2000$$

$$= 1450 \text{ lbs/suspension}$$

The required spring constant for the leaf is

$$k_{leaf} = \frac{P}{\delta} = \frac{1450}{2} = 725 \text{ lbs/in}$$

but

$$k = \frac{Ebt^3}{6l^3}$$

We know

$$E = 30 \times 10^6 \text{ psi}$$

$$l = 15''$$

$$t = 1/4''$$

Therefore,

$$b = \frac{k_{leaf}(6l^3)}{Et^3}$$

$$= \frac{725(6)(15)^3}{30 \times 10^6 (1/4)^3}$$

$$= 31.3''$$

Given that each leaf is 3″ wide, the maximum number of leafs are

$$N_L = \frac{31.32}{3} \approx 10 \text{ leafs or } 11 \text{ leafs.}$$

Now check the maximum stress in the leaf springs

$$\sigma = k_n(6Pl/bt^2)$$

where k_n is the notch factor, assumed to be 1.3, therefore

$$\sigma = 1.3 \left(\frac{6(1450)(15)}{31.32(1/4)^2} \right)$$

$$= 86666 \text{ psi}$$

Because of high stress value, we will use a surface-hardened leaf spring having a maximum allowable stress of 100,000 psi. Therefore, the Safety Factor is

$$SF = \frac{10000}{86666} \approx 1.15$$

Example 14-4

A leaf spring is subjected to fatigue loading. Using the Goodman criterion, design the leaf spring based on the following data:

$P_{max} = 4750$ lb, $P_{min} = 750$ lbs., $S_e = 90,000$ psi, $S_{ut} = 180,000$ psi, the length between shackles = 30″, $t = 0.3''$, $E = 30 \times 10^6$ psi, the safety factor = 1.3, and the number of leafs = 5.

Solution

Let us find an amplitude load

$$P_a = \frac{P_{max} - P_{min}}{2}$$

$$= \frac{4750 - 750}{2}$$

$$P_a = 2000 \text{ lbs}$$

Similarly, we can find the mean load

$$P_m = \frac{4750 + 750}{2}$$

$$= 2750 \text{ lbs}$$

Now mean and amplitude stresses are thus (assume notch factor $K_n = 1.4$)

$$\therefore \sigma_m = \frac{k_n 6 P_m l}{bh^2}$$

$$= \frac{1.4(6)(2750)(15)}{b(0.3)^2}$$

Similarly,

$$\sigma_a = \frac{k_n 6 P_a L}{bh^2}$$

$$= \frac{1.4(6)(2000)(15)}{b(0.3)^2}$$

but

$$\text{SF}\sigma_a = S_a$$

$$\text{SF}\sigma_m = S_m$$

and using the Goodman criterion, which is

$$\frac{\text{SF}\sigma_a}{S_e} + \frac{\text{SF}\sigma_m}{S_{ut}} = 1$$

where SF = 1.3, S_e = 90,000 psi, and S_{ut} = 180,000 psi.

Now, solving for b we get

$$b = 15.3426''$$

Therefore:

$$b' = \text{the width of the leaf} = \frac{15.34}{5} \approx 3''$$

Spring Constant

$$k = \frac{Ebt^3}{6L^3}$$

$$= \frac{30 \times 10^6 (15.34)(0.3)^3}{6(15)^3}$$

$$k = 600 \text{ lbs/in}$$

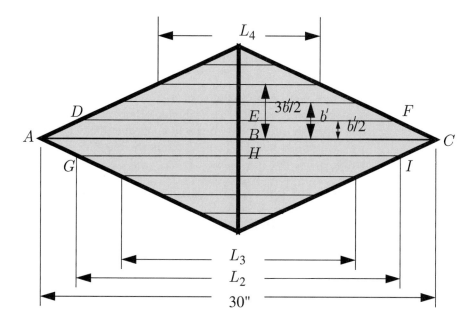

Figure 14–9

14.8 Method to Determine Leaf Length

We have seen that the design of a leaf spring was based on a triangular beam fixed at one end. In reality, the two triangular beams are placed back to back, as shown in Fig. 14–9, to come up with the length at each leaf. The total length of the beam is taken to be equal to the distance between two shackles, assuming zero camber. However the camber is always provided and the lengths will be more than just a distance between two shackles. For now we will assume that the leafs are straight. Such simplification will make the procedure to determine leaf lengths easier. This procedure can then be extended to account for camber. Minimum camber is equal to a maximum deflection expected plus some clash allowance, which can be as much as 50 percent of the maximum deflection expected. Let us now study the procedure used to find the length of each leaf spring designed in this way.

STEP 1.   Draw two triangular beams each having a length equal to the length given (15 inches in this case), as shown in Fig. 14–9. Note that the total length L_1 is 30 inches. This is the length of the first leaf.

STEP 2.   Join points ABC, the centerline points. Now draw two parallel lines DEF and GHI at a distance of $b'/2$ from centerline ABC

(where b' is the width of the leaf). Measure the length of one of the lines between the two points where it intersects the triangular beams. This length, L_2, will be the length of the second leaf.

STEP 3. Draw two more lines parallel to ABC but now at a distance of b' measured from ABC on each side. Measure the length of one of these lines between the two points where it intersects the triangular beams. This length is the length of the third leaf, L_3.

STEP 4. Repeat Step 3, but draw the parallel lines at a distance of $3b'/2$ from centerline ABC. Again measure the length of one of the lines. This length is L_4, the length of the fourth leaf.

This process is to be continued until the last length is obtained. Please realize that the best way to measure such lengths is to "draw" the figure, as shown in the example above, to scale. Then measure each length using an engineering scale. However, a mathematical model also can be developed to generate lengths L_1, L_2, L_3, etc. This concludes the design procedure for leaf springs.

PROBLEMS

14–1) A #13 gauge (0.091 in.) music wire helical compression spring has an outside coil diameter of 9/16 inches. Its free length is 3″, it has 21 active coils, and 2 square ground ends. It is subjected to $P_{min} = 5$ lb, and $P_{max} = 20$ lb. Determine the safety factor using both Goodman's criterion and the yield point shear stress. Also, determine the natural frequency f_n.

Assume $G = 11.5 \times 10^6$ psi, $S_{se} = 60{,}000$ psi, $S_{sut} = 130{,}000$ psi, $\rho = 0.282$ lb/in^3.

14–2) A helical compression spring is made of 0.046-in wire having a torsional yield strength of 108 kpsi. It has an outside coil diameter of 1/2 in and has 14 active coils.

    a. Find the maximum static load corresponding to the yield point of the material.

    b. Calculate the spring constant (scale) of the spring.

    c. What deflection would be caused by the load in (a)?

    d. If the spring has one dead turn at each end, what is the solid height?

e. What should be the free length of the spring so that when it is compressed solid, the stress will not exceed the yield point? (Provide ten percent clash allowance.)

14–3) A helical compression spring is made of 0.050-in diameter wire having a torsional yield strength of 110 kpsi. It has a mean diameter of 1/2 in and has 14 active coils.

a. Find the maximum static load corresponding to the yield point of the material.

b. Calculate the scale (spring constant k) of the spring ($G = 11.538 \times 10^6$ psi). What deflection would be caused by the load in (a)?

c. If the spring has one dead turn at each end, what is the solid height?

d. What should be the free length of the spring so that when it is compressed solid, the stress will not exceed the yield point? (Provide ten percent clash allowance.)

14–4) A helical steel compression spring is made of 0.5-inch diameter bar having a torsional yield strength of 110 kpsi. It has a coil mean diameter of 6 inches and has 10 active coils (2 dead coils).

a. Find the maximum static load corresponding to the yield point of the material.

b. Calculate the scale (spring constant k) of the spring ($G = 11.538 \times 10^6$ psi).

c. What deflection would be caused by the load in (a)?

d. If the spring has one dead turn at each end, what is the solid height?

e. What should be the free length of the spring so that when it is compressed solid, the stress will not exceed the yield point? (Provide ten percent clash allowance.)

14–5) Design a semi-elliptical spring to carry a dynamic load which varies from 1000 lb to 1800 lb. For $L = 36''$, $h = (1/80)\, b$, $K_n = 1.4$, $S_{ut} = 190$ ksi, $S_y = 170$ ksi, and $S_e = 85$ ksi.

a. Determine the total width b and the width of each leaf if the total number of leaves is 5. (Use both Goodman and Gerber criteria.)

      b.   Determine the spring rate.

      c.   Calculate the maximum deflection.

14–6) Design a semi-elliptical spring having length of 40 inches and an external load that varies between 6,000 lb to 8,400 lb cyclically. Assume the thickness of the leaf spring t to be equal to ($b/80$). Use the Gerber criterion and a safety factor equal to 1.3. The following material properties are to be used: S_{yp} = 180 ksi, S_e = 95 ksi, K_f = 1.3 and S_{yp} = 0.85 S_{ut}. (Design of the spring involves finding the width and thickness of each leaf and the length of each leaf and camber.) Use your engineering judgement for the camber.

14–7) Design a semi-elliptic leaf spring for a infinite fatigue life for a load that varies between 500 N and 1500 N. The maximum total length of the spring is to be in one meter made of steel having properties of S_{ut} = 1400 MPa, S_{yp} = 1100 MPa, and S_e = 600 MPa. Assume K_f = 1.2 (width b and thickness t have a ratio of 78).

      a.   Determine maximum deflection.

      b.   Determine the width of each leaf.

14–8) A steel alloy (S_u = 160 ksi, S_y = 110 ksi, S_n = 80 ksi) is used to make a five-leaf semi-elliptic leaf spring. (Each leaf is 0.125″ thickness by 2″ width.) K_f, due to stress concentration at the clips and the center hole, is 1.3. Use the simplified "triangular plate" model to design the spring.

      a.   What total spring length is needed to give a spring rate of 100 lb/in (i.e., 100 lb applied at the center causes a 1-inch deflection at the center)?

      b.   In service, the spring will carry a static load (applied to the center) of P, plus a superimposed dynamic load that varies from 0 to +P. What is the highest value of P that will give infinite life with a safety factor of 1.5?

Introduction to Gears and Design Summary for Spur Gears[1]

15.1 Introduction

Gears are wheels, made of various materials, on which teeth have been placed or machined to transmit rotational motion and force from one shaft to another. Today, most mechanical equipment incorporates some type of gearing. The object of a pair of gears is to cause rotation of the driven shaft due to rotation of the driving shaft. If the gears are the same size, the driven shaft rotates at the same speed as the driving shaft, but in the opposite direction. If the gears are of different sizes, the rotational speeds of the shafts are not the same. The shaft with the smaller gear will rotate faster and with less torque (rotational force). The smaller of the two gears (regardless of whether it is driver or driven) is called the *pinion*; the larger is called the *gear*. Gear sets can be designed to speed up the driven shaft (and decrease its torque) or to slow down the driven gear (and increase its torque). The former is called a *speed increaser* and the latter a *reduction gear*.

Gear drives are, for the most part, a "necessary evil" in equipment. They are used to transmit motion or force from one part of a mechanism to another, or because the speed of the power source and the driven equipment are selected to eliminate the use of gears if possible. This is done not because system engineers have an aversion to using gears, but because it is good engineering judgment to use as few parts as possible in a mechanism and still achieve the required objective. If cost and efficiency do not suffer too much, it is wise to match driver and driven and eliminate the use of gears.

---

[1] Part of the introduction to gears was provided by Universal Technical Systems, Inc., Rockford, IL.

It is usually desirable that the gears transmit motion uniformly from driver to driven. This means that if the driver is given uniform angular velocity, the driven will also have uniform angular velocity. Nonuniform motion is not welcome in most machinery. Gears that turn with a "jerky" motion will cause vibration and noise in precision equipment. When gears transmit uniform angular motion, the profiles on the teeth are said to be *conjugate*. In some cases, conjugate action is given up to obtain the highest possible efficiency. For instance, in a watch, the second hand may come to a complete stop five times per second. Watches and other clockworks are usually not constructed with conjugate gears.

Most any continuous curve can be used for the profile on one of the gears, and the profile on the other gear can be made conjugate to the first. This approach is not very efficient; the design of the gears and the tooling to manufacture them is much simplified if a standard form is used for the profiles. Thus, a tooth form is used which ensures that gears of all sizes in the system will be conjugate to each other.

One of the first tooth forms used in the gear industry was based on the cycloid. This form satisfies all the requirements of power gearing and was used successfully for many years. According to Wilfred Lewis, a pioneer in the development of gearing:

> *The practical consideration of cost demands the formation of gear teeth upon some interchangeable system. The cycloidal system cannot compete with the involute, because its cutters are formed with greater difficulty and with less accuracy, and a further expense is entailed by the necessity for more accurate center distances. Cycloidal teeth must not only be accurately spaced and shaped, but their wheel centers must also be fixed with equal care to obtain satisfactory results.*

And in the words of George B. Grant:

> *There is no more need of two kinds of tooth curves for gears of the same pitch than there is need for two different threads for standard screws, or two different coins of the same value, and the cycloidal tooth would never be missed if it were dropped altogether. But it was first in the field, it is simple in theory, and has the recommendation of many well-meaning teachers, and holds its position by means of "human inertia", or the natural reluctance of the average human mind to adopt a change, particularly a change for the better.* (Treatise on Gearing, 1890)

In the years since 1900, "human inertia" has been overcome, and the cycloidal tooth is now part of the history of gearing. Seldom used for power gearing, it has been replaced by the involute tooth form. (However, it is still used for special applications such as impellers and blowers.)

Many curves can have an involute. The *evolute* of a curve is the locus of its center of curvature. If a curve is an evolute of another, then the second curve is the *involute* of the first. We will investigate the involute of the circle for use as a gear tooth form. The involute of a circle can be imagined by envisioning the path of the end of a string as it is unwound from a circle. The radius of the involute is then the length of the unwound string at any position of the end of the string. The circle is the locus of the centers of curvature of the involute and is therefore its evolute.

The involute of a circle has many advantages as the form for gear tooth profiles. We will investigate how this form is used for gear teeth and some of the reasons for its use.

15.2 Belts and Friction Discs

Suppose that we have been assigned to design a system to connect two parallel shafts and cause the driven shaft to turn at one-half the speed of the driving shaft in the opposite direction.One way of connecting shafts to transmit rotation is with a belt running over a pulley on each shaft. If we use a pulley on the driven shaft that has twice the perimeter of the pulley on the driving shaft, then the driven shaft will turn at one-half the speed of the driving shaft. If we use a simple belt drive, both shafts will turn in the same direction. To change the direction of rotation of the pulley, we will need to cross the belt between the driving pulley and the driven pulley. Such a drive is sketched in Fig. 15–1 (I and III).

Another system we might use is a friction drive. We would put a disk on each shaft and transmit rotation from one shaft to the other by friction between the disks. If the diameter of the disk on the driven shaft is twice the diameter of the disk on the driving shaft, then the driven shaft will turn at one-half the speed of the driving shaft and in the opposite direction. This type of drive is sketched in Fig. 15–1 (II and IV).

Let us examine the relationship involved in these drives using some examples. In Fig. 15–1 (I), the driving pulley is labeled D_R and the driven D_N. Suppose we make the driving pulley (Dbl) 1" in diameter, the driven pulley (Db2) 2" in diameter and the distance between the shafts (the center distance CD) 2.25". Since we are connecting the pulleys with a belt, CD may be any value greater than 1.5". The belt will be "wound up" on Dbl at the same rate as it is "unwound" from Db2. Since the circumference of Dbl is one-half that of Db2, Db2 will turn at one-half the speed of Dbl. This *reduction ratio* (mg) is said to be (speed of driver)/(speed of driven), or 2.0 to 1.

Now let us replace our crossed belt drive with a friction drive. We know that the center distance here is to be 2.25" and the reduction ratio is to be 2.0 to 1. We can surmise that the driven disc (D_2) must be twice the size of the driving disc (D_1). Then D_2/D_1 must be equal to 2. We would soon find, by trial and error, that

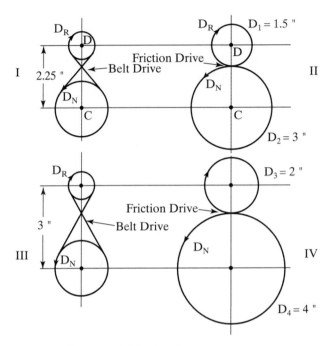

Figure 15-1   Belt drive and friction drives

$D_1=1.5"$ and $D_2=3"$ will meet our conditions. Two simple formulas give these results: $D_1=2 \times CD/(mg+1)$, $D_2=D_1 \times mg$.

Both drive I and drive II will give us the same motion on the driving and driven shafts. The rotation of the driven shaft will be uniform motion as long as the motion of the driving shaft is uniform. The point where the belt crosses in drive I is at the same location between the shafts as the contact point in drive II. We call this point the *pitch point* for reasons that will be obvious later in our discussion. The location of the pitch point can be found with the same formulas used to find the diameters of the friction disks.

Since we have two possibilities worked out for our assigned task, this is the point in the design process when the boss shows up and informs us that, due to factors beyond his or her control, the center distance must be increased to 3" for the good of the continued existence of the company.

So we go back to work. Let us try to salvage what we can from our efforts before we change the center distance.

To increase the center distance of our belt drive, it appears that all we need do is to use the pulleys we have already and get a longer belt. The driven pulley will still be 2" in diameter and the driving pulley 1" in diameter. Our reduction ratio is still $2/1=2.0$. All is well. This change is sketched in Fig. 15-1 (III). However, things are not so simple with our friction drive. If we just increase the center distance, the disks will no longer be in contact, and we no longer have any drive at all. Accordingly, we need to calculate new diameters for our disks. Now we have $CD=3''$, and mg is still equal to 2.0. Then $D_1=2 \times CD/(mg+1)=2''$ and $D_2=D_1 \times mg=4''$. (See Fig. 15–1 (IV).) We also have a new pitch point for drives III and IV. Notice that the change in center distance did *not* change the size of the pulleys, but did change the size of the friction disks and the location of the pitch point.

If we attached cardboard to the back of either of the belt pulleys and then attached a scribe to the crossed belt, the scribe would trace an involute on the cardboard with the pulley diameter as the base circle of the involute. This is obvious, since a point on the crossed belt is being unwound from the pulley the same as the string unwinding from a circle. The pulleys in our belt drives behave the same, geometrically, as the base circles in an involute gear system. Further, the friction disks in our friction drives behave the same as operating pitch diameters in an involute gear system. (The formulas for operating pitch diameters are the ones already given for friction disk diameters.)

15.3 The Involute Curve for a Spur Gear

Since the involute of the circle is the basis for the tooth profiles, and since we will be working with the design and analysis of gears and tooling, we will derive the basic equation of the involute for reference.

Fig. 15–2 is a sketch of an involute from the base circle D_B, which can have any diameter except zero. R_b is the radius of D_B, and angles are measured in radians. (An angle in radians is the arc length divided by the radius of the circle.) The shape of an involute is dependent only on the size of its base circle. The involute does not exist below its base circle.

From Fig. 15–2:

P is any point on the involute.

r is the radius to P from the center of the base circle.

R_c is the length of the "string" that has been "unwound" from the base circle, which is equal to the radius of curvature of the involute at point P.

$$dR_c = \sqrt{(r^2 - R_b^2)} \tag{15–1}$$

The length of the arc subtended by the angle ε is equal to R_c ;

then $\varepsilon = R_c/R_b$. (The angle ε is called the *roll angle* at P).

The tangent of the angle φ is R_c/R_b, so that $\phi = \arctan(R_c/R_b)$. (The angle φ is called the *pressure angle* at P.)

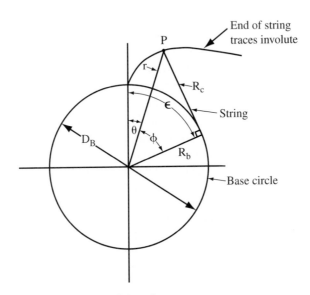

Figure 15-2   Base circle and involute

The angle

$$\theta = \varepsilon - \phi \tag{15-2}$$

is called the *involute* of the angle ϕ, sometimes written as inv ϕ.

We can then write the polar equation of the involute as

$$\theta = \frac{\sqrt{(r^2 - R_b^2)}}{R_b} - \arctan\left[\frac{\sqrt{(r^2 - R_b^2)}}{R_b}\right] \tag{15-3}$$

or

$$\theta = \text{inv } \phi = \tan \phi - \phi \tag{15-4}$$

15.4 The Involute Cam

Fig. 15–3 is a sketch of an involute cam. The cam follower starts in position #1 with the involute at position a. We then rotate the base circle from position a to position b, and the cam follower rises to position #2. The distance between positions #1 and #2 is equal to the length of the arc from a to b on the base circle. The cam follower always contacts the involute along a line tangent to the base circle and normal to the involute. (The line is our generating "string.")

If the base circle is turned at a uniform rate, the cam will rise at a uniform rate. The involute is a constant-rise cam.

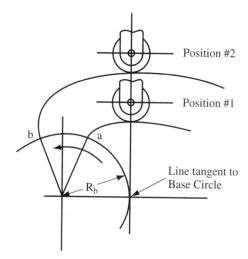

Figure 15-3 Involute cam

15.5 Action of One Involute Against Another

Now we will remove the cam follower and substitute another involute. This situation is represented in Fig. 15–4. Our original base circle and involute are labeled #1 and the new base circle and involute (replacing the cam follower) are labeled #2. In this configuration, if base circle #1 is rotated at a uniform rate, base circle #2 will rotate at a uniform rate. As the common tangent is "fed" off of base circle #1, it will be "fed" onto base circle #2. This is the same geometric case as our old crossed-belt drive, but with the pulleys replaced by base circles and the belt by the common tangent line. The speed ratio of the base circles will be inversely proportional to their diameters. The reduction ratio is R_{b2}/R_{b1}.

The *pitch point* is established by the intersection of the common tangent line and the line drawn between the centers of the base circles (the *line of centers*). Note that *a single involute (or a single gear) has no operating pitch diameter until it is put in contact with another involute.*

15.6 Center Distance, Operating Pitch Diameter (PD), and Pressure Angle

In Fig. 15–5, two pairs of involutes are in contact at different center distances. In fact, the only difference between the right and left sketches is the center distance: *The corresponding involutes and base circles are identical.* The common tangent to the base circles in both cases is the line of action and the path of contact between the involutes. The pitch point is, once again, the intersection of the

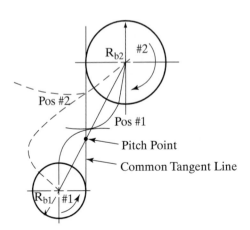

Figure 15–4

common tangent and the line of centers. All forces between the involutes are along the common tangent. The normal to the contact point *always* goes through the pitch point. This is the necessary and sufficient condition for conjugate action. Only with the involute tooth form are the path of contact and the line of action in the same straight line.

The angle between the common tangent and a line perpendicular to the line of centers is called the *operating pressure angle* and is denoted by ϕ. The operating pressure angle does *not* exist until two involutes are brought into contact. There is a definite relationship between the operating pitch diameters (OPDs) and the operating pressure angles (OPAs). Note that the sizes of the OPDs and the OPAs depend *only* upon the sizes of the base circles and the distance between them.

From the conditions in Fig. 15–5:

CD = center distance

R_1 = pitch radius of involute 1

R_2 = pitch radius of involute 2

R_{b1} = base radius of involute 1

R_{b2} = base radius of involute 2

ϕ = pressure angle

$CD = R_1 + R_2$

cosine $\phi = (R_{b1} + R_{b2})/CD$

$R_{b1} = R_1$ cosine ϕ

$R_{b2} = R_2$ cosine ϕ

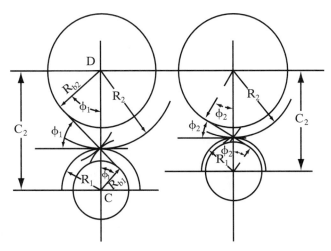

Figure 15–5 Center distance and pressure angles

15.7 Action of an Involute Against a Straight Line

Now let us look at the conditions obtained when an involute acts against a straight line. This set of circumstances is sketched in Fig. 15–6.

When an involute acts against a straight line, the line is tangent to the involute and perpendicular to the line of action (the line that generates the involute). If the straight line is constrained to move only in the direction of the line of action, it will be moved at a uniform rate that is the same as the cam follower in Fig. 15–3.

Now consider the motion of a straight line constrained to move in the direction given by A–A'. The distance the line travels in the direction A–A', from position #1 to position #2, is designated D_1, the distance along the line of action is designated D, and the angle between the line of action and A–A' is designated ϕ. It then follows that $D_1 = D/\text{cosine } \phi$. As the value of D changes uniformly, and as the value of ϕ is constant, the value of D_1 also changes uniformly. Since cosine ϕ is always less than 1, the value of D_1 will never be less than D.

Therefore, when the line against which the involute acts is constrained to move in the direction A–A', the distance it travels along A–A' will be greater than the distance along the line of action. However, its rates of motion will be uniform, as long as the rate of rotation of the involute is uniform. If the involute makes a complete revolution, the value of D would become $2\pi R_b$ and the value of D_1 would become $2\pi R_b/\text{cosine } \phi$, which represents the circumference of a pitch disk that drives a straightedge by friction when the straightedge is parallel to A–A'. The radius of this pitch disk, R_1, is equal to $R_b/\text{cosine } \phi$. In Fig. 15–6, this radius is

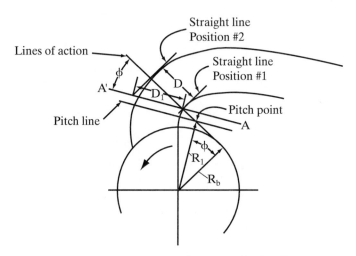

Figure 15–6   Action of an involute against a straight line

established by the intersection of the line of action with a line from the center of the base circle that is perpendicular to A–A'. This intersection is the pitch point; hence, *the form of the involute rack is a straight line.* Since the direction of A–A', and therefore ϕ, is arbitrary, any point of the rack may be used as a pitch point without affecting its value as the basic rack of an interchangeable gear system. (The rack for a gear system may also be thought of as the form of a gear with an infinite number of teeth and where the reference pitch diameter is a straight line.)

15.8 Summary of Basic Concepts

From what we have learned in our brief exploration of parallel axis drives and the involute of a circle, we can make a small summary of the properties of this gear tooth form:

1.   When we transmit rotary motion from one shaft to another, we usually want uniform motion (conjugate action).

2.   In theory, belt drives or friction disk drives will give us conjugate action. (The trouble is, however, that belts stretch and friction disks slip at relatively light loads, so we usually need something better—perhaps teeth on wheels that still give conjugate action.)

3.   We pick the involute of the circle for our gear tooth form.

Why do we pick the involute? Following are some reasons:

a.   Only the involute will be conjugate at various center distances. (Tight tolerance on CD not necessary.)

b.   Only with the involute will the path of contact and the line of action be the same straight line, because the pressure angle is constant.

c.   The form of the basic rack for the involute, and only for the involute, is a straight line. Any point on the rack can be used for the pitch point. (See later.)

d.   Since the basic rack is straight sided, so are generating-type tools based on the rack. These straight-sided are much easier to make and check than are curved tools.

e.   The involute system is simple in theory and *very* flexible (compared to other systems, regardless of what you are thinking right now).

f.   Gears are probably the most efficient means of transmitting power or motion between shafts. The involute gear is a good compromise between efficiency, flexibility, and load capacity.

g.   Most of the gear production machinery and tooling in the world is based on the involute system. (You can actually get your gears made and checked.)

15.9 Tooth Profiles Based on the Involute

Now that we have decided on a profile shape for our gear teeth, we need to develop this form into a gear. Starting with our base circle, we will develop a series of involute curves that will become part of the teeth on the gear. For now, we will look at only one side of the teeth, as gear teeth are usually symmetrical. Fig. 15–7 shows a base circle on which we have developed eight equally spaced involute curves. This figure might be constructed by winding a string around the circle. On the string, we have marked eight evenly spaced points. As we unwind the string, all eight points will describe involutes of the circle. The distance between involutes will be the same, measured on any tangent line to the circle (the string). The distance between successive involutes is also equal to one-eighth of the circumference of the base circle. This distance is called the *base pitch* of the gear. (Only the involute gear has such a base pitch.) The base pitch of two mating involute gears must be the same to obtain smooth, continuous action as contact is passed from one tooth to the next. The base pitch can be calculated by dividing the circumference of the base circle by the number of teeth on the

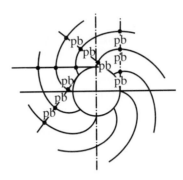

Figure 15–7　Base circle and involute curves

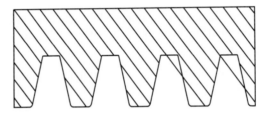

Figure 15–8　Involute rack

gear. The base pitch is also equal to the distance on the basic rack from one tooth to the next normal to the flanks of the straight rack teeth. Fig. 15–8 is a sketch of an involute rack.

Now we will add the other side of the teeth. It is usual practice with parallel-axis involute gears to make both sides of the teeth the same. (This, however, is not a requirement for continuous action.) Fig. 15–7 shows a base circle with eight involute teeth. Notice that, for a given tooth thickness, we are limited to the portion of the involute curve that we can use for our tooth flanks. However, since *any* point on the involute curve can be used for a pitch point (remember, the basic rack is a straight line), we are not limited by this consideration. Still, the figure does indicate two limitations. First, *there is no involute below the base circle*; any tooth action must take place above the base circle. This fact may seem obvious, but many gear sets do not operate properly because it was not considered. (When we examine the formulas normally used to calculate elements of the gear set, contact below the base circle can be overlooked.) Second, we cannot use any portion of the involute above the point where the left and right curves intersect. At this intersection the teeth would be pointed, and any material above the intersection would be cut away. This fact also is sometimes overlooked until the gear is cut, and then it becomes obvious (and somewhat uncomfortable for the designer).

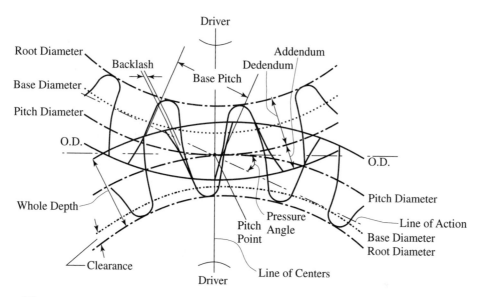

Figure 15–9 Gear nomenclature

The preceding are two of the limitations on correct gear tooth action. We will discuss some others as we proceed, but these two are fundamental.

15.10 Contact Ratio

To obtain smooth, continuous action from a pair of gears, we must be sure of two geometrical conditions. First, as we have discussed, the profiles must be conjugate. This is assured by using involute curves. Second, contact may not cease for any tooth before the next tooth has come into contact with it. This means that the working flanks of our teeth must be long enough to ensure a *contact ratio* greater than or equal to 1. Referring to Fig. 15–9, we see that, for the gears pictured, the active length of the contact on the line of action is bounded by the outside diameters of the gears. (It is *not* bounded by the base circles or the diameters of the pointed teeth.) To be assured of a contact ratio greater than 1, the active length of contact on the line of action must be longer than one base pitch. The length of contact divided by the base pitch is called the *contact ratio*. (It is actually the average number of teeth in contact through one mesh cycle.) Note that *the only requirements for smooth, continuous action of involute gears are that the base pitches be the same and the contact ratio be equal to or greater than 1.*

Table 15–1 presents terminology of gears, which we propose to use while analyzing and designing gears.

Table 15–1 Terminology of Gears

BASE DIAMETER: The diameter of the base circle from which the involute tooth profile is developed.

PITCH DIAMETER: The diameter of the pitch circle. In theory, it is the imaginary circle that rolls without slippage with a pitch circle of a mating gear.

OUTSIDE DIAMETER: The diameter of the addendum or outside circle.

ROOT DIAMETER: The diameter of the root circle, or the circle that is tangent to the bottoms of the spaces between the teeth.

LINE OF CENTERS: The line that connects the centers of the pitch circles of two mating gears.

PITCH POINT: The point of tangency of two pitch circles on the line of centers.

LINE OF ACTION: The straight line passing through the pitch point and tangent to the base circles. It is actually the path of contact of mating involutes.

BASE PITCH: The fundamental distance between adjacent curves along a common normal such as the line of action.

ADDENDUM: The radial distance between the pitch circle and the outside diameter.

DEDENDUM: The radial distance between the pitch circle and the root circle.

CLEARANCE: The amount by which the dedendum in a gear exceeds the addendum of its mating gear.

PRESSURE ANGLE: The angle between the line of action and the line tangent to the pitch circle at the pitch point. Since an involute has no specific pressure angle until it is brought into intimate contact with another involute, the operating pressure angle is determined by the center distance at which a pair of gears operates. For uniformity and various economies, standard pressure angles are established for standard gear tooth systems.

WORKING DEPTH: The depth of engagement of two mating gears—in effect, the sum of their addenda.

WHOLE DEPTH: The total depth of a space between two teeth, equal to the addendum plus dedendum, or the working depth plus clearance.

FACE WIDTH: The length of the teeth in an axial plane.

BACKLASH: The amount by which the width of a space between teeth exceeds the thickness of the engaging tooth on the pitch circles.

For further clarification of definitions of "Gearometry" elements, see AGMA Standard Number 112.03, entitled "Gear Nomenclature."

15.11 Standards

We now have an understanding of the form of the involute gear in terms of the base circle, the generation of the involute, and how involute gears mesh together to form a gear set. The definition and calculation of involute gear systems in terms of the base circles is difficult and rather clumsy. To facilitate the specification and calculation of gears and related tooling, a system of standards has been developed. *This system of standards, while making calculations and specifications easier, does much to mask the flexibility and adaptability of the involute system*. With this caveat in mind, we will again look at some basic elements of gears in terms of these standards.

The *diametrical pitch* is the number of teeth per inch of gear pitch diameter. The diametral pitch cannot be measured on a gear or a tool.

The *pitch diameter* of a gear is the diameter that is obtained by dividing the number of teeth by the diametral pitch. The pitch diameter cannot be found or measured on a gear; it will be referred to as the *reference pitch diameter* in our discussions.

The *pitch line* of the rack is a line parallel to the tips and roots of the rack teeth. (Usually, the pitch line is between the tips and roots of the teeth; sometimes, it is defined as the location where the tooth and space of the rack teeth are equal.)

The *circular pitch* is the distance from a point on the rack to the corresponding point on the next tooth or space parallel to the pitch line. The circular pitch is also the arc length from a point on a gear tooth or space to the corresponding point on the next tooth or space along the reference pitch diameter. The circular pitch is equal to π divided by the diametrical pitch.

These values are reference values defined by the standards. They all have corresponding operating values when two involute gears are mated with each other or with a rack:

1. *Involute gears do not have an operating pitch diameter until mated with another involute gear.*

2. *Involute gears do not have an operating pressure angle until mated with another involute gear.*

3. *Involute gears do not have an operating diametral pitch until mated with another involute gear.*

4. *Involute gears do not have an operating circular pitch until mated with another involute gear.*

The formula for the standard center distance is

$$C = \frac{(N_1 + N_2)}{(2P)} \tag{15-5}$$

$$P = \frac{N}{d_p}$$

The formula for the standard outside diameter is

$$OD = \frac{N}{P} + \frac{2}{P} \tag{15-6}$$

The formula for the standard circular pitch is

$$\text{CP} = \frac{\pi}{P} = \frac{\pi d_p}{N} \tag{15–7}$$

The formula for the standard zero-backlash tooth thickness on the reference PD is

$$\text{TT} + \frac{\text{CP}}{2} = 0 \tag{15–8}$$

In the preceding formulas, N_1, N_2 are the number of teeth in mating gears, P is the Diametrical pitch, and TT is the tooth thickness.

If a set of gears is made to these proportions, then the standard values will be the same as the operating values. There is *no* good reason, however, for using standard proportions. In fact, there are many good reasons for *not* using them. We will investigate many of these reasons as we proceed further.

Nonstandard proportion gears can be made with standard cutting tools with no sacrifice in cost or operation. (No change is necessary in tool specifications.) As a matter of fact, quite the contrary is true: Using standard cutting tools, we can obtain a significant improvement in operation and capacity of the gears by utilizing nonstandard proportions. The only real problem we will have is being confused by the standards until the flexibility of the system is overlooked. The advent of the computer has done much to help this situation, as the calculation and optimization of nonstandard gears can now be done quickly and accurately. Thus, *in the design of gears, the standards are helpful only in finding a starting place from which to optimize the design.*

15.12 Kinematics and Kinetics of Spur Gears

Gear kinematics is the study of the motion of gears without considering the cause of that motion. Gear motion includes *displacement, velocity, acceleration,* and, sometimes, the time rate of change of acceleration, called *jerk*. Kinetics is the study of cause of motion—forces, torques, or moments. First we will study the kinematics of gear trains and planetary gears. Later on, we will study the kinetics of gears.

Let us consider two spur gears (a pinion and a gear) that are in contact with each other, as shown in Fig. 15–10. One has a diametrical pitch P, and number of teeth N_1 and is rotating clockwise at a speed of n_1 revolutions per second (rps). We are interested in finding the speed n_2 of the gear in contact with the first gear and having the same diametrical pitch (this is a requirement for smooth operation of such a gear train), but a different number of teeth N_2. Let us look at such a problem from a fundamental point of view.

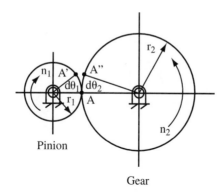

Figure 15-10   Pinion and gear working as gear train

Let the pinion (gear 1) rotate through an angle $d\theta_1$, thus moving point A to A' by $r_1 d\theta_1$. Of course, the gear (gear 2) now must rotate to A'' by an amount $r_2 d\theta_2$ that is the same as $r_1 d\theta_1$ in order to have no relative displacement during the motion from A to A', which is required for smooth operation. Thus,

$$r_1 d\theta_1 = r_2 d\theta_2 \tag{15-9}$$

If we divide this equation by dt, we obtain

$$r_1 \frac{d\theta_1}{dt} = r_2 \frac{d\theta_2}{dt} \tag{15-10}$$

But, by definition, we know that $\dfrac{d\theta}{dt} = \omega$, the angular velocity, and therefore,

$$r_1 \omega_1 = r_2 \omega_2 \tag{15-11}$$

But $\omega_1 = 2\pi n_1$ and $\omega_2 = 2\pi n_2$; also, $r_1 = \dfrac{N_1}{2P}$ and $r_2 = \dfrac{N_2}{P}$, where P is the diametrical pitch. Therefore, substituting the preceding values into Equation (15-11), we get

$$N_1 n_1 = N_2 n_2$$

or

$$n_2 = \frac{N_1 n_1}{N_2} \tag{15-12}$$

Thus, we can find the speed of gear 2 if the speed of gear 1 and the number of teeth in each gear are known. Such speed is in the direction opposite that of

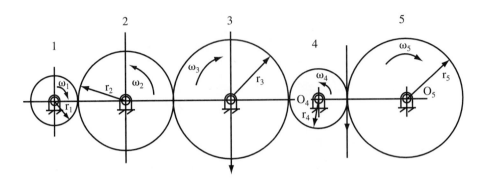

Figure 15–11   Spur gear train having different teeth for each gear

the speed of gear 1; that is, if gear 1 is rotating clockwise, gear 2 will rotate counterclockwise.

Now let us take a gear train such as that shown in Fig. 15–11.

We see that if we takes gear 1 and gear 2, then

$$n_2 = \frac{N_1}{N_2}n_1 \qquad\qquad (15\text{--}13)$$

If we take gear 2 and gear 3, then we can similarly show that

$$n_3 = \frac{N_2 n_2}{N_3} \qquad\qquad (15\text{--}14)$$

But $n_2 = \dfrac{N_1}{N_2}n_1$

Hence,

$$n_3 = \frac{N_2 N_1}{N_3 N_2}n_1 \qquad\qquad (15\text{--}15)$$

Now, by taking gear 3 and gear 4, we can show that

$$n_4 = \frac{N_3}{N_4}n_3$$

Substituting for n_3 from Equation (15–16), we get

$$n_4 = \frac{N_1 N_2 N_3}{N_2 N_3 N_4} n_1$$

Carrying the analysis further, we obtain

$$n_5 = \frac{N_1 N_2 N_3 N_4}{N_2 N_3 N_4 N_5} n_1$$

The preceding equation now shows a very important relationship between the rotational speed of the last gear if the speed of the initial gear is known, namely,

$$n_{last} = \frac{\text{Product of driving gears}}{\text{Product of driven gears}} n_{initial} \tag{15–16}$$

Clearly, gear 1 is driving gear 2, gear 2 is driving gear 3, gear 3 is driving gear 4, and gear 4 is driving gear 5. Thus, the product of the driving gears is $N_1 N_2 N_3 N_4$. Similarly, gear 2 is driven by gear 1, gear 3 is driven by gear 2, gear 4 is driven by gear 3, and gear 5 is driven by gear 4. Therefore, the product of the driven gears is $N_2 N_3 N_4 N_5$. Hence, Equation (15–17) can be used for any gear train. As far as the rotational direction of the last gear is concerned, it is easy to see that for an *even* number of gears in a train, the direction of the last gear is in the direction opposite that of gear 1. For a gear train with an *odd* number of gears, the last gear is rotating in the same direction as the first gear. In fact, if one wants to have a gear rotate in the same direction as the pinion, then a third gear called an idler is introduced between the pinion and the gear, as shown in Fig. 15–12.

15.13 Planetary Gear Kinetics

A planetary gear is shown in Fig. 15–12. Gear 1 is called a sun gear, gear 2 is called a planetary gear, and gear 4 is called a ring gear. Arm 3 connecting the sun gear hub and the planetary gear hub is like a gear, but has no teeth.

To understand the kinematics of a planetary gear, one must know the *speed of any two gears*. Only then, using the laws of consistency of motion, can one derive equations to obtain the motion of the other two gears.

Let us study the approach taken in relating the known speed of given gears to the unknown speed of the other gears, based on the geometry of the planetary gear train. Let n_1, n_2, n_3, and n_4 be the speeds (in rpm or rps) of gears 1, 2, 3, and 4, respectively. (Note that arm 3 can be treated as gear 3.) Also, let N_1, N_2, N_3, and N_4 be the number of teeth in gears 1, 2, 3, and 4, respectively. (Note that $N_3 = N_1$

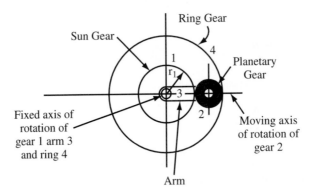

Figure 15–12 Sketch of planetary gear, showing sun gear, arm, planetary gear, and ring gear

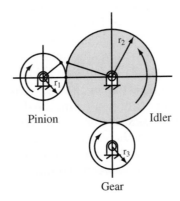

Figure 15–13 Gear train with an idler gear

+ N_2; we will see why later.) We do not, of course, have teeth in arm 3, but we know the arm radius, which is equal to the sum of the sun and planetary gear radii. Thus, we can assume that the arm has fictitious teeth. We will also assume that gears 1, 2, 3, and 4 have angular velocities ω_1, ω_2, ω_3, and ω_4, respectively, in radians/second, all assumed to be clockwise. Now if we take gear 1, the sun gear, then we see that at A we have $r_1\omega_1$ as velocity V_A. (See Fig. 15–14.) Similarly, at B gear 4 has velocity V_B, which is equal to $r_4\omega_4$, at C and the arm has velocity $r_3\omega_3$. Of course, if one takes gear 2, then the common points are A, B, and C. Then the velocity at A on the planetary gear is same as that on the sun gear, the velocity at B on the planetary gear is the same as that on the ring, and the velocity at C is the same as that on arm 3. Thus, if one isolates the planetary gear 2, which is a rigid body, and uses the velocities shown in Fig. 15–14, then, assuming that we

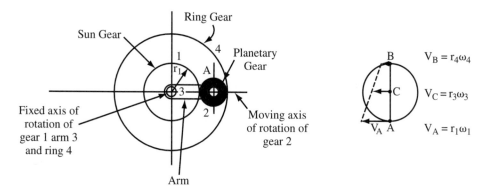

Figure 15–14 Planetary gear and isolated planetary gear showing velocity vectors at A, B, and C

know two velocities, say, the velocity at A and the velocity at B, we can find the velocity at C using relative velocity equations for rigid bodies. That is,

$$\bar{V}_C = \bar{V}_B + \bar{V}_{C/B} \tag{15–17}$$

or

$$\bar{V}_C = \bar{V}_A + \bar{V}_{C/A} \tag{15–18}$$

But $V_c = r_3\omega_3$, $V_B = r_4\omega_4$, $V_A = r_1\omega_1$, $V_{C/B} = r_{BC}\omega_2$, and $V_{C/A} = r_{AC}\omega_2$. Note, however, that if all the angular velocities of all the gears are assumed to be counterclockwise, then V_A, V_B, and V_C are to the left, and $V_{C/A}$ is also to the left, but $V_{C/B}$ is to the right if we let B be fixed and let ω_2 be counterclockwise with respect to B. Then $V_{C/B}$ is negative if V_A, V_B, V_C, and $V_{C/A}$ are assumed to be positive—that is, to the left. Accordingly, we can modify Equations (15–17) and (15–18) as the following two scalar equations:

$$r_3\omega_3 = r_4\omega_4 - r_{BC}\omega_2 \tag{15–19}$$

$$r_3\omega_3 = r_1\omega_1 + r_{AC}\omega_2 \tag{15–20}$$

Now we have two simultaneous linear equations and two unknowns, ω_2 and ω_3, that we can solve using the matrix equation.

$$\begin{bmatrix} r_{BC} & r_3 \\ -r_{AC} & r_3 \end{bmatrix} \begin{Bmatrix} \omega_2 \\ \omega_3 \end{Bmatrix} = \begin{Bmatrix} r_4\omega_4 \\ r_1\omega_1 \end{Bmatrix}$$

At times, only the number of teeth and the diametrical pitch (which is common for each gear) are given. In that case, one can use the following additional equations to obtain the radii:

$$r_1 = \frac{N_1}{2P} \qquad r_2 = \frac{N_2}{2P} \qquad r_4 = \frac{N_4}{2P} \qquad r_3 = r_1 + r_2$$

If the angular velocity in radians is not given, but the speed in revolutions per second is, then, since

$$\omega = \frac{n}{2\pi}$$

it follows that

$$\omega_1 = \frac{n_1}{2\pi} \qquad \omega_2 = \frac{n_2}{2\pi} \qquad \omega_3 = \frac{n_3}{2\pi} \qquad \omega_4 = \frac{n_4}{2\pi}$$

For ω to be in rad/sec, n must be revolutions per second (rps).

Let us study an example to understand the application of the equations we have derived.

Example 15-1

a.   For a specific overdrive, a ring is expected to rotate at 1.5 times the speed of an input arm. (The arm is an input.) The sun gear is locked and the planetary gear has 20 teeth; the angular velocity of the arm is 1 rad/sec. Determine:

  1.   The number of teeth required in the sun gear and the ring gear.

  2.   The angular speed of the planetary gear and its instantaneous axis of rotation. If needed, assume that $p = 0.7$, where p is the circular pitch, equal to $(\pi d_p)/N$, in which d_p is the pitch diameter of the gear and N is the number of teeth in the gear. Is it necessary to know p and ω_{arm} to solve the problem?

b.   What will be the speed of the output ring gear if the same planetary gear system as in part a is used, but now the sun gear is an input gear and the arm is locked. Can we use such a gear as a reverse mechanism? If so, what is the ratio of the output torque to the input torque?

Use the following diagram to answer the preceding questions:

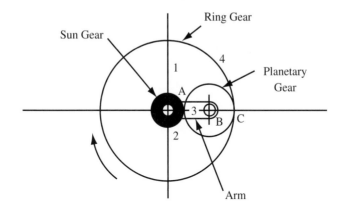

Solution

Since the sun gear is locked, the planetary gear has its instant center at the intersection of the sun and planetary gears. We are given that the angular velocity of the ring is 1.5 times the angular velocity of the arm; that is,

$$\omega_r = 1.5\omega_a$$

Now,

$$V_C = \omega_r \frac{N_r}{2} = 1.5\omega_a \frac{N_r}{2}$$

But

$$N_r = \frac{N_s}{2} + N_P$$

Therefore,

$$V_C = 1.5\omega_a \left(\frac{N_s}{4} + \frac{N_P}{2} \right)$$

But

$$V_B = \omega_a \frac{N_a}{2}$$

or, since $N_a = N_s + N_p$,

$$V_B = \omega_a \frac{N_s + N_P}{2}$$

Next, we look at the velocity diagram for the planetary gear:

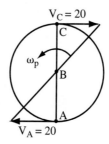

We see that

$$\frac{V_C}{N_p} = \frac{V_B}{N_p/2} \qquad (Note: d_P = \frac{N_P}{P})$$

Substituting for V_c and V_b, and knowing that $N_P = 20$, we obtain

$$1.5\omega_a\left(\frac{N_S}{4} + \frac{\omega}{2}\right)\left(\frac{1.5}{20}\right) = \left(\frac{N_S}{2} + \frac{20}{2}\right)\frac{1}{\left(\frac{20}{2}\right)}\omega_a$$

Canceling ω_a, we find that the number of teeth in the sun gear is

$$N_S = 5.73$$

(Naturally, one has to use six teeth or five teeth. This change will change kinematics of the gears.) To solve for the number of teeth in the ring gear, we note that

$$N_r = ((N_s/(2)) + Np)$$

$$= ((40/(2)) + 20) = 22.87$$

Since the instant center for the planetary gear is at A, the speed of the planetary gear is

$$\omega_p = \left(\frac{V_B}{(N_p/2)}\right)_n = \frac{(N_s + N_p)}{\frac{2}{N_P/2}}\omega_a$$

$$= \left(\frac{N_s + N_p}{N_p}\right)\omega_a \quad (since \ V_B = \left(\frac{N_s + N_p}{2}\right)\omega_a)$$

Thus,

$$\omega_p = \frac{(40 + 20)}{20}(1) = 3 \ rad/sec$$

These calculations show that a circular pitch was not necessary.

Now let us work on part b, in which the arm is locked and the instant center for the planetary gear is at its geometric center. Then

$$V_A = -V_C$$

But

$$V_C = \left(\frac{N_s}{2} + N_p\right)\omega_{ring}$$

so

$$V_A = \frac{N_s}{2}\omega_s = \frac{40}{2}(1) = 20$$

($V_B = 0$, as the arm is fixed.)

Now we draw a free-body diagram for the planetary gear, showing these velocities:

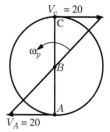

If the instant center is at B, then

$$V_C = -V_A = -20$$

But

$$V_C = \left(\frac{N_s}{2} + N_p\right)\omega_{ring}$$

So

$$\left(\frac{N_s}{2} + N_p\right)\omega_{ring} = -20$$

$$\left(\frac{40}{2} + 20\right)\omega_{ring} = -20$$

Therefore,

$$\omega_{ring} = \frac{-20}{40} = -0.5 \text{ rad/sec}$$

$$= 0.5 \text{ rad/sec clockwise}$$

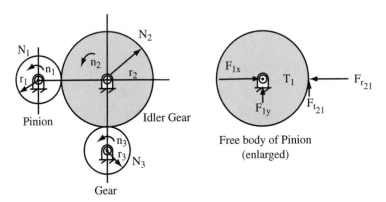

Figure 15–15 Gear train and free-body diagram showing forces acting on the pinion

This shows that the ring gear rotates at half the speed of the sun gear and in the opposite direction. Such a system can be used as a reverse gear.

The foregoing problem is solved using lead model *ex15-1.tk*, or *planetory.tk*.

15.14 Kinetics of Spur Gears

Thus far, we have studied the kinematics of spur gears. Now let us examine the kinetics of a gear train having a pinion, an idler gear, and a gear, as shown in Fig. 15–15. We assume that the pinion has N_1 teeth and is rotating at speed n_1. We also assume that the idler gear has N_2 teeth and is rotating at speed n_2. Gear 1 has N_3 teeth and is rotating at speed n_3. If the pinion is driven by a motor having a specified horsepower, then the torque transmitted to the pinion by the motor is

$$T_1 = \frac{(\text{HP})}{2\pi n_1}(33,000) \tag{15–21}$$

where T_1 is the torque in ft-lb. (*Note*: If n_1 is in rps, then multiply by 550; if n_1 is in rpm, then multiply by 33,000.)

In SI units, the power is given in kW, so that

$$T_1 = \frac{\text{kW}(1,000)}{2\pi n_1} \tag{15–22}$$

where T_1 is the torque in Nm, n_1 is the speed in rps, and kW is kilowatts (1,000 watts of power).

Now, knowing the torque T_1, we can find the forces acting at the hub of the idler and the intersection of the pinion and idler gear. Since the pinion or gear is rotating at a constant speed, we can use equations of equilibrium, i.e., $\Sigma \overline{F} = 0$ and $\Sigma \overline{M} = 0$. Let us draw a free-body diagram of the pinion, as in Fig. 15–15.

Then we can define F_{1x} and F_{1y} as reaction forces at the hub of the pinion. Also, let F_{r21} and F_{t21} be the components of the gear force in the radial and tangential directions, respectively. (The subscript 21 stands for the force coming from gear 2 to gear 1.) Finally, since F is acting at a pressure angle ϕ, it follows that $F = F_{t21}/\cos \phi$ and $F_{r21} = F\sin\phi$.

Now let T_1, n_1, and r_1 be the applied torque, rotational speed, and radius of the pinion, respectively. Then, applying the equations of equilibrium, we get (see the free-body diagram of the pinion in Fig. 15–15).

$$F_{1x} - F_{r21} = 0 \quad \text{for } \Sigma F_x = 0$$

$$F_{1y} + F_{t21} = 0 \quad \text{for } \Sigma F_y = 0$$

$$F_{t21}(r_1) - T_0 = 0 \quad \text{for } \Sigma M_0 = 0$$

(The last equation will yield the value of F_{t21}.)

Now, we know that $F = F_{t21}/\cos\phi$ and $F_{r21} = F\sin\phi$. Thus, the foregoing equations are sufficient to find the reaction and applied forces at the intersection of the pinion and the idler gear.

Next, if one is required to find reaction forces on an idler gear, that can be done by drawing a free-body diagram such as Fig. 15–16 and then using the equations of equilibrium.

Since action and reaction forces are equal in magnitude, but opposite in direction, we can show that $F_{r12} = -F_{r21}$ and $F_{t12} = -F_{t21}$, the forces at the interface of gears 1 and 2. Let F_{x2} and F_{y2} be the reaction forces at the hub of gear 2 (the idler). Let F_{t32} and F_{r32} be the interface forces from gear 3. (Note that $F_{t32} = -F_{t23}$ and $F_{r32} = -F_{r23}$.) Let n_2 be the rotational speed of gear 2, having a radius r_2. We can assume that no external torque is applied or transferred through an idler gear. Therefore,

$$\text{for } \Sigma F_x = 0 \qquad F_{x2} + F_{r12} + F_{t32} = 0 \tag{15–23}$$

$$\text{for } \Sigma F_y = 0 \qquad F_{y2} - F_{t12} + F_{r32} = 0 \tag{15–24}$$

$$\text{for } \Sigma M_0 = 0 \qquad F_{t12}r_2 + F_{t12}r_2 + F_{t32}r_2 = 0 \tag{15–25}$$

These three equations can be solved for the three unknowns F_{x2}, F_{y2}, and F_{t32}. Note that $F_{r32} = F_{t32} \tan \phi$, where ϕ is the pressure angle. A similar approach can be used to find the forces acting on an output gear, as well as its hub reaction forces.

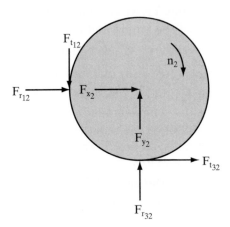

Figure 15–16   Free-body diagram for an idler gear

15.15 Design of Spur Gears

Design of Gear Tooth for Bending Fatigue Stress

In spur gear design, one needs to know the *bending stress* and *pitting resistance stress*. So far we had assumed that the tangential and radial force was acting at the base circle radius of a gear. Moreover, for design of the gear tooth we will neglect the radial load and assume the tangential load to be acting at the tip of the gear tooth. Such an assumption will yield a conservative design. As a review and to show the point of application of the tangential force, Fig. 15–17 is drawn. Fig. 15–17 also shows other parameters required in design for example the width b. The bending stress in FPS units due to the tangential force is given by formula

$$\sigma = \frac{F_t P}{bJ} K_v K_o K_m \qquad (15\text{–}26)$$

With SI units, we use the formula

$$\sigma = K_v K_o K_m \frac{F_t}{bJm} \qquad (15\text{–}27)$$

The meaning of the variables in either formula is as follows:

J: geometry factor (includes the Lewis form factor, which is not discussed in this book)

K_v: velocity or dynamic factor

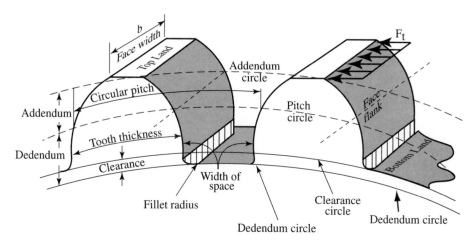

Figure 15–17 Assumed location of F_t on gear tooth

K_o: overload factor (reflects nonuniformity of driving and load torques)

K_m: mounting factor

(All of the preceding factors are shown in Example 15-2.)

F_t: transverse load in lb or newtons F_t (in FPS) $= \dfrac{(HP)}{V} \times 550$
  or F_t (in newtons) $= (kW)\, 1{,}000/V$

V: pitch line velocity ($V = 2\pi rn$) in ft/sec or m/sec if r is in ft or r is in meters, respectively, and n is in rps

b: face width in inches or meters

r: radius of gear in feet or meters

n: rps

P: diametrical pitch (N/d_p), where d_p is in inches

N: number of teeth in the gear

m: module $\dfrac{d_p}{N}$, where d is in meters

σ: effective bending fatigue stress

Factors such as J, K_v, etc., are given in the solution of Example 15-2.

The stress obtained using Equations (15–26) or (15–27) must be equal to or less than the material's fatigue strength for a certain life, or it must be equal to or less than the endurance limit if a life greater than 10^6 cycles is desired. The endurance limit S_e, for gear material is obtained using the equation

$$S_e = S_{e'} C_L C_G C_S k_r k_t k_{ms}$$

where S_e is the endurance limit for a machine element, $S_{e'}$ is the standard R. Moore endurance limit, C_L is the load factor (1 for bending loads), C_G is the gradient factor (1 for $P > 5$ and 0.85 for $P \leq 5$), C_S is the surface factor (see Fig. 15–22), k_r is the reliability factor (see Table 15–4), k_t is the temperature factor $= \dfrac{620}{(460 + T)}$ for $T > 160°F$ (for $T < 160°F$, $k_t = 1$), and k_{ms} is the mean stress factor (1 for an idler gear and 1.4 for a pinion).

Once the bending stress and allowable strength for the gear are determined, the safety factor based on bending fatigue is calculated using the equation

$$SF)_b = \frac{S_e}{\sigma}$$

The recommended value of $SF)_b$ is between 1.6 and 2.

15.16 Design of Tooth for Surface-Wear Fatigue Stress

To design a gear tooth against pitting, we use the fundamental American Gear Manufacturing Association (AGMA) formula for surface wear fatigue (i.e., pitting resistance), which is based on the Hertz stress formula for two cylinders loaded as shown in Fig. 15–18.

$$\sigma_H = C_p \sqrt{\frac{F_t}{b d_p I} K_v K_o K_m} \qquad (15\text{–}28)$$

where b is the width of a tooth, d_p is the pitch circle diameter of the pinion, d_g is the pitch circle diameter of the gear, and

$$I = \frac{\sin\phi \cos\phi}{2} \frac{R}{R + 1} \qquad (15\text{–}29)$$

in which ϕ is the pressure angle and $R = d_g/d_p$.

(*Note:* The subscript p denotes the pinion and g the gear.)

Now only one variable needs to be defined, namely,

$$C_p = 0.564 \sqrt{\frac{1}{\dfrac{1 - v_p^2}{E_p} + \dfrac{1 - v_g^2}{E_g}}} \qquad (15\text{–}30)$$

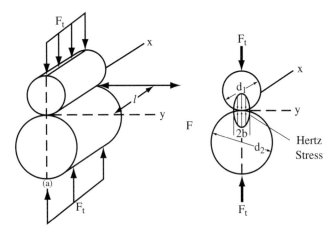

Figure 15–18   Hertz stress due to F_t

where v is Poisson's ratio and E is the modulus of elasticity. The effective pitting stress σ_H must be equal to or less than the pitting strength S_H, which is given by the equation

$$S_H = S_{fe}C_{Li}C_R \tag{15–31}$$

where S_{fe} is the surface fatigue strength for a gear material, C_{Li} is the life factor (for a lifetime greater than 10^4 and less than 10^{11} cycles, see Fig. 15–24), and C_R is the reliability factor. The safety factor based on the load should fall somewhere between 1.3 and 1.6.

15.17 Design Hints

For a face width b,

$$p \leq b \leq 3p$$

where $p = (\pi d)/N = \pi/P$ is the circular pitch. We take values of p on the basis of an inventory of gear-cutting tools and N on the basis of space requirements. Next, we find

$$V = \frac{\pi d n}{12}$$

where n is in rpm, d is in inches, and V is in feet per minute. Then we find

$$F_t = \frac{33{,}000\ (\text{HP})}{V}$$

or, in SI units,

$$F_t = \frac{kW\,60{,}000}{V}$$

where V is in m/min. Finally, we find σ or σ_H and do the comparison by way of the actual load. (See Examples 15-2 and 15-3.)

Make sure that you check for interference and the contact ratio. The equation for interference is, of course,

$$r_{a(max)} = [r_b^2 + c^2(\sin^2\phi)]^{0.5}$$

where $r_{a(max)}$ is the radius of the maximum noninterfering addendum circle of the gear or pinion, c is the center distance P_1P_2, ϕ is the actual pressure angle, and $r_b = r\cos\phi$.

As stated earlier, the contact ratio, which should be greater than 1 (1.5 is a good number), is

$$CR = \frac{\left(r_{ap}^2 + r_{bp}^2\right)^{0.5} + \left(r_{ag}^2 + r_{bg}^2\right)^{0.5} - c\sin\phi}{p_b} \tag{15-32}$$

where r_{ap}, and r_{ag} are the addendum radii of the mating pinion and gear, r_{bp}, and r_{bg} are the base-circle radii of the mating pinon and gear, and p_b is the base-circle pitch (i.e., $p_b = \pi d_b/N$, where $d_b = d\cos\phi$). Consequently, $p_b = p\cos\phi$.

Example 15-2

Fig. 15-19 shows a power-transmitting device used in the landing gear of an aircraft. The device transmits $\pi/2$ HP at 300 rpm from the pinion shaft to the gear shaft. If the pinion has 20 teeth and the gear has 60 teeth with a diametrical pitch equal to 10, find the bending stress and the safety factor based on bending if the fatigue strength (endurance limit) for the steel used in the pinion is 64 ksi and in the gear is 50 ksi. The width of the gear cannot exceed 1 in.

This example is based on bending fatigue.

Solution

In the design of a gear train (2 or more gears), it is required that for smooth operations, the diametrical pitch is the same for every gear in the train. Let us first design the pinion and then we will check the gear requirements.

The circular pitch $p = \dfrac{\pi}{P} = \dfrac{\pi}{10} = 0.314$

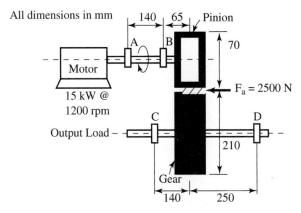

All dimensions in mm

Figure 15–19 Pinion and gear for landing gear

Since $0.314 \leq b \leq 3$, we use b ≈ 1. Now

$$d_p = \frac{N_p}{P} = \frac{20}{10} = 2'' \text{ and } d_G = \frac{N_G}{P} = \frac{60}{10} = 6''$$

The pitch line velocity

$$V = \frac{\pi d_p n_p}{12} = \frac{\pi(2)(300)}{12} = 50\pi \text{ ft/min}$$

$$F_t = \frac{(33,000)\,(HP)}{V} = \frac{33,000\left(\frac{\pi}{2}\right)}{50\pi} = 230 \text{ lb}$$

Assuming a precision gear that is shaved and ground for $V = 50\pi$, the

Table 15–2 Overload Correction Factor K_o

	Driven machinery shock		
Source of power	**Medium**	**Light**	**Uniform**
Medium shock	2.25	1.75	1.50
Light shock	2.00	1.50	1.25
Uniform	1.50	1.25	1.0

velocity or dynamic factor K_v is approximately 1.1. (See Fig. 15–20.)

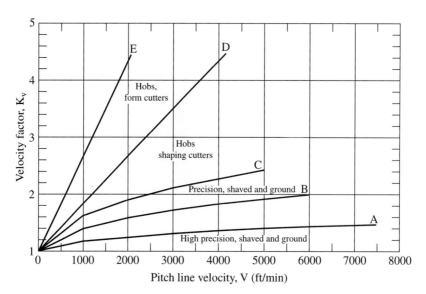

Figure 15–20   Graph showing pitch line velocity vs. velocity factor for various types of gears

The geometry factor J is found from Fig. 15–21. (It is assumed that the fillet radius is $\dfrac{0.35}{P} = 0.035''$ and that no sharing takes place, so as to obtain maximum stress.) From the figure, $J = 0.24$ (assuming a pressure angle $\phi = 20°$; if not, then use $\phi = 25$ (a different graph will be needed)).

Table 15–3 Reliability Correction Factor k_r

Reliability	99.999	99.99	99	90	50
k_r	0.66	0.7	0.81	0.9	1.0

The value of K_o, the overload factor, is obtained from Table 15–2, which is based on experimental evidence as per AGMA recommendations. We choose $K_o = 1.50$ based on light–light shock.

The mounting correction factor K_m is based on Table 15–4. If we assume that we have less rigid mounting, then $K_m = 1.6$, and we can calculate the bending stress:

$$\sigma = \frac{F_t P}{bJ} K_v K_o K_m$$

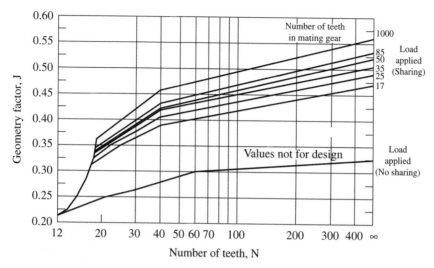

Figure 15–21 Geometry factor J vs. number of teeth for $\phi = 20$ and fillet radius > .03"

$$= \frac{230(10)(1.1)(1.50)(1.6)}{(1)(0.24)}$$

$$= 25{,}300 \text{ psi}$$

Now we can find the fatigue strength of the pinion using the formula

$$S_e = S_{e\Delta}C_L C_G C_s k_r k_t k_{ms}$$

where $C_L = 1$, $C_G = 1$ for $P > 5$, and C_S is obtained using Fig. 15–22 for $S_{ut} = 2 S_{e'}$ $= 2 (64) = 128$ ksi and the fine-ground commercially polished gear. The result is $C_S = 0.9$. Similarly, from Table 15–3, $k_r = 0.81$ for 99% reliability.

Table 15–4 Mounting Correction Factor K_m

Type of mounting	Face width			
	0–2 in	6 in	9 in	16 in and up
No rigid mounting	2.2	2.2	2.2	2.2
Less rigid mounting	1.6	1.7	1.8	2.2
Accurate mounting	1.3	1.4	1.5	1.8

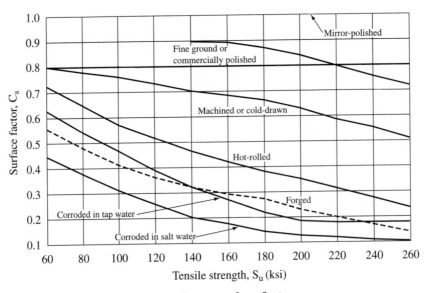

Figure 15–22   Tensile strength vs. surface factor

Since $k_t = \dfrac{620}{(620 + T)}$ for $T > 160°$, we assume that $T < 160°$; therefore, $k_t = 1$ and $k_m = 1.4$ for the pinion. Thus,

$$S_e = 64(1)(1)(0.9)(0.81)(1)(1.4) = 65.32 \text{ ksi}$$

The safety factor based on bending stress is $\text{SF})_b = \dfrac{S_e}{\sigma}$

$$= \dfrac{65.32}{25.3} = 2.58$$

which is greater than the recommended 2, and therefore, we can reduce the width of the gear if a lower safety factor is desired.

The last thing we need to do is check the safety factor for the gear. We leave this task as an exercise for the reader. (Take $S_{e'} = 50$ ksi for the gear material.)

Example 15-3

This example is based on surface fatigue. Fig. 15–23 shows a motor-driven system driving an external load rotating at 100 rpm. The machine requires minimum shock, uniform driving and driven system, has accurate mounting. If the number of teeth in the pinion is 16 and the number of teeth in the gear is 48 with a diametrical pitch of $P = 8$, and pressure angle $\phi = 20°$. If the HP transmitted is 3, determine the surface fatigue stress and find the safety factor based on pitting if the surface fatigue strength of the pinion is 100 ksi and the surface

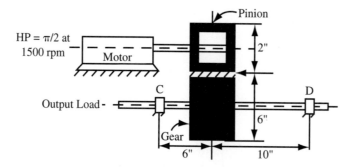

Figure 15–23 Pinion and gear for landing gear

fatigue strength of the gear is 90 ksi. Assume that the gear and pinion are made of steel with $E = 30 \times 10^6$ psi and $v = 0.30$.

Solution

We will design the pinion first.

$$\text{HP} = \frac{F_t V_p}{33,000}$$

where $V_p = 2\pi r_p n_p$

$$= 2\pi\left(\frac{N}{(2)P}\right) n_p$$

$$= 2\pi\left(\frac{16}{2(8)}\right)(100)$$

$$V_p = 200\pi \text{ ft/min}$$

Therefore,

$$F_t = \frac{\text{HP}(33,000)}{V}$$

$$= \frac{3(33,000)}{200\pi}$$

$$\approx 176 \text{ lb}$$

To find σ_H, we need C_p and I, in addition to R, d_p, d_z, K_v, K_o, and K_m. We have

$$R = \frac{d_G}{d_p} = \frac{N_G}{N_p}$$

$$= \frac{48}{16} = 3$$

$$I = \frac{\sin\phi\cos\phi}{2}\left(\frac{R}{R+1}\right)$$

$$= \frac{\sin 20°\cos 20°}{2}\left(\frac{3}{3+1}\right)$$

$$= 0.1205$$

$$C_p = 0.564\sqrt{\frac{1}{\frac{1-0.3^2}{30\times10^6}+\frac{1-0.3^2}{30\times10^6}}}$$

$$= 2{,}289.8$$

$$\approx 2300$$

For uniform speed, we use $K_o = 1$, for the dynamic factor, we use $K_v = 1.1$ (from Fig. 15–20), and for the mounting factor, we use $K_m = 1.6$ (see Table 15–4). Therefore,

$$\sigma_H = C_P\sqrt{\frac{F_t(SF)}{bd_pI}}(K_rK_oK_m)$$

$$\sigma_H = 2{,}300\sqrt{\frac{176(1.1)(1)(1.6)}{(1)(2)(0.1205)}}(\sqrt{SF})$$

and

$$\sigma_H = 82{,}458\sqrt{SF}\text{ psi}$$

Finally, to calculate the surface strength, we use

$$S_H = S_{fe}C_{Li}C_R$$

Assuming 90% reliability, we have $C_R = 0.9$ and $C_{Li} = 1$. If the lifetime is $= 10\times10^6$ cycles (see Fig. 15–24), then

$$S_H = 100(1)(0.9) = 90\text{ ksi}$$

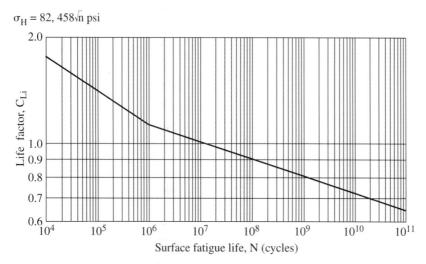

$\sigma_H = 82, 458\sqrt{n}$ psi

Figure 15–24 Graph showing surface fatigue life vs. life factor C_{Li} for a motor-driven system

$$S_H = 90 = \sqrt{SF}(82.46)$$

and it follows that

$$SF = \left(\frac{90}{82.46}\right)^2 = 1.20$$

which is much less than the safety factor based on bending. You may want to use a higher strength pinion material or the next width size, say, $b = 1.5$, to obtain a reasonable safety factor.

We leave it as an exercise for the reader to check the safety factor of the gear, using the data given.

PROBLEMS

15–1) A pair of mating gears has $20°$ full-depth teeth with a diametric pitch of 8. Both gear and pinion are made of steel heat treated to 350 Bhn (Brinell hardness number), and both have a face width of 1.0 in. The teeth are cut with a top-quality hobbing operation. The pinion has 20 teeth and rotates at 1,100 rpm. It is mounted outboard on the shaft of an electric motor and drives a 40-tooth gear that is positioned inboard on an accurately mounted blower shaft. The design life corresponds to five years of operation at 60 hours per week, 50 weeks per year. Using a reliability of 99% and a safety factor of 1.5, estimate the horsepower that can be transmitted, based only on bending fatigue. Use the following diagram:

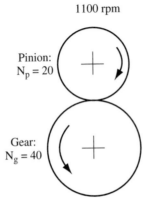

1100 rpm

Pinion:
$N_p = 20$

Gear:
$N_g = 40$

15–2) For the gear train shown in the following diagram, the pinion shaft transmits 20 hp at 2,500 rpm.

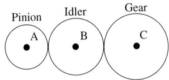

Pinion   Idler   Gear

A   B   C

The overall speed reduction ratio (from pinion to gear) is 3.5:1. The pinion has 14 teeth, a 25° pressure angle, and a diametric pitch of 6. The idler has 17 teeth. Assume that the pinion meshes with the idler and the idler meshes with the gear.

    a.   Find the diameters of the pinion, idler, and gear, as well as the mean and alternating components of the loads transmitted on each gear.

    b.   Draw free-body diagrams of the pinion, idler, and gear, indicating the forces acting on them.

    c.   Determine the torques and transmitted loads of the pinion idler and the gear.

    d.   Determine the reactions at B.

15–3)   a.   For the gear train of Problem 15–2, assume that the power transmitted is increased in such a way that the tangential component of the load transmitted on a tooth is 500 lb, all other conditions being the same. Determine the suitable face width and the idler-tooth stress. The teeth have AGMA full-depth profiles AISI1010 (Bhn = 56). Assume that the load and the source are both uniform in nature. The gears are high precision, accurately mounted, and shaved and ground. (*Hint*: You may assume a face width in the middle of the recommended range for your initial calculations. Provide a safety factor of 1.5 with 99% reliability. (S_u = 0.5 Bhn ksi, S'_n = 0.5S_u) Assume that C_L = 1, C_G = 1, C_S = 0.9, k_r = 0.814, k_t = 1, k_{ms} = 1, k_o = 1, k_m = 1.3, k_v = 1.225, and S_H = (0.48 Bhn − 10) ksi

b. Check the gear train for surface fatigue. Comment.

Clutches and Brakes

16.1 Introduction

Clutches are used to engage or disengage the power source from power-carrying devices in aircrafts, automobiles, and various types of machinery. They provide a smooth, gradual connection and disconnection of power sources and power-driven systems having a common axis of rotation.

16.2 Design of Clutch Based on Uniform Wear

One of the most common types of clutches is the *axial clutch* or *disk clutch*, shown in Fig. 16–1. With the aid of an external force, this device utilizes the frictional force developed between two surfaces when they are brought in contact with each other. The higher the force, the higher is the resistive torque developed, as the friction force developed is proportional to an applied normal force.

Disk clutches usually have a large frictional area and thus develop high torque. They also have effective heat dissipation capacity and favorable pressure distribution. The capacity of a clutch is determined by its torque-transmitting capacity. Whenever two clutch plates come in contact due to an applied force F, a resistive torque is developed because of the frictional properties of the clutch pads. The mathematical analysis that follows will help us in determining the torque capacity and other properties of clutches and brakes.

The nomenclature used in developing a mathematical model for a clutch is as follows:

a: inner radius of clutch pad $= \dfrac{d}{2}$

b: outer radius of clutch pad $= \dfrac{D}{2}$

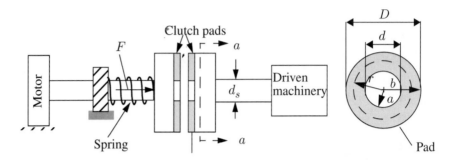

Figure 16–1   Clutch showing shoe internal diameter d and external diameter D

f: coefficient of friction for pads

P_a: pressure at radius a which is the maximum pressure

T: torque developed (using the maximum allowable pressure)

P: pressure at a distance r

F: applied force

The torque T that is developed is an important design variable. It can be determined as follows.

First, we assume that the pads wear out uniformly under an external force F. For example, the work done by the maximum allowable pressure P_a at $d/2$ is the same as the work done by P at some radius r (see Fig. 16–1) and can be expressed in equation form by using nomenclature developed earlier. We have

$$P_a(\Delta A)ad\theta = P(\Delta A)rd\theta$$

But

$$a = d/2$$

Therefore, after cancelling ΔA term we get

$$P_a(d/2) = Pr$$

$$\therefore P = \frac{P_a d}{2r}$$

for $d/2 \le r \le D/2$. (See Fig. 16–1.)

The force required to create frictional torque is

$$F = \int_{d/2}^{D/2} P(2\pi r dr) \qquad (16\text{--}1)$$

but $P = \dfrac{P_a d}{2r}$, substituting and integrating Equation (16–1), we get,

$$F = \frac{\pi P_a d}{2}(D-d) \qquad (16\text{--}2)$$

where P_a is the maximum allowable pressure, which is a material property of the clutch.

The force F is the maximum force required, which must be provided externally to create the maximum pressue P_a at a radial distance a in the clutch pad. (The spring provides the force F.)

Similarly, the maximum torque developed due to coefficient of friction, f, is given by

$$T = \int_{d/2}^{D/2} f(2\pi r dr P)r$$

Rewriting P in terms of P_a and integrating gives

$$T = \frac{\pi f P_a d}{2}(D^2 - d^2) \qquad (16\text{--}3)$$

For the maximum torque, d is related to D. The exact relationship can be obtained by differentiating Equation (16–3) with respect to the inner diameter d and setting the resulting partial derivative equal to zero; that is,

$$\frac{\partial T}{\partial d} = 0 = D^2 - 3d^2$$

so that

$$d^2 = D^2/3$$

Substituting for d^2 in Equation (16–3), we get

$$T_{max} = (\pi f P_a D / 2\sqrt{3})\left(D^2 - \frac{D^2}{3}\right)$$

or

$$T_{max} = \pi f P_a \frac{D^3}{3\sqrt{3}}$$

Rearranging Equation (16–2) we obtain

$$P_a = \frac{2F}{\pi d(D-d)}$$

Now we substitute for Pa in Equation (16–3). The torque, in terms of the force F, is given by

$$T = \frac{Ff}{4}(D+d) \tag{16–4}$$

But for maximum torque $d = D/\sqrt{3}$; therefore,

$$T_{max} = \frac{Ff}{4}\left(D + \frac{D}{\sqrt{3}}\right)$$

Thus, we see that the torque is proportional to both the force F and the coefficient of friction, f.

The value of the maximum allowable pressure P_a is a material property based on the strength and heat dissipation capacity of a material. Table 16–1 gives approximate values of the maximum allowable pressure for various types of commonly used materials.

Table 16–1 Maximum allowable values for clutch materials

Material	Maximum allowable pressure (psi)	Coefficient of friction	Maximum allowable temperature	Maximum speed (fpm)
Molded lining (asbestos)	95	.4 to .45	500	5,000
Woven lining (asbestos)	95	.2 to .45	500	4,000 to 7,000
Brake blocks (asbestos)	160	.4 to .6	700	7,000
Metal (copper or iron)	500	.1 to .45	1,000	7,000

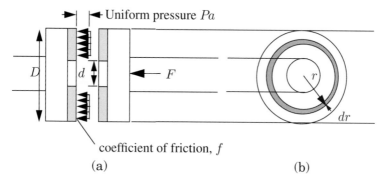

Figure 16–2   Clutch showing uniform pressure applied by force F

16.3 Clutch with Uniform Pressure

If we assume that the maximum allowable pressure P_a is applied uniformly throughout the contacting surface of the clutch, as shown in Fig. 16–2 (which is possible if the external diameter D is not much larger than internal diameter d (for example, $D/d < 2$), then it can be shown that the external force F required to create such pressure is

$$F = \int_{d/2}^{D/2} P_a 2\pi r\, dr \tag{16-5}$$

$$= 2\pi P_a \left| \frac{r^2}{2} \right|_{d/2}^{D/2}$$

so that

$$F = \frac{\pi P_a}{4}(D^2 - d^2) \tag{16-6}$$

The torque developed is obtained by integrating the moment developed by force F (see Fig. 16–2(b)); that is,

$$T = \int_{d/2}^{D/2} P_a (2\pi r\, dr) f(r)$$

$$= 2\pi P_a f \left| \frac{r^3}{3} \right|_{d/2}^{D/2}$$

$$= \frac{\pi P_a f}{12}[D^3 - d^3]$$

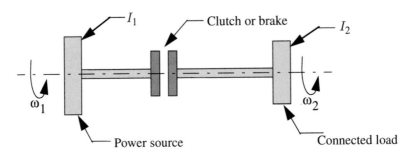

Figure 16–3   Power source and connected load

But

$$P_a = \frac{4F}{\pi(D^2 - d^2)}$$

Therefore,

$$T = \frac{Ff}{3}\left(\frac{D^3 - d^3}{D^2 - d^2}\right) \tag{16–7}$$

16.4 Energy Considerations

Suppose we have a clutch plate connected to a power source, as shown in Fig. 16–3. The power source, which is a rotating system, has a mass moment of intertia I_1 and is rotating at an angular velocity of ω_1. The connecting system has a mass moment of inertia I_2, and is rotating at an angular velocity ω_2. Dynamic theory tells us that $I_1\ddot{\theta}_1 = -T$ and $I_2\ddot{\theta}_2 = T$. (Note that resting torque creates a positive acceleration in a driven system.) Integrating both equations, we get

$$I_1\dot{\theta}_1 = -Tt + C' \qquad I_2\dot{\theta}_2 = Tt + C \tag{16–8}$$

Rearranging yields

$$\dot{\theta}_1 = \frac{-Tt}{I_1} + \frac{C'}{I_1} \qquad \dot{\theta}_2 = \frac{Tt}{I_2} + \frac{C}{I_2}$$

Let

$$C_1 = \frac{C'}{I_1} \qquad C_2 = \frac{C}{I_2}$$

Then

$$\dot{\theta}_1 = \frac{-Tt}{I_1} + C_1 \qquad \dot{\theta}_2 = \frac{Tt}{I_2} + C_2$$

But at $t = 0$, $\dot{\theta}_1 = \omega_1$ and $\dot{\theta}_2 = \omega_2$.

The preceding initial conditions require $C_1 = \omega_1$ and $C_2 = \omega_2$. Rearranging the foregoing equations and subtracting, we get

$$\dot{\theta}_1 - \dot{\theta}_2 = \omega_1 - \omega_2 - T\left(\frac{I_1 + I_2}{I_1 I_2}\right)t \qquad (16\text{--}9)$$

Now let us find the time t_1 taken for $\dot{\theta}_1 - \dot{\theta}_2 = 0$ (i.e., both disks are rotating with the same angular velocity). Setting Equation (16–9) equal to zero and setting $t = t_1$, we get, upon rearranging,

$$t_1 = \frac{I_1 I_2 (\omega_1 - \omega_2)}{(I_1 + I_2)T} \qquad (16\text{--}10)$$

Equation (16–10) shows that the time required for the two angular velocities to become equal is proportional to the difference of the initial angular velocities (ω_1–ω_2) and inversely proportional to the torque T (clutch capacity).

If we assume that the torque developed during the clutching operation is constant, then the energy developed per unit time is given by

$$u = T\dot{\theta}_r$$

where $\dot{\theta}_r = \dot{\theta}_1 - \dot{\theta}_2$. Substituting Equation (16–9) for $\dot{\theta}_1 - \dot{\theta}_2$ yields

$$u = T\left[(\omega_1 - \omega_2) - T\left(\frac{I_1 + I_2}{I_1 I_2}\right)t\right]$$

Therefore, the total energy over the period t_1 is

$$E = \int_0^{t_1} u\,dt = T\int_0^{t_1}\left[(\omega_1 - \omega_2) - T\left(\frac{I_1 + I_2}{I_1 I_2}\right)t\right]dt$$

Upon integrating and rearranging, we get

$$E = \frac{I_1 I_2}{2(I_1 + I_2)}(\omega_1 - \omega_2)^2 \qquad (16\text{--}11)$$

Let us now determine the temperature rise in a clutch pad. If we assume that the heat is absorbed by a clutch system of weight W, then we can show that

$$\Delta T = E/(9,336CW) \tag{16–12}$$

where ΔT is the rise in temperature, E is the energy dissipated, W is the weight of the clutch system that absorbs heat, C is the specific heat of the clutch material (use .12 for steel or cast iron), and 9,336 is an amount of BTUs required to raise the temperature 1°F per unit weight of the clutch. It is important to make sure that the initial temperature of the brake system plus ΔT, the temperature rise due to relative velocity $(\dot\theta_2 - \dot\theta_1)$, is not higher than that allowed for the clutch material. Such maximum allowable temperature values for various materials are provided by the manufacturer or can be found in Table 16–1 for commonly used materials.

16.5 Brakes

Brakes are very similar to clutches, except that their role is reversed; that is, one has ω_2 or $\omega_1 = 0$. Ultimately, the external braking action makes ω_1 or $\omega_2 = 0$.

Caliper Disk Brakes

A sketch of a typical caliper disk brake is shown in Fig. 16–4. Caliper disk brakes are popular on passenger cars. They have a greater cooling capacity and therefore are resistant to fading. An antilocking system can also be fitted more easily to caliper-type brakes. The figure shows the caliper pads, which are on both sides of the disk at a radius r_{pc}. The pressure Pa is applied through the pressure cylinder to the brake pads (calipers). This pressure is based on the force F applied by the cylinder, the area of the pad, and the number of pads.

The force required to create an allowable pressure on the pads is given by

$$F = NA_pP_a \tag{16–13}$$

where F is the force required for N pads, N is the number of pads, A_p is the cross-sectional area of the pad, and P_a is the allowable pressure. The resisting torque generated is

$$T_r = fF(r_{pc})$$

where f is the coefficient of friction and r_{pc} is the radius to the center of gravity of the pads.

If the mass moment of inertia of a disk and the mass connected to it is I, then the time required to stop the mass from rotating with an initial angular velocity ω is given by

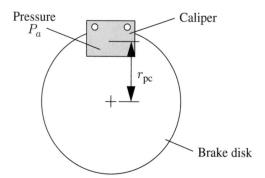

Figure 16–4 Sketch of caliper-type brake

$$t = \frac{I\omega}{T_r} \qquad\qquad (16\text{–}14)$$

The temperature rise is given by an equation similar to the one derived in the analysis of the clutch; that is,

$$\Delta T = \frac{E}{9336(C)W} \qquad\qquad (16\text{–}15)$$

where ΔT is the rise in temperature in degrees Fahrenheit, and E is the total energy and is given by an equation

$$E = I_1 I_2 (\omega_1 - \omega_2)^2 / 2(I_1 + I_2)$$

W is the weight of the brake system that absorbs heat, and C is the specific heat of the brake material (use .12 for steel or cast iron).

Make sure that the rise in temperature is not greater than that allowed for the material used in making the pads. (See Table 16–1 for the maximum allowable temperature.)

Disk brakes are getting to be a very common type of brake used in automobiles. Even more common is an antilock braking system. The advantage of antilock brakes is to prevent the wheel from getting locked. This is invaluable on slippery roads, as keeping the wheels from locking helps the driver to control the vehicle during emergency stops. One is cautioned not to pump the brake pedal when operating a vehicle equipped with antilock brakes. Instead, one should continuously press the brake pedal. It is important to recognize, however, that even under ideal conditions, a steady pressure on the brake pedal will require a slightly longer stopping distance, compared with a vehicle without an antilock braking system. It is considered important that the disk of the antilock brakes should contain holes at intervals to prevent overheating during the braking operation.

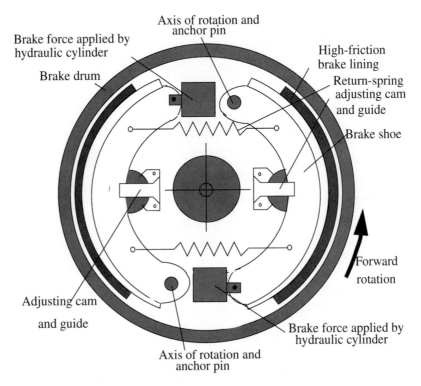

Figure 16–5   General arrangement of internal drum brake

Internally Expanding, Long or Short Drum Brakes and Clutches

Expanding clutches and brakes are used in automobiles, trucks, textile machinery, excavators, airplanes, motorcycles, etc. The drum clutches or brakes have shoes, one or more pistons, anchor pins, springs, and adjusting cams as major components. A sketch of such a brake is shown in Fig. 16–5.

Mathematical modeling of drum brakes requires that the pressure developed between a brake shoe and drum be proportional to the vertical distance from the pin. Since this distance is itself proportional to $\sin \theta$ (see Fig. 16–6), it is clear that the vertical distance is maximal at $\sin \theta_a = 90$ if θ_2, the angle at which the brake pad ends, is greater than $90°$. However, if θ_2 is less than $90°$, then $\sin \theta_a = \sin \theta_2$. Using this assumption, we can show that

$$\frac{P}{\sin \theta} = \frac{P_a}{\sin \theta_a} \qquad (\theta_a = 90° \text{ if } \theta_2 > 90 \text{ and } \theta_a = \theta_2 \text{ if } \theta_2 < 90°) \qquad (16\text{–}16)$$

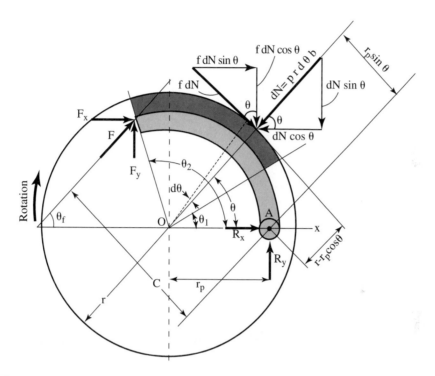

Figure 16–6   Forces developed on brake shoe due to applied force F

(Note that θ_2 is measured from an axis passing through the center of the drum and pin, as shown in Fig. 16–6. Therefore, always let the x-axis pass through the pin axis.)

If $\theta_1 = 0$, then $P = \left(\dfrac{P_a}{\sin \theta_a}\right) \sin \theta_1 = 0$. Hence, to have a value of pressure

$P > 0$, θ_1 (the angle at which the brake shoe starts) should be greater than 20° for efficient brake design.   If $\theta_1 < 10^\circ$, the brake lining will be underutilized.

Now let us analyze the internal forces developed at the interface of the brake lining and the drum due to an external force F applied by a hydraulic piston. Let the x-axis pass through the center of the pin, P and the center of the drum, O. (See Fig. 16–6.) Following is the nomenclature we shall use:

θ_1: angle where the brake lining starts

θ_2: angle where brake lining ends

C: perpendicular distance from P to force F

r: radius of the drum (approximately equal to radius of the brake lining)

r_p: radius to the pin

M: moment due to friction force

M_n: moment due to normal force

F: external force applied to shoe

T: torque developed due to friction force

R: pin reactions

P: interface pressure at angle θ

b: width of brake lining

C: perpendicular distance to force F, measured from pin axis

θ_f: angle made by force F with respect to pin (x-) axis

The friction force is a function of the normal force. Therefore, at θ, let the normal force $dN = Pr d\theta b$. Such a force produces a friction force $dF_r = dN\,f$ where f is the coefficient of friction. Now, the geometry shows that $d_n = r_p \sin\theta$ and $d_f = r - r_p \cos\theta$, where d_n and d_f are the perpendicular distances from the pin at A to the forces dN and dFr, respectively. Hence, the moment of friction about A is

$$M_f = \int_{\theta_1}^{\theta_2} dF_r d_f$$

where

$$dF_r = dN_f$$

and

$$d_f = r - r_p \cos\theta \quad \text{(see Fig. 16–6)}$$

But

$$dN = Pbrd\theta$$

However, it is assumed that

$$\frac{P}{\sin\theta} = \frac{P_a}{\sin\theta_a}.$$

Therefore,

$$dN = (P_a \sin\theta / \sin\theta_a)\, brd\theta$$

Therefore,

$$M_f = \int_{\theta_1}^{\theta_2} \frac{fP_a br}{\sin \theta_a} \sin \theta \, (r - r_p \cos \theta) \, d\theta$$

or

$$M_f = \frac{fP_a br}{\sin \theta_a} C_{11} \tag{16–17}$$

where

$$C_{11} = \int_{\theta_1}^{\theta_2} \sin \theta (r - r_p \cos \theta) d\theta \tag{16–18}$$

Similarly, the moment of the normal force dN about the pin axis at A is

$$M_n = \int_{\theta_1}^{\theta_2} dN \, r_p \sin \theta$$

But

$$dN = \frac{P_a \sin \theta}{\sin \theta_a} br d\theta$$

Thus,

$$M_n = \frac{P_a brr_p}{\sin \theta_a} \int_{\theta_1}^{\theta_2} \sin^2 \theta d\theta$$

or

$$M_n = \frac{P_a brr_p}{\sin \theta_a} C_2 \tag{16–19}$$

where

$$C_2 = \int_{\theta_1}^{\theta_2} \sin^2 \theta d\theta \text{ (constant for a given case and yet to be determined)}$$

Since the system is in equilibrium, the moments due to the force F, M_f, and M_n must add up to zero:

$$FC + M_f - M_n = 0$$

Therefore,

$$F = \frac{M_n - M_f}{C} \tag{16--20}$$

For self-locking (i.e., for $F = 0$), we have $M_n = M_f$. Thus, to prevent self-locking, M_n must be greater than M_f.

Because a minimum force is required to energize such a pad, the right side of the brake lining is called self-energizing, and the left side is called non-self-energizing if we have clockwise rotation. For counterclockwise rotation, the reverse is true.

Now we can find the resisting torque on the drum by realizing that only the friction force dF_r produces torque about the center of the drum. The torque developed on the right shoe is called T_r, where the subscript r describes the right side. We have

$$T_r = \int_{\theta_1}^{\theta_2} dF_r r$$

But

$$dF_r = fdN = f \frac{P_a br \sin\theta d\theta}{\sin\theta_a}.$$

Substituting for dF_r and integrating this equation, we get

$$T_r = \frac{fP_a br^2}{\sin\theta_a} (\cos\theta_1 - \cos\theta_2) \tag{16--21}$$

Thus, the braking torque T_r is directly proportional to the coefficient of friction, the maximum pressure, the width of the brake lining and the radius of the drum square. For an ideal design, it is expected that the value of T_r is the maximum.

Similarly, to find the pin reactions, we make use of the fact that the system is in static equilibrium. That is, for $\Sigma F x = 0$, we get

$$R_x - \int_{\theta_1}^{\theta_2} dN \cos\theta + \int_{\theta_1}^{\theta_2} dF_r \sin\theta + F_x = 0 \tag{16--22}$$

where $F_x = F \cos \theta_f$ and θ_f is the angle between the force F and x-axis. (Remember again, the x-axis always passes through the pin axis.) Similarly, for $\Sigma F_y = 0$, we get

$$R_y - \int_{\theta_1}^{\theta_2} dN \sin \theta - \int_{\theta_1}^{\theta_2} dF_r \cos \theta + F_y = 0 \qquad (16\text{--}23)$$

where $F_y = F \sin \theta_f$. Simplification of Equations (16–23) and (16–24) then yields

$$R_x = \frac{P_a br}{\sin \theta_a} [C_1 - f C_2] - F \cos \theta_f \qquad (16\text{--}24)$$

and

$$R_y = \frac{P_a br}{\sin \theta_a} [C_2 + f C_1] - F \sin \theta_f \qquad (16\text{--}25)$$

where

$$C_1 = \int_{\theta_1}^{\theta_2} \sin \theta \cos \theta \, d\theta = \left[\frac{1}{2} \sin^2 \theta \right]_{\theta_1}^{\theta_2}$$

or

$$C_1 = \frac{1}{2} [\sin^2 \theta_2 - \sin^2 \theta_1] \qquad (16\text{--}26)$$

and

$$C_2 = \int_{\theta_1}^{\theta_2} \sin^2 \theta \, d\theta$$

$$= \left[\frac{\theta}{2} - \frac{1}{4} \sin 2\theta \right]_{\theta_1}^{\theta_2}$$

or

$$C_2 = \left[\left(\frac{\theta_2 - \theta_1}{2} \right) - \frac{1}{4} (\sin 2\theta_2 - \sin 2\theta_1) \right] \qquad (16\text{--}27)$$

Remember that θ_1 and θ_2 must be in radians in these equations.

We now have a mathematical model to determine the resisting torque, pin reactions, and force F required to develop the maximum allowable pressure in

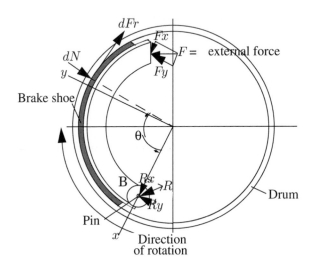

Figure 16–7   Directions of normal force dN and friction force dFr at an
angle θ with respect to the x-axis (right side of shoe is not
shown)

the brake shoe if the geometrical dimensions and allowable shoe pressure are
given.

Let us now analyze the left-side brake lining as the pressure developed on
the left side is not the same as on the right side. Fig. 16–7 shows the dN and dF_r
directions for clockwise drum motion, as was the case for the right-side brake lining. Clearly, the force F is common to both the linings, assuming that we use the
same pressure cylinder to energize the left and right linings. Thereafter, by taking moments of friction force and normal force about the left pin, we get, after
rearranging,

$$F = \frac{M_{nl} + M_{fl}}{C} \text{ (since } M_{nl} \text{ and } M_{fl} \text{ produce clockwise moment about } B\text{)}$$

It is now obvious that the left-side lining has no self-energizing effect, as
$MN + Mf$ is always greater than zero. The condition for self-energizing was that
$F = 0$, i.e., $MN - Mf = 0$. This is not possible in the case of the left lining if the
drum is rotating in a clockwise direction. However, for counterclockwise rotation,
the left lining will become self-energizing, and the right-side lining will be non-
self-energizing, as expected.

Returning to our discussion concerning the left lining, recall that M_n and
M_f were proportional to an external maximum pressure P_a for the right lining.
However, it is not known how much maximum pressure (say, P_{al}) is developed on
the left-side brake lining, but we can see that Equations (16–18) and (16–20)

indicate that M_{nr} and M_{fr} are proportional to the maximum pressure on the right side. Therefore, we can conclude that the moments on the left side are proportional to the pressure on the left side, and we have

$$\frac{M_{nr}}{P_{ar}} = k = \frac{M_{nl}}{P_{al}}, \text{ where } k \text{ is a proportionality constant. Solving for } M_{nl}, \text{ we}$$

obtain

$$M_{nl} = \frac{M_{nr}}{P_{ar}}P_{al} \tag{16–28}$$

Similarly,

$$\frac{M_{fl}}{P_{al}} = \frac{M_{fr}}{P_{ar}}$$

and it follows that

$$M_{fl} = \frac{M_{fr}}{P_{ar}}P_{al} \tag{16–29}$$

Now,

$$F = \frac{M_{nl} + M_{fl}}{C}$$

Substituting for M_{nl} and M_{fr} from Equations (16–28) and (16–29), we get

$$F = P_{al}\frac{M_{nr}}{P_{ar}} + P_{al}\frac{M_{fr}}{P_{ar}} \tag{16–30}$$

We can solve this equation for P_{al}, as all the other values are known. Once P_{al} is known, we can use the same torque equation that was derived for the right-side brake lining, with P_{ar} replaced by P_{al}; that is,

$$T_l = \frac{fP_{al}br^2}{\sin\theta_a}(\cos\theta_1 - \cos\theta_2) \tag{16–31}$$

where T_l is the resisting torque developed by the left lining.

Thus, the total resistance torque capacity (called the braking torque) of the brake linings in a drum is the sum of the two torques obtained, i.e.,

$$T = T_r + T_1 \tag{16–32}$$

The torque T is called the *braking torque capacity* of the brake system. We can prove that, for a drum having the mass moment of inertia about its centroidal axis equal to I_g,

$$T = I_g \alpha$$

where α is the angular acceleration of the drum. Then, knowing α, we can find the time required to stop the drum from rotating at an angular velocity ω

The left-side pin reactions are not as high as the right-side pin reactions for clockwise rotation. In fact, it can be proven that the left-side pin reactions are

$$R_x = \frac{P_{al} br}{\sin \theta a}(C_1 + f C_2) - F_x$$

and

$$R_y = \frac{P_{al} br}{\sin \theta_a}(C_2 - f C_1) - F_y$$

where $F_x = F\cos \theta_f$, $F_y = F\sin \theta_f$, and C_1 and C_2 are as defined earlier. However, remember that the left-side pin reactions will be the same as the right-side pin reactions if the rotation of the wheel is reversed. Therefore, one should always design a pin for the maximum reaction value and use the same pin on both sides. The same applies for the rest of the system if one wants to develop a conservative design. However, all automobile manufacturers use a left shoe that is different from the right shoe to provide an economical design.

Example 16-1

The brake shown in the following figure exerts a force $F = 700$ lb on each shoe. The drum diameter is 13 in, and the brake shoes have a face width of 1.5 in. The coefficient of friction of the lining is 0.32, and the maximum allowable pressure is 175 psi. Assume that the pin radius is 5 in and the brake lining starts at $\theta_1 = 5°$ and ends at $\theta_2 = 120°$. The pin is located 30° from the vertical axis, and $C = 8.66$ in.

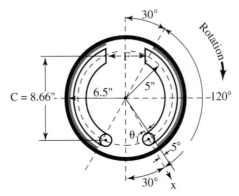

a. Determine the maximum pressure developed, and indicate on which shoe it occurs.

b. Calculate the braking torque seen by each shoe, and find the total braking torque.

c. Find the resulting hinge-pin reactions.

Solution a. We see that $\theta_a = 90°$. Integrating Equation (16–17) from θ_1 to θ_2 yields

$$M_f = \frac{fP_{ar}br}{\sin\theta_a} r\cos\theta \Big|_{\theta_1}^{\theta_2} - r_p\frac{1}{2}\sin 2\theta \Big|_{\theta_1}^{\theta_2}$$

where P_{ar} is the pressure developed on the right-side brake pad. Thus,

$$M_f = \frac{0.32P_{ar}(1.5)(6.5)}{\sin 90°}(+6.5\cos(120°)-6.5\cos(5°))$$

$$-\frac{5}{2}(\sin 2(120)-\sin 2(5))$$

$$= 24.55P_{ar} \text{ in-lb}$$

We will use this relation later.

The moment caused by the normal forces is calculated by integrating Equation (16–19) from θ_1 to θ_2, resulting in

$$M_n = \frac{P_{ar}brr_p}{\sin\theta_a}\left[\left(\frac{\theta}{2}-\frac{1}{4}\sin^2\theta\right)\Big|_{\theta_1}^{\theta_2}\right]$$

$$= \frac{P_a(1.5)(6.5)(5)}{\sin(90°)}\left[\left(\frac{120\pi}{2(180)}-\frac{1}{4}\sin^2(120°)\right)-\left(+\frac{0.0873}{2}-\frac{1}{4}\sin^2(5°)\right)\right]$$

$$= 61.605P_{ar} \text{ in-lb}$$

We will also use this relation later.

Next we use Equation (16–20), since the actuating force is given. We can solve for P_{ar} in the right shoe:

$$F = \frac{M_N - M_f}{C} = \frac{61.605P_{ar}-24.55P_{ar}}{8.66} = 700 \text{ lb}$$

It follows that $P_{ar} = 163.65$ psi, which is less than the allowed $P_{amax} = 200$ psi. Therefore, the value of P_{ar} is acceptable.

For the left shoe, (*Note:* $M_n/P_{ar} = 61.605$ etc.)

$$F = \frac{M_{nl} + M_{fl}}{C} = \frac{61.605 P_{al} + 24.55 P_{al}}{8.66} = 700 \text{ lb}$$

Hence, $P_{al} = 70.37$ psi

This pressure on the left shoe is much less than that on the right shoe, so at times the left shoe is made of a weaker material.

b. Next, let us determine the torque developed by the right and left shoes. From Equation (16–21), the resisting torque developed by the right shoe is

$$T_r = \frac{f P_{ar} b r^2 (\cos\theta_1 - \cos\theta_2)}{\sin\theta_a}$$

$$= \frac{0.32(163.6)(1.5)(6.5)^2 (\cos(5°) - \cos(120°))}{\sin(90°)}$$

$$= 4,965.6 \text{ in-lb}$$

The torque developed by the left shoe is (Equation (16–21))

$$T_l = \frac{f P_{al} b r^2 (\cos\theta_1 - \cos\theta_2)}{\sin\theta_a}$$

$$T_l = \frac{0.32(70.40)(1.5)(6.5)^2 (\cos(5°) - \cos(120°))}{\sin(90°)}$$

$$= 2,135.17 \text{ in-lb}$$

So the total torque is

$$T_{TOTAL} = T_r + T_1$$

$$= 4,965.6 + 2,135.6$$

$$\approx 7,100 \text{ in-lb}$$

c. Now let us calculate the C_i values of $C{i,i} = 1, 2$:

$$C_1 = \frac{1}{2}\sin^2\theta_2 - \frac{1}{2}\sin^2\theta_1 = 0.371$$

$$C_2 = \left(\frac{\theta_2}{2} - \frac{1}{4}\sin 2\theta_2\right) - \left(\frac{\theta_1}{2} - \frac{1}{4}\sin 2\theta_1\right) = 1.263$$

Now the reactions at the right-shoe pin are

$$R_{xr} = \frac{P_{ar}br}{\sin\theta_a}(C_1 - fC_2) - F_x$$

$$= -402.83 \text{ lb (check for accuracy) and}$$

$$R_{yr} = \frac{P_{ar}br}{\sin\theta_a}(C_2 + fC_1) - F_y$$

$$= 1,599.30 \text{ lb (check for accuracy)}$$

Therefore, the resultant reaction at the right-shoe pin is

$$R_r = [(R_{xr})^2 + (R_{yr})^2]^{\frac{1}{2}}$$

$$= \sqrt{(402.83)^2 + (1,599.33)^2} = 1,649 \text{ lb}$$

Similarly, the reactions at the left-shoe pin are

$$R_{xl} = \frac{P_{al}br}{\sin\theta_a}(C_1 + fC_2) - F_x = 182 \text{ lb}$$

$$R_{yl} = \frac{P_{al}br}{\sin\theta_a}(C_2 - fC_1) - F_y = 179 \text{ lb}$$

Thus, the resultant reaction at the left-shoe pin is

$$R_l = [(R_{xl})^2 + (R_{yl})^2]^{\frac{1}{2}} = 255 \text{ lb}$$

Note that the magnitude of the left reaction is less than that of the right. Therefore, we design a pin for the maximum reaction and use it for both of the shoes. For this example, we design the pin to carry 1,649 lb. Auto manufacturers however, use two different pins to carry design loads.

The preceding problem is solved using the lead model *ex16-1.tk*, or *brake.tk*. The model and its results are shown on pages 368 and 369. You can now try to solve homework problems by using this lead model. Note that the mathematical model for internal drum brake design is extensive. Therefore, use the lead model to check your hand calculations.

RULES SHEET

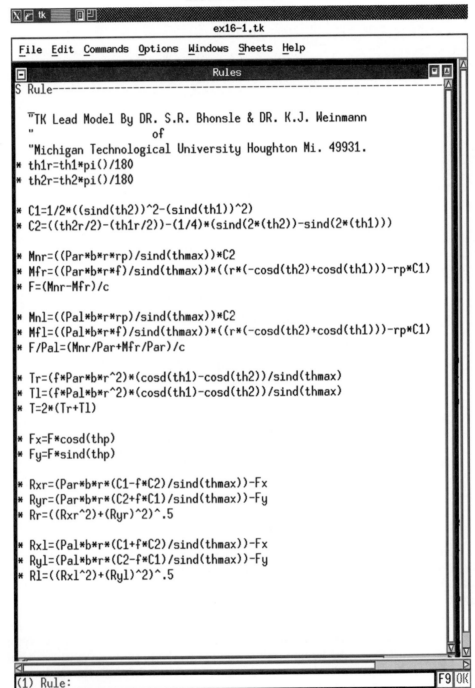

```
X  tk      
                              ex16-1.tk

 File  Edit  Commands  Options  Windows  Sheets  Help

                              Rules
S Rule------------------------------------------------------

  "TK Lead Model By DR. S.R. Bhonsle & DR. K.J. Weinmann
  "                    of
  "Michigan Technological University Houghton Mi. 49931.
* th1r=th1*pi()/180
* th2r=th2*pi()/180

* C1=1/2*((sind(th2))^2-(sind(th1))^2)
* C2=((th2r/2)-(th1r/2))-(1/4)*(sind(2*(th2))-sind(2*(th1)))

* Mnr=((Par*b*r*rp)/sind(thmax))*C2
* Mfr=((Par*b*r*f)/sind(thmax))*((r*(-cosd(th2)+cosd(th1)))-rp*C1)
* F=(Mnr-Mfr)/c

* Mnl=((Pal*b*r*rp)/sind(thmax))*C2
* Mfl=((Pal*b*r*f)/sind(thmax))*((r*(-cosd(th2)+cosd(th1)))-rp*C1)
* F/Pal=(Mnr/Par+Mfr/Par)/c

* Tr=(f*Par*b*r^2)*(cosd(th1)-cosd(th2))/sind(thmax)
* Tl=(f*Pal*b*r^2)*(cosd(th1)-cosd(th2))/sind(thmax)
* T=2*(Tr+Tl)

* Fx=F*cosd(thp)
* Fy=F*sind(thp)

* Rxr=(Par*b*r*(C1-f*C2)/sind(thmax))-Fx
* Ryr=(Par*b*r*(C2+f*C1)/sind(thmax))-Fy
* Rr=((Rxr^2)+(Ryr)^2)^.5

* Rxl=(Pal*b*r*(C1+f*C2)/sind(thmax))-Fx
* Ryl=(Pal*b*r*(C2-f*C1)/sind(thmax))-Fy
* Rl=((Rxl^2)+(Ryl)^2)^.5




(1) Rule:                                                   F9  OK
```

VARIABLES SHEET

```
┌─────────────────────────────────────────────────────────────────────┐
│ X  tk        ⬚⬚                                                        │
├─────────────────────────────────────────────────────────────────────┤
│                            ex16-1.tk                                  │
├─────────────────────────────────────────────────────────────────────┤
│ File  Edit  Commands  Options  Windows  Sheets  Help                  │
├─────────────────────────────────────────────────────────────────────┤
```

St	Input	Name	Output	Unit	Comment
	5	th1		deg	Angle from pin axis to start of shoe
		th1r	.08726646	rad	th1 in radians
	120	th2		deg	Angle from pin axis to end of shoe
		th2r	2.0943951	rad	th2 in radians
	60	thp		deg	Angle from pin axis to horizontal axis
	90	thmax		deg	thmax=90deg if th2 > 90
					thmax=th2 if th2 < 90
		C1	.37120194		Constant for calculations
		C2	1.2634827		Constant for calculations
	8.66	c		in	Perp. distance to force F from pin
	1.5	b		in	Brake face width (mm or in)
	6.5	r		in	Radius of the drum(in)
	5	rp		in	Radius of pin (mm or in)
	.32	f			Coefficient of friction
	700	F		lb	Actuating force
		Par	163.64896	psi	Max. pressure developed in a shoe (Self-Energizing)
		Pal	70.368206	psi	Max. pre. developed in opposite shoe (Nonself-Energizing)

```
(18) Status:                                                    F9 OK
```

St	Input	Name	Output	Unit	Comment
		Pal	70.368206	psi	Max. pre. developed in opposite shoe (Nonself-Energizing)
		Tr	4966.2391	lb-in	Torque developed in right shoe
		Tl	2135.4571	lb-in	Torque developed in left shoe
		T	14203.392	lb-in	Resultant torque developed
		Mnr	10081.276	lb-in	Mom. due to normal force on right shoe
		Mfr	4018.4616	lb-in	Moment due to friction on right shoe
		Mnl	4334.8963	lb-in	Moment due to friction on left shoe
		Mfl	1727.9177	lb-in	Mom. due to normal force on left shoe
		Fx	350	lb	Reaction force in x direction
		Fy	606.21778	lb	Reaction force in y direction
		Rxr	-402.8336	lb	Right pin reaction in x direction
		Ryr	1599.2967	lb	Right pin reaction in y direction
		Rxl	182.07406	lb	Left pin reaction in x direction
		Ryl	179.14814	lb	Left pin reaction in y direction
		Rr	1649.2498	lb	Resultant force in right pin
		Rl	255.43104	lb	Resultant force in left pin

PROBLEMS

16–1) Determine a suitable size and required force for an axial disk clutch to transmit 7.5 hp at 1,725 rpm. Assume a safety factor of 2 and a uniform wear model. Assume also that the clutch has a single dry disk for which the maximum allowable pressure is 225 psi and the friction coefficient is 0.35. Design for the maximum torque-transmitting capacity for the clutch.

16–2) Given the following diagram of a cone clutch, use your understanding of a plate clutch to

    a.   Derive the equation for F in terms of p and r. (*Note*: $P_a r_i = pr$ for uniform wear, where p_a is the allowable pressure at r_i)

    b.   Derive the equation for the torque T developed in terms of a, P_a, r_i, r_o, f, and b, where f is the coefficient of friction and b is the width of the cone.

    c.   Find the time required for stopping the cup if it is rotating at 900 rpm, assuming that the cone is stationary, that $I_{cup} = 4I_{cone}$, $ri = 10$, $r_o = 16$, and $f = 35$, and that the cup is made of steel.

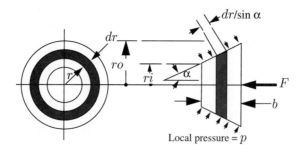

Local pressure $= p$

16–3) An internal drum brake has an inside rim diameter of 14" and a pin radius of 6". The shoe has a face width of 2.5". If an actuating force $F = 800$ lb is applied and the mean coefficient of friction is 0.32,

    a.   Determine the maximum pressure developed and its location if $\theta_1 = 0$ and $\theta_2 = 120°$.

    b.   Determine the total torque developed due to the force F.

c. Determine the pin reactions. (Do we need to find pin reactions at both A and B?)

Use the following diagram to aid in your calculations:

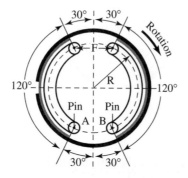

Use the lead model to solve the Problems 16–4 and 16–5.

16–4) A 420-mm-diameter brake drum has four internally expanding shoes. Each of the hinge pins A and B supports a pair of shoes and has a radius of 150 mm. The face width of the shoes is 100 mm. A coefficient of friction of 0.28 is to be used and a maximum pressure of 1250 kPa is allowed.

a. Determine the actuating force F required to develop maximum allowable pressure ($\theta_1 = 10°$ and $\theta_2 = 65°$).

b. Calculate the braking capacity (total torque developed).

c. Determine the maximum pin reactions.

Use the following diagram to aid in your calculations.

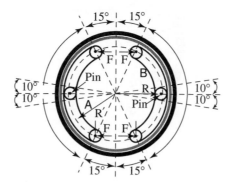

16–5) The brake drum shown in the following diagram is 320 mm in diameter.

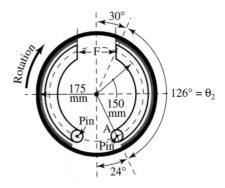

A force F on each shoe is applied as shown. A face width of 35 mm is used. The lining has a coefficient of friction of 0.3 and a pressure not to exceed 1,200 kPa.

    a.   Determine the actuating force F.

    b.   Find the breaking capacity.

    c.   Calculate the hinge-pin reaction on the right side.

16–6) A force F develops maximum pressure on the lining of a brake equal to 0.80 MPa. The coefficient of fraction between the drum and the lining is 0.40. The width of the lining is 80 mm. Refer to the diagram that follows, and

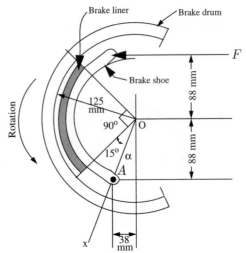

a. For the given rotation, determine whether the brake is self-energizing.

b. Find the force F required to develop the maximum pressure.

c. Indicate the location at which the maximum pressure will occur.

d. Find the braking torque developed due to left shoe.

e. After the force is applied, if the drum makes 20 revolutions uniformly in 5 seconds before coming to a complete stop, estimate the braking power.

f. Calculate the diameter of the pin required at A if the yield strength of the pin material is 500 MPa.

16–7) a. Develop or modify the existing lead model to solve Problem 16–6.

The maximum pressure developed due to the force F on the lining of the brake shoe in Problem 16–6 is equal to 0.75 MPa. The friction coefficient between the drum and the lining is 0.25. The width of the lining is 50 mm. Using the TK model, answer the following questions:

b. For the given rotation, determine whether the brake is self-energizing.

c. Find the force F required to develop the maximum pressure.

d. Indicate the location at which the maximum pressure will occur.

e. Find the torque developed about the center of the brake.

f. After the force is applied, if the drum makes 20 revolutions uniformly in 5 seconds before coming to a complete stop, estimate the braking power.

16–8) The maximum pressure developed due to the force F on the lining of the brake shoe in Problem 16–5 is equal to 0.85 MPa. The friction coefficient between the drum and the lining is 0.35. The width of the lining is 150 mm. Use the TK lead model to answer the following questions:

a. For the given rotation, determine whether the brake is self-energizing.

b. Find the force F required to develop the maximum pressure.

c. Indicate the location at which the maximum pressure will occur.

d.   Find the torque developed about the center of the brake.

e.   After the force is applied, if the drum makes 20 revolutions uniformly in 5 seconds before coming to a complete stop, estimate the braking power.

f.   Calculate the diameter of the pin required at A if the yield strength of the pin material is 400 MPa.

g.   Calculate the diameter of the pin required at A if the yield strength of the pin material is 500 MPa.

16–9)   a.   For the following figure, find the equations that relate the torque developed to an external force F.

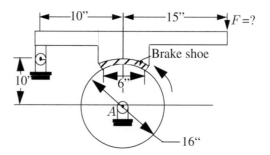

16–10)   A short-shoe and external brake-drum-arrangement is shown in Problem 16–9. The projected size of the shoe is 6" × 15". The shoe has a coefficient of friction of 0.45 and permits an average pressure of 18,000 psi, based on the projected area of contact. The initial speed of the drum is 1,000 rpm.

a.   Draw a free-body diagram of the brake shoe, showing all the forces acting on it.

b.   What value of external force F can be applied without exceeding the allowable contact pressure?

c.   What is the resulting brake torque?

d.   Is the brake self-energizing or de-energizing for the direction of rotation?

e.   What resultant force is applied to the pivot bearing at point A?

f.   If the full application of the force F brings the drum from 1,200 rpm to a stop in 4 seconds, how much heat is generated?

g.   What is the average power developed by the brake during the stop?

Shafts and Axles

S hafts are rotating members, usually circular in cross section, that are used to transmit power from one location on the shaft to another location or another shaft with the aid of gears, pulleys, etc. A spindle is a short shaft. An axle is a shaft that does not rotate.

To design a shaft, it is necessary to understand the theory behind the causes of deflection and its relation to geometrical and mechanical properties, including buckling.

Shafts are subjected to bending moments, torques, shear forces, and axial loads. Previously, we learned how to draw bending, torque, shear, and axial load diagrams. From such diagrams, we can locate critical cross sections and determine geometrical requirements for a given shaft if the allowable stresses are given; or we can find the stresses if the geometry of the cross section, locations of the load, and length of the shaft are given. Also, using the energy approach (Castigliano's theorem), we can determine the critical angular deflections, linear deflections, or slopes wherever such values are needed.

The types of deflections (linear or angular) in a shaft are due to either the bending moment M, the torque T, or the shear force V. Deflections due to the bending moment or shear force create slopes at the bearing support. It is critical that one have minimum slope at bearings, as the excessive deflection at the bearings reduces their life.

The bending moment, torsion, and shear cause stresses in the shaft that may result in failure. The design of a shaft is therefore mainly based on static or fatigue stresses, deflections, and slopes at bearings.

It is necessary that one be knowledgeable in regard to stresses in the shaft and the strength of the shaft material. Also, one must know the static and fatigue strengths of the material—that is, the life of a machine element under dynamic loads. This added knowledge will help us manufacture shafts that will work for their entire designed life.

We first develop the stress formulas, using the same notation as in Chapter 3. The axial stress is

$$\sigma_{P_{max}} = V_{xx}/A \tag{17-1}$$

where V_{xx} is the maximum axial load.

The bending stress is

$$\sigma_{M_{max}} = \frac{M_{max}C}{I} \tag{17-2}$$

where

$$M_{max} = [M_{xy}^2 + M_{xz}^2]_{max}^{1/2} \tag{17-3}$$

The shear stress due to the torque load is

$$\tau_{T_{max}} = \frac{TC}{J} \tag{17-4}$$

where $C = d_o/2$ in which d_o is the outside diameter of the shaft, and $J = 2I$, in which

$$I = \frac{\pi(d_o^4 - d_i^4)}{64}$$

where d_o is the outside diameter and d_i is the inside diameter of the shaft if it is hollow. The shear stress due to the transverse shear load is

$$\tau_{T_{max}} = \frac{VQ}{It} \tag{17-5}$$

where

$$V = [V_{xy}^2 + V_{xz}^2]^{1/2}$$

$$Q = \frac{\pi}{8}(d)^2\left(\frac{2d}{3\pi}\right) \text{ (for a solid shaft)}$$

and $t = d$ for a solid shaft or $t = d_o - d_i$ for a hollow shaft.

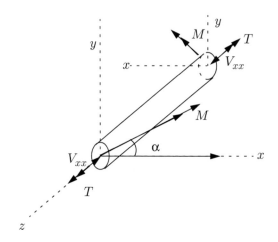

Figure 17–1 Shaft subjected to moment, axial, and torque loads

If we take an element at the location where the effect of bending, axial, and shear stress (torque) is critical, then we can show that the maximum shear stress in a solid shaft is given by

$$\tau_{max_{ME}} = \frac{2}{\pi d^3}[(8M + V_{xx}d)^2 + (8T)^2]^{1/2} \tag{17–6}$$

where the subscript $_{ME}$ denotes the machine element, M is the resultant critical moment, V_{xx} is the critical axial load, and T is the critical torque. (Note that the effect of the shear stress V is not shown, as V has no effect where the effect of M, V_{xx}, and T is maximum.)

The preceding loads are shown in Fig. 17–1. We can also prove equations such as Equation (17–6) for a hollow shaft.

If the shaft is subjected to dynamic loads, then it is recommended that one should follow the steps given below.

1.   Find the axial stress σ_{xx} and note that this is a mean stress.

2.   Find the maximum and minimum bending stress at the critical location, and then find the amplitude and mean stress using the formulas derived earlier.

3.   Find the maximum and minimum (torque) shear stress, and then find the amplitude and mean (torque) shear stress. (In this case, amplitude shear stress = 0.)

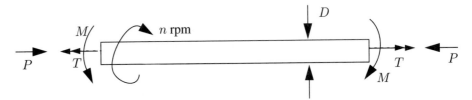

Figure 17–2   Shaft subjected to M, T, and P loads

4.   Knowing the total amplitude, mean normal stress, and amplitude and mean shear stress, find the principal amplitude and principal mean stresses.

5.   Using such principal stresses, find the equivalent amplitude and mean stresses.

6.   Using the Goodman or Gerber criterion, relate stresses in 5 to the fatigue strength and ultimate stress of the material.

The equations found will determine the required diameter of the shaft. The example that follows shows the proper procedure for determining the required diameter of a shaft subjected to complicated loads. In it, we assume that we have amplitude and mean moments, amplitude and mean axial forces, and amplitude and mean torques. We can then find $\sigma_{ea_{ME}}$ and $\sigma_{em_{ME}}$, the amplitude and mean stresses, respectively, (see Example 17-1), by properly substituting the amplitude and mean moment, axial load, and torque. Once the equivalent von Mises, σ_{ea}, and σ_{em} stresses in a machine element are determined, then using either the Goodman or Gerber criterion, we can find additional mathematical equations to solve for, say, the safety factor or the diameter of the shaft.

Example 17-1

A rotating shaft subjected to various loads is shown in Fig. 17–2. Determine the safety factor if the Goodman criterion is used. The load and associated data are as follows:

$$T = 12 \text{ Nm}, P = -200 \text{ N}, M = 20.9 \text{ Nm}, D = 0.016 \text{ m},$$

$$S_e = 291 \text{ MPa}, S_{ut} = 900 \text{ MPa}, K_f = 1.09 \text{ torsion},$$

$$K_f = 1.25 \text{ axial}, K_f = 1.25 \text{ bending}$$

(*Note:* K_f is the stress concentration factor for the respective loadings.)

Solution

Notice that P and T are always present. Therefore, they produce no amplitude stress but they produce a mean stress. M, however, produces alternating stress, as the shaft is rotating. Now,

$$\sigma_{xp})_m = K_f P/A = \frac{(1.25)(-200)}{(\pi/4)(0.016)^2} = -1.24 \text{ MPa}$$

and

$$\sigma_{xb})_a = K_f \frac{MC}{I} = \frac{(1.25)(20.9)(0.016/2)}{\dfrac{\pi(0.016)^4}{64}} = 65.0 \text{ MPa}$$

(Note that $\sigma_m = 0$ for bending.) Also,

$$\tau_{xy})_T = K_f \frac{TC}{J} = \frac{12\left(\dfrac{0.016}{2}\right)}{\dfrac{\pi(0.016)^4}{32}}(1.09) = 16.3 \text{ MPa}$$

The von Mises stress equation for 2-D plane stress is

$$\sigma_e = (\sigma_x^2 + \sigma_y^2 - \sigma_x\sigma_y + 3\tau_{xy}^2)^{1/2}$$

Therefore,

$$\sigma_{ea} = (\sigma_{xa}^2 + \sigma_{ya}^2 - \sigma_{xa}\sigma_{ya} + 3\tau_{xya}^2)^{1/2}$$

But $\sigma_{xa} = \sigma_{xb} = 65.0$ MPa, $\sigma_{ya} = 0$, and $\tau_{xya} = 0$. Thus,

$$\sigma_{ea} = (65.0^2)^{1/2} = 65.0 \text{ MPa}$$

Similarly,

$$\sigma_{em} = (\sigma_{xm}^2 + \sigma_{ym}^2 - \sigma_{xm}\sigma_{ym} + 3\tau_{xym}^2)^{1/2}$$

But $\sigma_{xP} = -1.24$ MPa, $\sigma_{ym} = 0$ MPa, and $\tau_{xyT} = 16.3$ MPa. Hence,

$$\sigma_{em} = ((-1.24)^2 + 0 - 0 + 3(16.3)^2)^{1/2} = 28.1968$$

Now we plot the Goodman diagram and draw the line through σ_{ea} and σ_{em}:

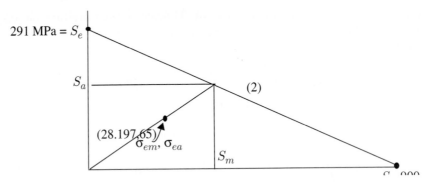

Next, we use the Goodman diagram to develop the equations

$$\frac{S_a}{S_m} = \frac{\sigma_{ea}}{\sigma_{em}} = \frac{65}{28} = 2.304$$

and

$$\frac{S_a}{S_e} + \frac{S_m}{S_{ut}} = 1$$

Solving these equations simultaneously, we get

$$S_a \approx 260 \text{ MPa}$$
$$S_m \approx 112 \text{ MPa}$$

Therefore,

$$S_f = \frac{S_a}{\sigma_a}$$

$$= \frac{260}{65}$$

so that

$$S_f = 4$$

Now do the problem using the Gerber criterion.

The lead model *shaft1.tk,* or *ex17-1.tk,*was developed to analyze or design a shaft subjected to static or fatigue loads. The solution to the preceding problem, using this lead model, is on pages 381 and 382.

RULES SHEET

```
X  tk          
                        ex17-1.tk
 File  Edit  Commands  Options  Windows  Sheets  Help

┌─────────────────────── Rules ───────────────────────┐
Rule─────────────────────────────────────────────
    "TK Lead Model By DR. S.R. Bhonsle & DR. K.J. Weinmann
    "                    of
    "Michigan Technological University Houghton Mi. 49931.


" Failure theories
" Modified Goodman equation
if (T=1) then (1/Sut)+(r/Se)=(1/Sm)
" Soderberg equation
if (T=2) then (1/Sy)+(r/Se)=(1/Sm)
" Gerber parabolic equation
if (T=3) then (1/Sut)^2+(r/Se)=1/Sm
" Quadratic equation
if (T=4) then (1/Sut)^2+(r/Se)^2=(1/Sm)^2

" Mean and alternating stress
Sigem=((Sigxm^2)+(Sigym^2)-(Sigxm*Sigym)+3*Tauxym^2)^.5
Sigea=((Sigxa^2)+(Sigya^2)-(Sigxa*Sigya)+3*Tauxya^2)^.5
Sa/Sm=r
n=Sm/Sigem
r=Sigea/Sigem

" Properties of circular cross section
A=pi()/4*(Do^2-Di^2)
I=pi()/64*(Do^4-Di^4)
c=Do/2
Sigxm=Kfa*(P/A)
Sigxa=Kfb*(32*M/(pi()*Do^3))
Tauxym=Kft*(c*Tor/((1/32)*pi()*Do^4))
_


(31) Rule:                                        F9 OK
```

VARIABLES SHEET

```
ex17-1.tk
```

File Edit Commands Options Windows Sheets Help

St	Input	Name	Output	Unit	Comment
	1	T			Failure equation number, where
					T=1 refers to Goodman
					T=2 refers to Soderberg
					T=3 refers to Geber parabolic
					T=4 refers to Quadratic
	900	Sut		MPa	Ultimate tensile strength
	291	Se		MPa	Endurance limit
		Sm	110.75583	MPa	Mean strength
	700	Sy		MPa	Yield strength
	-200	P		N	Axial load
		A	.00020106	m^2	Cross sectional area
	20.9	M		Nm	Moment
		c	.008	m	Distance from neutral surface extreme fiber
		I	3.217E-9	m^4	Second moment of inertia
		n	3.9279451		Safety factor
	.016	Do		m	Outer diameter
	0	Di		m	Inner diameter
		Sigem	28.196889	MPa	VonMises Mean Stress
		Sigea	64.967545	MPa	VonMises Alternating Stress
		Sigxm	-1.243398	MPa	Mean Axial Stress due to P
	0	Sigym		MPa	
		Tauxym	16.263646	MPa	Mean Shear Stress
		Sigxa	64.967545	MPa	
	0	Sigya		MPa	
	0	Tauxya		MPa	
	1.25	Kfa			Axial Stress Concentration Fac
	1.25	Kfb			Bending Stress Concentration F
	1.09	Kft			Torsional Stress Concentration
	12	Tor		Nm	Axially Applied Torque

(35) Name: F9 OK

PROBLEMS

17–1)   A bevel pinion and shaft are shown in the following figure:

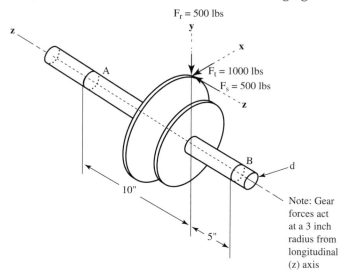

$F_r = 500$ lbs

$F_t = 1000$ lbs

$F_s = 500$ lbs

10"

5"

Note: Gear
forces act
at a 3 inch
radius from
longitudinal
(z) axis

Bearing A takes thrust. The left end of the shaft is coupled to an electric motor, and the right end is free. The load components applied by the mating bevel gear are shown in the figure.

    a.   Find the reactions at A and B.

    b.   Draw the axial load, torsional load, bending load, and shear load diagrams for the range A–B.

    c.   Determine the critical section, and find the maximum bending moment, shear, and axial loads at that section.

    d.   Find the diameter of the shaft required if $\tau_{yp} = 40{,}000$ psi and the safety factor $= 1.6$. (Assume that K_f, the stress concentration factor, equals 1.3 for each type of load.)

        1.   Use the maximum shear stress design criterion to solve for shaft diameter.

        2.   Use the Goodman criterion and design for a fatigue life of 10^6 cycles ($S_{se} \approx 0.50\tau_{yp}$, $S_{sut} = 100{,}000$ psi) to solve for half diameter.

17–2)   Solve Problem 17–1 if the endurance limit in shear $S_{se} = 40{,}000$ psi and the ultimate shear strength $S_{sut} = 90{,}000$ psi. Use the Gerber criterion.

17–3) Develop a TK lead model for Problem 17–1, and determine the required diameter for the shaft, using:

    a.   The maximum-shear-stress failure criterion

    b.   The Gerber criterion for a fatigue life of 10^6 cycles

    c.   The Goodman criterion for a fatigue life of 10^4 cycles Assume that all necessary strengths for the shaft material are given.

17–4) If S_{ut} = 60,000 psi and S_e = 30,000 psi, determine the shaft diameters in Problem 17–1, using the three criteria suggested in Problem 17–3. (*Note:* S_{sut} = 0.85S_{ut}/2 and S_{se} = 0.57(S_e).)

17–5) Draw bending moment, shear, torque, and axial diagrams for the shaft shown in the following diagram:

    a.   Determine the critical cross section and find the principal stresses and the maximum shear stress in terms of the diameter of shaft.

    b.   Using the maximum shear stress failure criterion, determine the required diameter of the shaft if S_{syp} = 20,000 psi and the safety factor is 1.6. Take P = 10,000 lb, W = 1,000 lb/ft, and T = 40,000 in-lb.

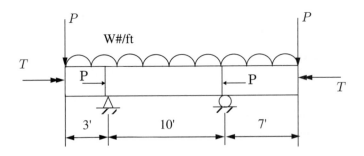

CHAPTER 18

Roller Contact Bearings

18.1 Introduction

The final two chapters in the book present the study of bearings. This chapter introduces the method used to *select* ball or roller contact bearings, and Chapter 19 presents the methods used to *analyze, design, and/or optimize* journal bearings. The *design* of ball or roller bearings is not presented, since most of the design data are obtained experimentally, based on fatigue failure. Such design data are presented in the form of equations, which were developed using regression analysis.

18.2 Bearings

More than a hundred types of bearings are available from various manufacturers. However, we will cover only a very limited number and learn how to select them. Broadly speaking, there are two types of bearings:

1. Ball bearings (used in high-speed applications)
2. Roller bearings (used in high-load applications)

Ball Bearings

The major parts of a typical bearing, shown in Fig. 18–1, are the outer ring, the inner ring, the balls or rollers, and a separator. Separators keep the balls or rollers from rubbing against each other. Separators are also known collectively as a *cage*. The inner and outer ring surfaces, which come in contact with the balls or rollers have curvilinear grooves to let the balls or rollers, have a larger contact area and thus a larger load-carrying capacity. It is possible to allow the inner

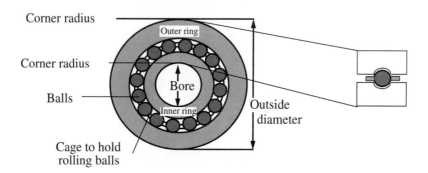

Figure 18–1 Typical ball bearing

ring to rotate and fix the outer ring, or allow the outer ring to rotate and fix the inner ring. It is advisable, however, to fix the outer ring and allow the inner ring to rotate for a longer life of the bearing.

Four types of ball bearings are produced: heavy, medium, light, and extra light. The last three are designated the 300, 200, and 100 series, respectively. There are also other types of bearings, such as axial and thrust bearings, angular contact bearings, self-aligning bearings, etc. These are shown in Fig. 18–2. The design (configuration) of a bearing allows it to carry some amount of axial thrust, as we will see when the load equations are presented later. The equations to be used if excessive thrust exists are given still later.

The basis for selecting a bearing for a desired application is as follows:

1. Static loading
2. Speed (linear surface speed)
3. Lubrication
4. Sealing (to prevent dirt from entering)
5. Size
6. Life
7. Reliability

One should also consider the following when selecting a bearing for a certain application:

a. Material properties of the bearing (high-carbon chromium steel (52100) hardened to Rockwell 58–65 is primarily used in making balls and rings)

b. Type of loads

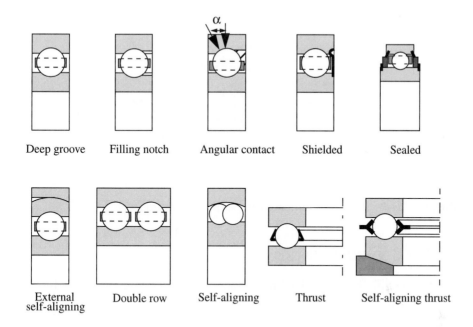

Figure 18–2 Types of ball bearings

    c.   Friction

    d.   Heat

    e.   Resistance to corrosion

    f.   Kinematic problem

    g.   Lubrication

    h.   Machine tolerance

    i.   Assembly

    j.   Use

    k.   Cost

    l.   Size.

Bearings are designed to carry radial loads and axial loads, which are shown in Fig. 18–3.

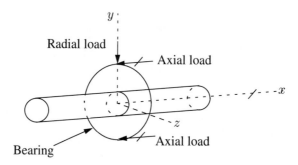

Figure 18–3   Bearing subjected to radial and axial loads

Roller Contact Bearings

Roller contact bearings are known chiefly by three names, all of which refer to the same type of bearing:

1.   Rolling contact bearing

2.   Antifriction bearing (There is some friction in a bearing!)

3.   Roller bearing

Roller bearings are capable of carrying shock or impact loads, both radially and axially. The axial load is the load in the longitudinal direction of the bearing or the shaft on which the bearings are mounted. Fig. 18–4 shows the various types of roller bearings.

18.3 Advantages of Bearings

Ball or roller bearings have the following advantages:

a.   bearings are capable of carrying radial and thrust loads with low friction

b.   low maintenance

c.   easy to replace when worn

d.   minimal space requirements

e.   good lubrication, as many bearings come with sealed lubricant.

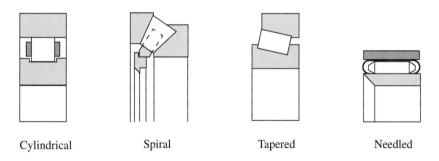

Cylindrical   Spiral   Tapered   Needled

Figure 18–4 Types of roller bearings

Bearing Life

Experimental results are used to ascertain a bearing's life under a given radial load. A bearing is considered to have failed if a certain amount of pitting or scaling (1 mm in size) takes place. If 100 bearings were tested for a given *radial load*, and if the failure life is recorded in ascending order, then the life of the 10th-ranked bearing that failed during testing is taken as the *rated life*, or L_{10} *life*. Such life is expressed in terms of the total number of revolutions (or total number of hours at constant speed) undergone to develop a failure criterion. Simply put, the life at 10% failure is considered the rated life for a given load.

The *basic load rating C* (also called the rated capacity, dynamic load rating, basic load capacity, or specific dynamic capacity) is defined as a constant radial load that a group of bearings can endure for a *specified* life—for example, 90 million cycles (90×10^6) of the inner ring, or 3,000 hours of operation at 500 cycles per minute. Such life is also called the *minimum life*, and is nothing more than the L_{10} life. If one requires a median life, it is usually between four and five times the L_{10} life; however, most bearings are designed for 90 to 99.99% reliability.

Charts based on experimental data provide us with the rated life and the corresponding rated capacity C. In the actual selection of a bearing, however, one may be required to choose a bearing to have an actual life of, say, L_a (for 90% reliability) and a radial load (actual load) F_a. It is found experimentally that the following relationship exists between the rated life, rated load capacity, actual bearing life required, and actual radial load on the bearing:

$$\frac{L_r}{L_a}(C/F_a)^{3.33} = 1 \tag{18-1}$$

Here, L_a is the required (actual) life corresponding to the actual load F_a, C is the rated load capacity (obtained from tables; check the catalog for the rated life and the rated load), F_a is the (actual) radial load, and L_r is the rated life (for example some manufacturers state related load for life of 90×10^6 for 90% reliability) this value may change, depending upon the manufacturer. Most of the time L_a and F_a are known or given, and if we know, for example, $L_r = (90 \times 10^6)$, we can obtain the required rated load capacity by rearranging Equation (18–1) to the form

$$C = F_a (L_a/L_r)^{0.3} \qquad\qquad (18\text{–}2)$$

Equation (18–1) shows that, if the load F_a is doubled relative to C, the life of the bearing is diminished by a factor of 10.

The foregoing equations naturally are based on 90% reliability, as that is what the L_{10} life means. However, most designs are based on a reliability factor greater then 90%; that is, if a greater reliability is desired, the equations are modified to account for such a change. Also, sometimes it is required to have bearings in a group of say two, three, or more. Then the reliability of the group is related to that of an individual bearing by

$$R_n = (R)^n \qquad\qquad (18\text{–}3)$$

where R_n is the group reliability, n is the number of bearings, and R is the reliability of one bearing. Thus, Equation (18–3) states that the reliability of a group of, say, four bearings reduces to $(0.9)^4 = 65.60\%$, which is much less than the 90% reliability of one bearing. Therefore, for example, if a group of bearings is used, the reliability factor of each bearing must be much greater than 90% to obtain 90% group reliability.

To accommodate reliability, Equation (18–2) is modified to the form

$$L_a = k_r L_r (C/F_a)^{3.33} \qquad\qquad (18\text{–}4)$$

where k_r is a reliability factor that depends upon the reliability required. Values of k_r are given in Table 18–1 for various reliabilities.

Again, remember that if a group of bearings is used, then the reliability of the group is less than the reliability of any single member of the group. Equation (18–3) suggests that, to have the same load-carrying capacity as any of its members, the life of a group of bearings will be decreased.

18.4 Influence of Axial Load

Bearings are designed for both radial and axial loads. An axial load is parallel to a shaft's longitudinal axis. Most of the radial load-carrying bearings can carry some axial (thrust) loads. However, if an axial load is more than 35 percent of the radial load (or more than 68 percent, depending upon the type of bearing), one

Table 18–1 Reliability vs. k_r

Reliability	k_r
99%	0.22
98%	0.33
97%	0.44
96%	0.53
95%	0.62
92%	0.75
90%	1.00
50%	4.00

must account for such a load by increasing the actual radial load, calling it F_e (the equivalent load), and using F_e in place of, F_a, the actual load, in Equations (18–1) through (18–3).

Equations to determine the equivalent load F_e are given for two types of bearings, namely, for $\alpha = 0$ and $\alpha = 25°$, where α is the angle the load makes with respect to the radius (vertical line). (See Fig. 18–2 for α.)

Equations for Radial Ball Bearings ($\alpha = 0$)

The following procedure is used to derive the equivalent load if F_a, the actual radial load, and F_T, the actual thrust load, are given:

Determine the ratio F_T/F_a. Then

1. If $0 \le \dfrac{F_T}{F_a} < 0.35$, then $F_e = F_a$.

2. If $0.35 \le \dfrac{F_T}{F_a} < 10$, then $F_e = F_a[1 + 1.115(F_T/F_a - 0.35)]$.

3. If $\dfrac{F_T}{F_a} > 10$, then $F_e = 1.176\, F_T$.

Equations for Angular Ball Bearings ($\alpha = 25°$)

1.   If $0 \le F_T/F_a < 0.68$, then $F_e = F_a$.

2.   If $0.68 \le F_T/F_a < 10$, then $F_e = F_a \left[1 + 0.870 \left(\dfrac{F_T}{F_a} - 0.68 \right) \right]$.

3.   If $F_T/F_a > 10$, then, $F_e = 0.911 F_T$

Naturally, if a bearing is subjected to heavy shock loads, then its life is going to be less compared to a bearing that is subjected to moderate or light shock loads. Therefore, to account for shock, the following equation is used (the equations also account for thrust load and reliability):

$$L_a = k_r L_R \left(\frac{C}{F_e k_a} \right)^{3.33} \tag{18–5}$$

where k_a is the shock factor that accounts for the type of shock load.

Solving for C, we obtain

$$C = F_e k_a \left(\frac{L_a}{k_r L_R} \right)^{0.3} \tag{18–6}$$

Table 18.2 gives values of k_a depending upon the type of shock.

The values in Table 18–2 are merely suggestive for solving problems in this chapter and may change from manufacturer to manufacturer. It is strongly recommended that one study the manufacturer's catalog before selecting a bearing for a desired application. Shafts with gears having excellent mounting procedures typically have values of k_a that fit between the light-impact and no-impact values.

In sum, one can select a bearing on the basis of the following procedure:

1.   Determine the load on the bearing, i.e., F_a and F_T.

2.   Determine the ratio of F_T/F_a.

3.   Decide on the value of α.

4.   Find the equivalent load F_e.

5. From the required reliability, determine k_r from Table 18–1.

6. Depending upon the type of shock, select k_a from Table 18–2.

7. Use Equation (18–5) or Equation (18–6) to determine the actual life or the rated capacity required.

It is common practice for a bearing manufacturer to provide tables which show various types of bearings, their rated load capacity, etc. The tables have to be carefully studied as different manufacturers present different types of tables. Again, it is recommended that one call for catalogs from a manufacturer if one is to select bearings from such manufacturer for actual use.

Table 18–2 Shock factor for various types of bearings

Type of shock	k_a **for ball bearing**	k_a **for roller bearing**
No shock	1	1
Heavy impact	2 to 3	1.5 to 2.0
Medium impact	1.5 to 2.0	1.1 to 1.5
Light impact	1.2 to 1.5	1.0

Example 18-1

Select a ball bearing ($\alpha = 0$) required for a shaft in Problem 4–11 (see page 77) for the left support (O) and the right support (B) if the reliability required is 99% and $L_a = 90 \times 10^6$ cycles.

Solution

The solution to Problem 4–11 was obtained by the authors and the reactions at the left support are

$L_x = 2.42$ kN

$L_y = -2.09$ kN

$L_z = 3.06$ kN

(Note that the x-axis is along the longitudinal axis of the shaft.)

Therefore,

$$F_a = [L_y^2 + L_z^2]^{0.5}$$

$$= [(-2.09)^2 + (3.06)^2]^{0.5}$$

$$= 3.70 \text{ kN}$$

$$F_T = L_x = 2.42 \text{ kN}$$

$$r = \frac{F_T}{F_a} = \frac{2.42}{3.70} = 0.653$$

Now let us find the required capacity of the bearing if $k_a = 1.2$, $k_r = 0.22$ (for 99% reliability), and $\alpha = 0$. Since ratio $r = 0.653$ and $L_a = 90 \times 10^6$ cycles, we use

$$F_e = F_r(1 + 1.115((F_T/F_a) - 0.35))$$

$$= 3.70(1 + 1.115(0.653 - 0.35))$$

$$= 5.10 \text{ kN}$$

We know from Equation (18–6) that the required capacity of the bearing is

$$C = F_e k_a \left(\frac{L_a}{k_r L_R}\right)^{0.3}$$

$$= 5.10(1.2)\left(\frac{90 \times 10^6}{0.22(90 \times 10^6)}\right)^{0.3}$$

$$= 9.70 \text{ kN}$$

So we select a suitable bearing from the catalog to carry a 9.70–kN load. The different equation is used for $\alpha = 25°$. For this value of α, $c = 7.04$ kN.

Now as a practice, it is advised to select bearing at right support.

Lead Model

The lead model called *ex18-1.tk*, or *bearing.tk*, was used to solve Example 18-1. The Rules Sheet shows the equations used and the Variables Sheet shows the input and output (results) for $\alpha = 0$ and $\alpha = 25°$. There are no new commands in the Rules Sheet, and the model is easy to use or modify to suit your requirements.

RULES SHEET

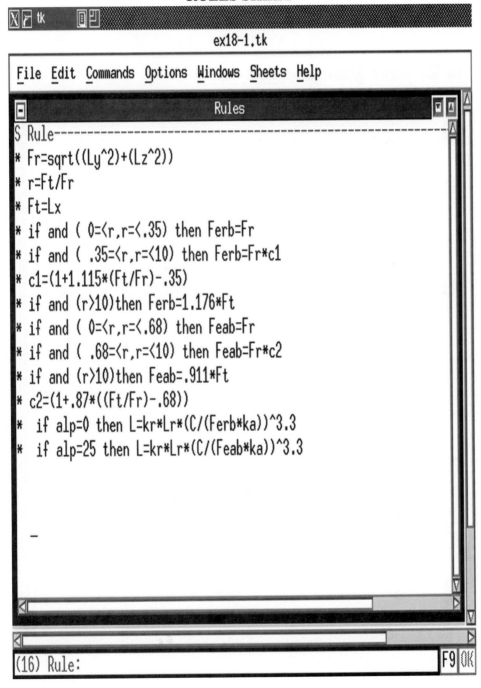

ex18-1.tk

File Edit Commands Options Windows Sheets Help

Rules

```
S Rule-------------------------------------------
* Fr=sqrt((Ly^2)+(Lz^2))
* r=Ft/Fr
* Ft=Lx
* if and ( 0=<r,r=<.35) then Ferb=Fr
* if and ( .35=<r,r=<10) then Ferb=Fr*c1
* c1=(1+1.115*(Ft/Fr)-.35)
* if and (r>10)then Ferb=1.176*Ft
* if and ( 0=<r,r=<.68) then Feab=Fr
* if and ( .68=<r,r=<10) then Feab=Fr*c2
* if and (r>10)then Feab=.911*Ft
* c2=(1+.87*((Ft/Fr)-.68))
*   if alp=0 then L=kr*Lr*(C/(Ferb*ka))^3.3
*   if alp=25 then L=kr*Lr*(C/(Feab*ka))^3.3



    -
```

(16) Rule: F9 OK

VARIABLES SHEET

```
XF tk      回凹
```

ex18-1.tk

```
File Edit Commands Options Windows Sheets Help
```

```
─                          Variables                        回回
St Input---- Name--- Output--- Unit----- Comment------------------------
              Fr      3.7056309 kN      'resultant radial force on a bearing
    -2.09     Ly              kN        'reaction on bearing in y direction
    2.42      Lx              kN        'reaction on bearing in x direction
              r       .65306019         'ratio of Ft/Fr
              Ft      2.42      kN       'force in axial( thrust) direction
    3.06      Lz              kN        'reaction on bearing in z direction


              c1      1.3781621         'constant
    _         Ferb    5.1069601 kN      'equivalent force for radial bearing(a
              Feab    3.7056309 kN      'equivalent force for axial bearing(al
              c2      .97656237         'constant
    25        alp                       'radial angle zero if radial bearing 2
    1E8       L                         'life of bearing in cycles
    .22       kr                        'realiability factor
    9E7       Lr                        'rated life of a bearing(90e6 cycles)
              C       7.2639887         'capacity of bearing in lbs or kN or N
    1.2       ka                        ' impact factor
```

```
◁                                                              ▷
(9) Input:                                                  F9 OK
```

PROBLEMS

18–1) What is the rated life of a bearing? A bearing manufacturer prepared a catalogue, based on a rated life of 10^5 cycles. If all other factors are the same, by what value should these ratings be multiplied when one compares them with the rating based on 10^6 cycles?

18–2) Determine the rated load, C, for an angular contact ball bearing for a shaft involving moderate shock and running at 500 rpm. The shaft is subjected to a 5–kN radial and 3–kN thrust load. The expected life of the shaft is three years operating 50 weeks per year and 40 hours per week. Due to the nature of work, the failure rate cannot be more than 2 percent.

18–3) Determine C for a bearing mounted on a shaft, using the following data: $F_r = 3.00$ kN, $F_T = 2$ kN, 95% reliability, heavy shock conditions, and a life requirement of 120×10^6 cycles. (Ans.: $C = 13.18$ kN.)

18–4) Determine the rated capacity of a bearing subjected to heavy impact and a load of 4.5 long tons, (1 long ton = 2,240 lb). The diameter of the shaft is 2 inches, and 95% reliability is required. The bearing must last for 80,000 miles of running if it is placed between a shaft and a wheel with a diameter of 30 inches.

18–5) For a bearing having $L_{10} = 90 \times 10^6$ cycles and $L_a = 1.267 \times 10^6$ cycles, estimate the life and rated load if $F_a = 1,000$ N and $F_t = 250$ N. The bearing is subjected to heavy impact, and 90% reliability is required.

18–6) A suitable ball bearing is used in a gearing application considered to be average with respect to shock loading. The shaft rotates at 3,000 rpm, and the bearing is subjected to a radial load of 800 N and a thrust load of 200 N. Estimate the bearing life in hours for 90% reliability if the rated load for the bearing is 1,000 N.

18–7)   A helical pinion is mounted on a shaft that is overhung from two support
bearings A and B, as shown in the following figure (not to scale).

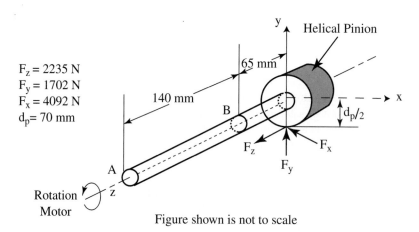

$F_z = 2235$ N
$F_y = 1702$ N
$F_x = 4092$ N
$d_p = 70$ mm

Figure shown is not to scale

The left end of the shaft is driven by an electric motor, while the right
end, closer to bearing B, is attached to the gear. The force components
acting on the pinion are as indicated. Assume that the entire thrust load
is taken by the bearing B.

a.   Sketch the load, shear force, and bending moment diagrams for
the shaft in the horizontal and vertical planes. Also, sketch the
shaft torsional and axial force diagrams. Mark the values with
the correct sign, using the appropriate sign convention.

b.   Determine the radial and axial loads applied to the bearings.

c.   If the rated load is 3,000 N at L_{10} life of 90×10^6 cycles, determine the life of the critical bearing.

18–8) The spur reducer shown in the following diagram (not to scale) is driven by a 15–kW motor at 1,200 rpm.

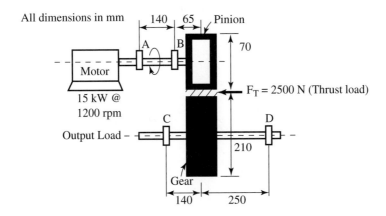

The pitch diameters of the pinion and gear are 70 mm and 210 mm, respectively. If the pressure angle is 0.35 radian, determine the tangential (F_T) and the radial (F_R) components of the load on the gear. Due to external loading conditions, it is estimated that the axial component of the force F_T on the gear is 2,500 N. If a bearing having a rated load of 4,000 N at $L_{10} = 90 \times 10^6$ is used, determine the life of the critical bearing; that is, calculate the least L_{10} life in hours of the bearings at the locations C and D. Assume an application factor of 1.2 on the load. (Bearing at C takes all the thrust load.)

18–9) Determine the rated load of a bearing with $L_{10} = 90 \times 10^6$ cycles if it supports a radial load of 7,500 N and a thrust load of 4,500 N. Assume that the inner ring rotates at 2,000 rpm. The bearing must have a life of 2.5×10^8 revolutions with 95% reliability.

18–10) A landing gear has a motor driving the pinion to turn the gear into the required position. A bevel pinion and shaft are shown in the following figure:

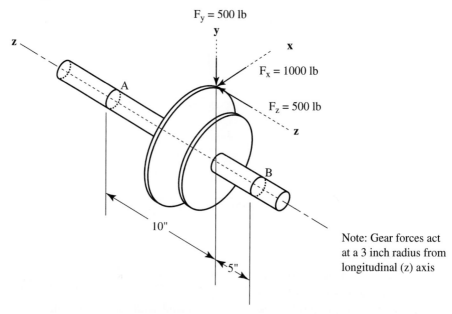

Note: Gear forces act at a 3 inch radius from longitudinal (z) axis

Bearing A takes thrust. The left end of the shaft is coupled to an electric motor, and the right end is free. Load components applied by the mating bevel gear are also shown.

    a.   Draw the axial load, torsional load, bending load, and shear load diagrams.

    b.   Determine the critical section and find the maximum shear stress at that section.

    c.   Determine C if a bearing with $L_{10} = 90 \times 10^6$ cycles is to be used. Assume moderate shock, 99% reliability, and an expected life of 10^6 cycles.

Lubrication Journal Bearings

19.1 Introduction

A bearing without balls or rollers, but to minimize friction if one uses lubricant between outer and inner ring then we have crude bearing called journal bearing. Journal bearings have many advantages over ball or roller bearings. A journal bearing is very reliable, has a minimum number of parts, requires less material than ball bearings, and is inexpensive to make. Also, journal bearings require less radial space, are well suited for shock and overload conditions, and are quieter than rolling bearings. The fatigue life of a journal bearing is much greater than that of a roller or ball bearing. Damage due to foreign bodies is less in journal bearings than in ball or roller bearings. However, journal bearings require lubrication. Still, the only condition required for the long life of a journal bearing is that the clean lubricating fluid be present at all times between the bearing and the shaft and that the fluid be at or below a certain temperature. Lubrication is a substance introduced between two moving surfaces so as to minimize friction and thus minimize wear and heating and maximize the life of the mating parts. Journal bearings are called 360° (full) bearings if they cover the full circumference of a journal (shaft) for a certain length. If, however, the bearing is made less than 360° in circumference, then it is called a partial bearing. In this chapter, we will cover only full journal bearings.

19.2 Lubrication

Since journal bearings require lubrication, one needs to understand the various types of lubricating films that may be used between two mating surfaces. Lubricating films are classified according to the degree with which they separate the sliding surfaces. If one has two rough mating surfaces, as shown in Fig. 19–1(a),

401

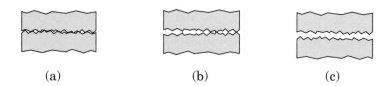

(a)   (b)   (c)

Figure 19–1   Three basic types of lubrication (a) Boundary (continuous
and extensive local contact) (b) Mixed film (intermittent
local contact) and (c) Hydrodynamic (surface separated)

and also has a lubricating substance between the two surfaces, such that the surfaces are in contact with each other, the lubricating film is known as *boundary lubricant*, and it has a coefficient of friction between 0.1 and 0.40 (high friction). The second type of lubricating condition between two surfaces does not allow continuous contact of the mating surfaces, as shown in Fig. 19–1(b). This type of lubrication is called *mixed-film lubrication*, and its coefficient of friction is between 0.02 and 0.10, which is much less than for boundary lubrication. The third type—the type we shall study in this chapter—is called *hydrodynamic lubrication* and is shown in Fig. 19–1(c). This type of lubrication, which maintains complete separation of the mating surfaces by means of a thick film, has a coefficient of friction in the range of 0.001 to 0.020, much less than the first two types of lubrication.

Studies of the behavior of journal bearings show that the basic parameters which influence the type of lubrication are the following:

1.   The viscosity μ (the lower the speed, the higher the viscosity of the lubricant is required)

2.   The rotating speed (the higher the speed, the lower the viscosity of the lubricant is required)

3.   The bearing unit load $P = W/Ld$, where W is the load on the bearing, L is the length of the bearing, d is the diameter of the bearing.

Still other parameters play a role, but are not as influential as the preceding ones. They will be discussed later.

Experimental studies show that if two dimensionless terms—the coefficient of friction f and $\mu n/P$ (where n is the speed)—are plotted against each other, then the relationship shown in Fig. 19–2 results. The figure shows that even with hydrodynamic lubrication, friction goes up as the speed goes up. This simply translates into requiring more horsepower to drive a bearing at higher speeds to overcome the friction between journal and hydrodynamic bearings. The plotted curve is known as the *Stribeck curve*.

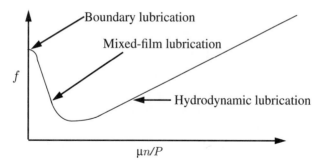

Figure 19-2   Plot of $\mu n/P$ vs. f (friction) (Stribeck curve)

19.3 Hydrodynamic Lubrication

Hydrodynamic lubrication can be achieved if:

1.   A suitable clean lubricant is present between two separating surfaces.

2.   Relative motion exists between two surfaces to be separated.

3.   Wedging action exists, provided by the eccentricity of the shaft. (Wedging action is described later in the chapter.)

Knowledge of certain lubrication properties and the laws governing lubrication behavior is necessary to develop the mathematical model needed to design, analyze, and/or optimize journal bearings. We will begin with Newton's law and then study lubrication properties and relate them to journal bearing behavior.

19.4 Newton's Law of Viscous Flow

Newton was the first scientist to explain the viscous behavior of a lubricating fluid. He conducted a simple experiment in which he introduced lubricating fluid at a certain temperature between one fixed and one moving plate (to create relative motion). He pulled the free plate with a velocity U. He found that the force required to pull the free plate depends upon the velocity gradient, i.e., $F\alpha(U/h)$. (See Fig. 19-3.) Upon dividing F by A and plotting F/A vs. U/h (see Fig. 19-4), Newton discovered that the slope of the curve is a constant that depends only on the type of fluid. Such a slope is called the *viscosity* of the fluid, μ.

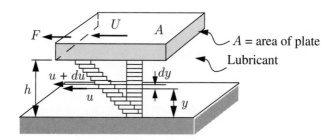

Figure 19–3   Two plates with relative motion and lubricant

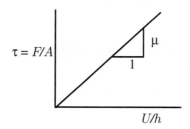

Figure 19–4   Plot of U/h vs. F/A

If A is the area of the free-plate surface that is in contact with the lubricant, then the shear stress between the plate and the lubricating fluid is given by

$$\tau = F/A = \frac{\mu U}{h}$$

This relationship between shear stress and the viscosity of a fluid is called *Newton's law for viscous fluids*.

Solving for μ, we get

$$\mu = \left(\frac{F}{A}\right)\left(\frac{h}{U}\right)$$

But $U/h = du/dh = du/dy$; therefore,

$$\mu = \tau \, dy/du$$

Let us now decide the units for μ in fps and SI units:

$$\mu = \left(\frac{lb}{in^2}\right)\left(\frac{in}{in/sec}\right)$$

$$\mu = \text{psi seconds}$$

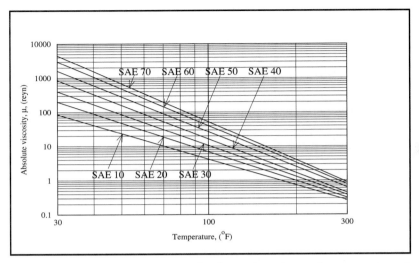

Figure 19–5   Relation between absolute viscosity and temperature in °F
for various types of SAE oils

Thus, μ has units of psi seconds in the fps system. A common unit for μ is reyns, where

1 reyn = 1 psi second

The viscosity, being small, is expressed as

1 microreyn = 10^{-6} reyn

Similarly, it can be shown that in the SI system μ has units of Pa seconds, also known as poise, where

1 poise = 1 Pa second

Since the viscosity has a very small value, it is expressed as millipoise or centipoise, where

1 millipoise = 10^{-3} PaS

1 centipoise = 10^{-2} PaS

The conversion from SI units to fps units for viscosity are 1 N sec/mm^2 = 145 lb sec/in^2 = 145 psi sec.

Another important relation worth studying is that between the viscosity and temperature of a lubricant. Such a relationship for a standard lubricant is shown in Figs. 19–5 and 19.6. The plots show that the viscosity decreases as the temperature increases.

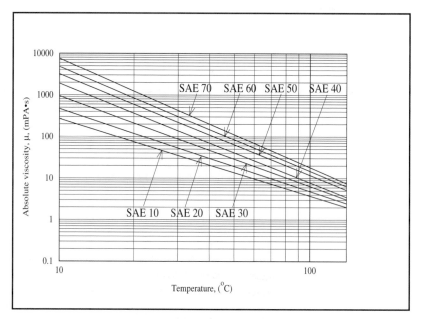

Figure 19–6   Relation between absolute viscosity and temperature in °C
for various types of SAE oils

The material we have studied so far is sufficient to develop a mathematical model for the analysis or design of a journal bearing. Now we turn to the theory that will be used in analysis or design of a journal bearing.

19.5 Hydrodynamic Lubrication Theory

The theory required to understand lubricant behavior was first developed by Reynolds, who suggested that if a lubricant is adhering to two surfaces, creating a fluid pressure of sufficient intensity to support an external load, then the *velocity, friction*, and *pressure distribution* in the thickness of the lubricant film can be determined, which in turn can help determine other unknown parameters required to design of a journal bearing.

Let us first examine the various positions of a journal bearing as the shaft starts from rest and increases its velocity to a constant angular velocity ω. We start with a shaft and journal bearing that are at rest. The cross section is as shown in Fig. 19–7(a.) As the shaft starts rotating with the bearing still stationary, the slow rotation will create *boundary lubrication*, as shown in Fig. 19–7(b). A further increase in speed will result in the *hydrodynamic condition* of Fig. 19–7(c). This condition creates enough internal pressure to support an external force W that is transferred from the journal to the bearing. At the same time, since the

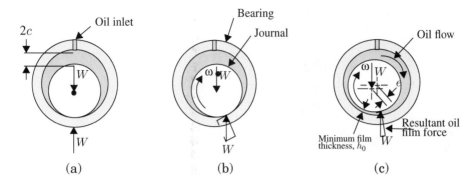

Figure 19–7 Various types of positions as a bearing picks up angular velocity (a) At rest, (b) Slow rotation (boundary lubrication), and (c) Fast rotation (hydrodynamic lubrication)

journal is rotating and the bearing is stationary, boundary conditions create velocity and velocity gradients throughout the thickness of the lubricant, due to the viscosity of the lubricant. Note that the initial concentric clearance c, which is defined as the difference between the radius of the journal bearing and the radius of the journal (shaft) is converted into a location where film thickness is minimal due to eccentricity developed during high-speed rotation.

The fast rotation during the hydrodynamic condition is of the most interest to us. However, note that the most damage to the bearings is done during slow speed—that is, during boundary lubrication—if the lubricant contains erosive particles (i.e., if the oil is dirty).

We shall develop the theory for one-dimensional flow—flow in the x-direction only. Then we will extend the theory to account for flow in the x- and z-directions, or two-dimensional flow (see Fig. 19–8). This extension is based on flat-plate flow, as shown in Fig. 19–3. Assume that the top plate is part of a journal and the bottom plate is a part of the bearing, with a lubricant between the two plates of viscosity μ. Assume also that the journal is rotating with velocity U and the bearing is stationary, as shown in Fig. 19–8.

In the development of the theory, Reynolds made the following simplifying assumptions:

1. The fluid film is thin enough to replace curved partial bearings with flat bearings (plane slider bearings).

2. The lubricants obey Newton's law of viscous fluids (i.e., $\tau = \mu \, du/dy$).

3. The force due to acceleration of the lubricant is neglected (i.e., $F = ma = 0$ is assumed).

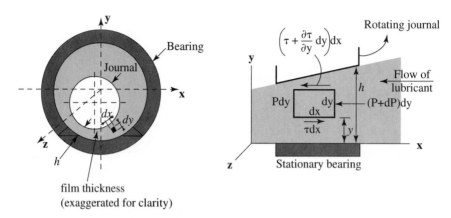

Figure 19–8   Cross section of journal bearing showing all applicable forces

4. The lubricant fluid is incompressible. (The mass flow in equals the mass flow out.)

5. The viscosity of the fluid is assumed to remain constant, even though the inlet temperature and outlet temperature of the lubricant are different. (Therefore, an average temperature value is used in any analysis or design.)

6. The pressure distribution throughout the film is not a function of z (the z-direction being along the longitudinal axis of the shaft).

7. There is no flow of lubricant in the z-direction. (In other words, the bearing and journal extend sufficiently long in the z-direction.)

8. The velocity distribution or the value of the velocity is a function of x and y, but not of z. That is, $U = f(x,y)$ (See Fig. 19–8.)

With these simplifying assumptions, Reynolds developed his theory, which is based upon the concept of equilibrium when the hydrodynamic condition is achieved.

Let us now select an element from Fig. 19–8 of width dx and height dy and of unit thickness in the z-direction, perpendicular to the page, as shown in Fig. 19–8. Since the pressure is assumed to change in the x-direction, but is constant in the y- and z-directions, we have a pressure P acting on the left side of the element and $P + dP$ on the right side. Similarly, due to the velocity gradient and the viscosity of the lubricant, the shear stress τ acts on the bottom surface. However, it changes on the top surface to $\tau + d\tau$ due to a change in velocity gradient.

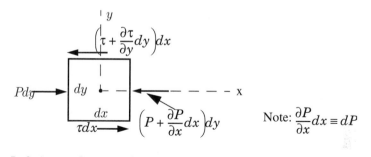

Figure 19–9 Lubricant element showing pressure and shear stresses

The isolated element is in static equilibrium, and therefore, it is necessary to multiply the stresses acting on it by the corresponding areas to obtain forces. It is further assumed that the element has unit thickness in the z direction. Summing the forces in the x-direction and setting them equal to zero, we get

$$Pdy + \tau dx - \left(P + \frac{\partial P}{\partial x}dx\right)dy - \left(\tau + \frac{\partial \tau}{\partial y}dy\right)dx = 0 \qquad (19\text{–}1)$$

Canceling terms and neglecting higher order terms, we get

$$\frac{dP}{dx} = \frac{\partial \tau}{\partial y} \qquad \left(\textit{Note:}\ \frac{\partial P}{\partial x} = \frac{dP}{dx}, \text{ since } P \neq f(z,y)\right)$$

But

$$\tau = \mu\frac{du}{dy} \qquad \text{(Newton's law for viscous flow)}$$

Hence,

$$\frac{\partial \tau}{\partial y} = \mu\frac{d^2 u}{dy^2} \qquad \text{(assuming that } \mu \text{ is a constant)}$$

Substituting for $\frac{\partial \tau}{\partial y}$ in the previous equation, we obtain

$$\frac{dP}{dx} = \mu\frac{d^2 u}{dy^2} \qquad (19\text{–}2)$$

Equation (19–2) is the fundamental differential equation developed by Reynolds for one-dimensional flow; therefore, it is called *Reynolds's one-dimensional differential equation*.

We can solve Equation (19–2) for the velocity distribution U throughout the thickness of the lubricant. Integrating the equation twice will result in

$$u = \frac{1}{\mu}\left(\frac{dP}{dx}\frac{y^2}{2} + c_1 y + c_2\right)$$

where $\dfrac{dP}{dx}$ is a constant. To evaluate the constants c_1 and c_2, let us satisfy the boundary conditions, namely,

$u = 0$ at $y = 0$ and $u = U$ at $y = h$ (see Fig. 19–3)

Substituting the boundary conditions, we get

$$u = \frac{1}{2}\mu\frac{dP}{dx}(y^2 - hy) + \frac{U}{h}y \tag{19–3}$$

Knowing the velocity distribution, we can determine the fluid flow (Q) in the x-direction. This is given by

$$Q = \int_0^h u\,dy$$

or, upon integration,

$$Q = \frac{Uh}{2} - \frac{h^3}{12\mu}\frac{dP}{dx} \tag{19–4}$$

Now we have an equation to determine the total flow in the x-direction for an incompressible fluid. We assume, naturally, that flow is in one direction. For incompressible flow, we have

$$Q_{x_1} = Q_{x_2} \qquad \text{(flow in equals flow out)}$$

Therefore,

$$\frac{dQ}{dx} = 0$$

Differentiating Equation (19–4) with respect to x, we get

$$\frac{dQ}{dx} = \frac{U\,dh}{2\,dx} - \frac{d}{dx}\left(\frac{h^3}{12\mu}\frac{dP}{dx}\right) = 0$$

or

$$\frac{d}{dx}\left(\frac{h^3}{\mu}\frac{dP}{dx}\right) = 6\left(U\frac{dh}{dx}\right) \tag{19–5}$$

Equation (19–5) is the classical Reynolds equation for one-dimensional flow based upon the assumptions that:

1. The fluid is Newtonian; that is, it obeys relation $\tau = \mu \, du/dy$.
2. The fluid is incompressible (i.e., flow in equals flow out).
3. The velocity U of the shaft is a constant. (There is no angular acceleration.)
4. The fluid has laminar flow (i.e., the velocity gradient is continuous).
5. There is no gravitational effect (i.e., $F = ma = 0$).
6. There is no slip at the boundary surfaces (i.e., $u = 0$ at the stationary bearing and $u = U$ at the rotating shaft).
7. The thickness of the lubricant is quite small compared to the size of the journal.

In sum, Reynolds developed two differential equations for one-dimensional flow [Equations (19–2) and (19–5)]. These equations have no direct solution, and therefore, one has to resort to numerical analysis to solve them. Since, in reality, one has to deal with two-dimensional flow, one can use a similar approach to that just presented. Extending the approach, Reynolds developed the following differential equation for two-dimensional flow (i.e., flow in the x- and z-directions):

$$\frac{\partial}{\partial x}\left(\frac{h^3}{\mu}\frac{\partial P}{\partial x}\right) + \frac{\partial}{\partial z}\left(\frac{h^3}{\mu}\frac{\partial P}{\partial z}\right) = 6U\frac{dh}{dx} \tag{19–6}$$

This differential equation has no analytical solution; hence, the only solution that exists is a numerical one. One of the empirical solutions is

$$(r/c)f = \phi\left[\left(\frac{r}{c}\right)^2\left(\frac{\mu n}{P}\right)\right]$$

where the variables are defined in Fig. 19–10.

A computerized numerical solution of Equation (19–6) by *Raimondi* and *Boyd* was developed and is presented in graphical form in Figs. 19–11 through 19–18. We have also taken the original data developed by Raimondi and Boyd, digitized it, and saved it in a lead model for efficient use. More information is given on the lead model later; now we will study the actual solution developed by Raimondi and Boyd.

These two researchers' obtained results using a variable called the characteristic number S (also called Sommerfeld's number). The characteristic number S is given by an equation

$$S = \frac{\mu n}{P}(r/c)^2$$

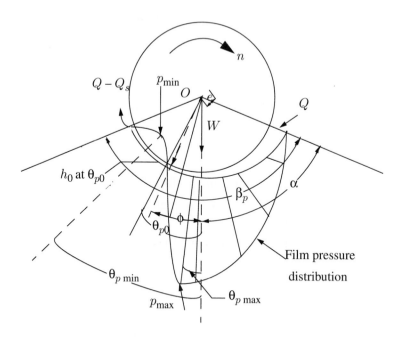

Figure 19–10   Sketch of journal bearing with variables

where μ is the viscosity of the lubricant, r is the radius of the bearing ($d = 2r$), c is the concentric clearance, and $P = \dfrac{W}{(Ld)}$, in which P is the unit load, W is the external load on the bearing, L is the length of the bearing, d is the diameter of the bearing. The Sommerfeld number is plotted on the x-axis of each of the graphs in Figs. 19–11 through 19–18.[1] The y-axis has variables such as h_0/c, where h_0 is the minimum film thickness (see Fig. 19–10) and c is the concentric clearance. Also plotted are variables such as $P/P_{\max}$, Q_s/Q, θ_p, θ_{P0}, e, Q, and others. Once again, it is recommended that one not read values off these graphs to get desired values; the lead model on journal bearings has them stored in the form of digitized data, which can be used in calculations. The interpolation is done automatically using the cubic mapping technique. The numerical solution is of great assistance in solving analysis and design problems involving journal bearings.

The next two examples are based on the graphs presented. The examples, however, are based on the average temperature of the lubricant, which is not

---

[1] Figs. 19–11 through 19–18 are for academic use. Use the lead model for correct analysis, design, or optimization of a journal bearing.

Table 19–1   Nomenclature

W = load, lb or N

n = speed, rps

r = journal radius, in or meters

D = journal diameter, in or meters

L = axial length of bearing, in or meters

$z = L/D$

P = load per unit projected bearing area

 = $W/2rL$, psi or N/m^2

μ = viscosity in reyns, lb sec/in^2, or N sec/m^2

h_0 = minimum film thickness, in or meters

c = radial clearance, in or meters

e = journal displacement or eccentricity, in or meters

e/c = eccentricity ratio, dimensionless

F = friction forces on journal, lb or N

$f = F/W$ = coefficient of friction

HP = power consumed in friction, horsepower

U = journal peripheral speed, in/sec or m/sec

Q = flow of lubricant drawn into clearance space by journal, in^3/sec, or m^3/sec

Q_s = side flow of lubricant, in^3/sec, or m^3/sec

β = angular length of bearing arc, deg (in this chapter, 360°)

α = leading angle, extending from beginning of bearing arc to the line of action load, de

ϕ = position of minimum film thickness, deg

$\theta_{p\,max}$ = position of maximum film pressure, deg

$\theta_{p\,min}$ = position of minimum film pressure, deg

p = film pressure, psi or N/m^2

p_{max} = maximum pressure developed in lubricant

θ_{po} = position at which film thickness becomes h_0

p_{min} = minimum pressure developed in lubricant

p_o = ambient pressure, psi or N/m^2

$S = (r/C)^2\mu N/P$ = bearing characteristic number or Sommerfeld number (dimensionless

t_1 = lubricant inlet temperature, °F or °C

t_2 = lubricant outlet temperature, °F or °C

Δ_t = temperature rise of lubricant = $t_2 + t_1$, °F or °C

t_a = average film temperature = $t_1 - \Delta_t/2$, °F or °C, or $t_{av} = (t_1 + t_2)/2$

c = specific heat of lubricant, BTU/lb °F

γ = weight per unit volume of lubricant, lb/in^3 or N/m^3

f_m = mechanical equivalent of heat

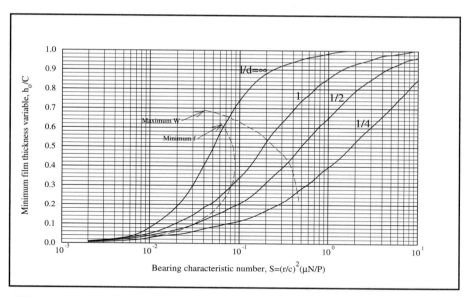

Figure 19–11   Graph of characteristic number versus minimum film thickness. (Dotted lines are optimum lines, left for minimum friction and right for maximum W.)

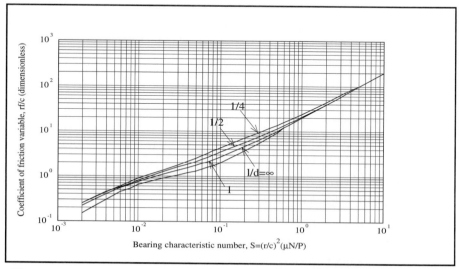

Figure 19–12   Graph of characteristic number versus coefficient of friction variable

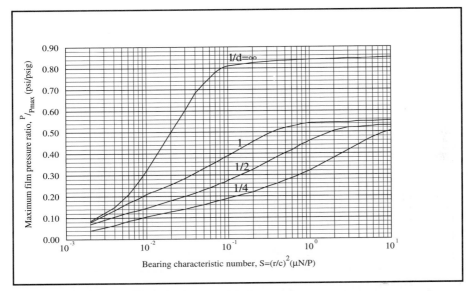

Figure 19–13 Graph of characteristic number versus pressure and maximum film pressure ratio

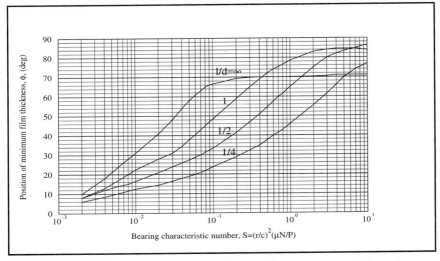

Figure 19–14 Graph of characteristic number versus position of minimum film thickness

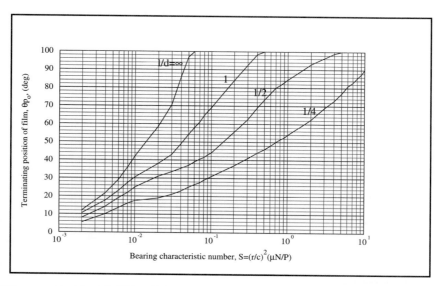

Figure 19–15   Graph of characteristic number versus terminating
position of film

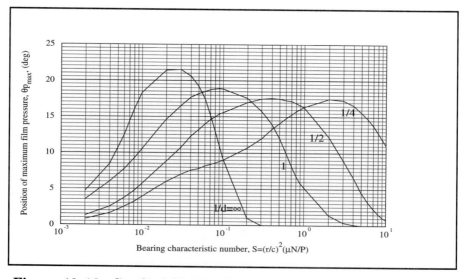

Figure 19–16   Graph of characteristic number versus position of
maximum film pressure

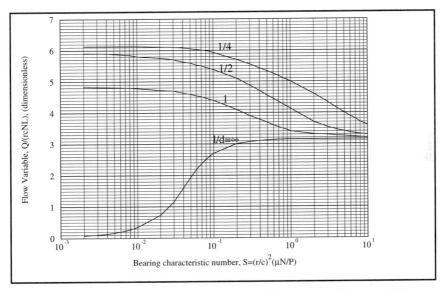

Figure 19–17 Graph of characteristic number versus flow variable

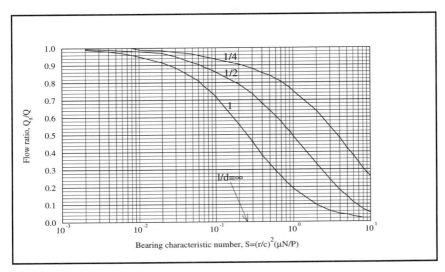

Figure 19–18 Graph of characteristic number versus flow ratio

known in general. Later on, we will determine the temperature at the exit if the entrance temperature is given. The temperature at the exit will allow us to determine the average temperature, which will be simply the average of the entrance and exit temperatures and can be used in the analysis or design of a journal bearing. To understand the graphs and example problems, let us first develop nomenclature (as shown in Table 19–1).

Example 19-1

A journal bearing transmits a radial load $W = 2.0$ kN. For the journal, $L/D = 1$, and SAE40 oil is used. $T_{av} = 65°$ C, and the shaft is rotating at 1,800 rpm. If the allowable average pressure = 1.5 MPa, then:

        a.   Determine L and D.

        b.   Determine values of c (the concentric clearance corresponding to two edges of the optimum zone).

        c.   Does c for minimum friction satisfy Trupler's criterion (explained shortly)?

Solution

We have

$$P = \frac{W}{LD}$$

$$P = 1.5 \times 10^6 = \frac{2 \times 10^3}{LD}$$

but $L = D$; therefore, solving for L we get

$$L = 36.5 \text{ mm} \approx 37 \text{ mm}$$

Now,

$$n = \frac{1,800}{60} = 30 \text{ rps}$$

Fig. 19–6 shows that $\mu = 22 \times 10^{-3}$ PaS at 65°C for SAE40 oil. The optimum $S = 0.082$ or 0.21. (See Fig. 19–11 for two optimum curves.) Thus, for $S = .082$ (this value is for minimum friction), we have

$$0.082 = (r/c)^2 \left(\frac{\mu n}{P}\right) = \left(\frac{18.5}{c}\right)^2 \left(\frac{0.022 \times 30}{1.5 \times 10^6}\right)$$

Solving for c, we get

$$c = 0.043 \text{ mm}$$

For $S = 0.21$ (this value is for maximum W), using the procedure as above but $S = 0.21$, we get $c = 0.026$ mm.

Now for maximum load, we use Trupler's criterion, viz.,

$h_0 = 0.0002 + 0.00004D$ (in FPS units, with h_0 and D in inches)

or

$h_0 = 0.005 + 0.00004D$ (in SI units, with h_0 and D in mm)

In this example,

$h_0 = 0.005 + 0.00004(37) = 0.0065$ mm

But $\dfrac{h_0}{c} = 0.19$. (See Fig. 19–11.) Therefore,

$h_0 = 0.19(0.029) = 0.00571 < 0.0065$ recommended

Hence, the minimum h_0 criterion is not satisfied. One may increase c value to get desired h_0 value.

The preceding example is solved using the lead model called *ex19-1.tk,* or *jbr.tk*. The Rules Sheet and the Variables Sheet are shown with the results on the pages 420 through 423. You may use this model to change the concentric clearance to satisfy Trupler's criterion. Note that the model does not have equations required for that criterion. Thus, you may add those in the Rules Sheet and solve for the value of c that will satisfy Trupler's criterion.

The lead model *ex19-1.tk*, or *jbr.tk*, is quite extensive in nature. Therefore, it may only be modified, and not replaced.

The Rules Sheet is also extensive but is self-explanatory and must be studied again, after one reads Sections 19.6 and 19.7. The Variables Sheet explains the variables used under the heading "Comment." You will see that, just as in FORTRAN, in TK Solver one can use the "call" command to go to other subroutines to get the desired results.

To make better use of the power of this lead model, it is strongly recommended that you read Sections 19.6 and 19.7. In these sections, you will see how to determine the average temperature of a fluid if the inlet temperature is known. Let us take one more example.

Example 19-2

A journal bearing is to be designed for minimum friction. The following data is given: $L = 50$ mm, $D = 50$ mm, $W = 5$kN, $n = 1,200$ rpm, $h_0 = 0.025$ mm. For *minimum friction*, find c and μ. Use these values to find f and the power loss.

Solution

From Fig. 19–11, for *minimum friction* and $L/D = 1$, the characteristic number is $S = 0.08$. Also, $h_0/c = 0.3$, from the same figure. Therefore,

$$c = h_0/0.3 = \frac{0.025}{0.3} = 0.083 \text{ mm}$$

RULES SHEET

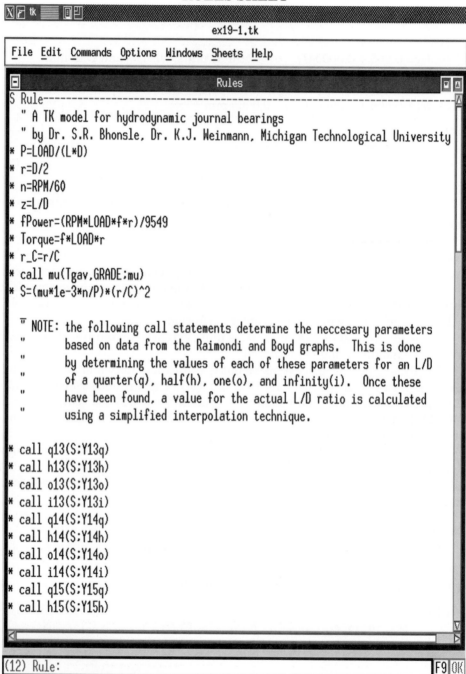

```
ex19-1.tk
```

File Edit Commands Options Windows Sheets Help

```
                               Rules
S Rule-----------------------------------------------------------------------
  " A TK model for hydrodynamic journal bearings
  " by Dr. S.R. Bhonsle, Dr. K.J. Weinmann, Michigan Technological University
* P=LOAD/(L*D)
* r=D/2
* n=RPM/60
* z=L/D
* fPower=(RPM*LOAD*f*r)/9549
* Torque=f*LOAD*r
* r_C=r/C
* call mu(Tgav,GRADE;mu)
* S=(mu*1e-3*n/P)*(r/C)^2


  " NOTE: the following call statements determine the neccesary parameters
  "        based on data from the Raimondi and Boyd graphs.  This is done
  "        by determining the values of each of these parameters for an L/D
  "        of a quarter(q), half(h), one(o), and infinity(i).  Once these
  "        have been found, a value for the actual L/D ratio is calculated
  "        using a simplified interpolation technique.

* call q13(S;Y13q)
* call h13(S;Y13h)
* call o13(S;Y13o)
* call i13(S;Y13i)
* call q14(S;Y14q)
* call h14(S;Y14h)
* call o14(S;Y14o)
* call i14(S;Y14i)
* call q15(S;Y15q)
* call h15(S;Y15h)

(12) Rule:                                                              F9 OK
```

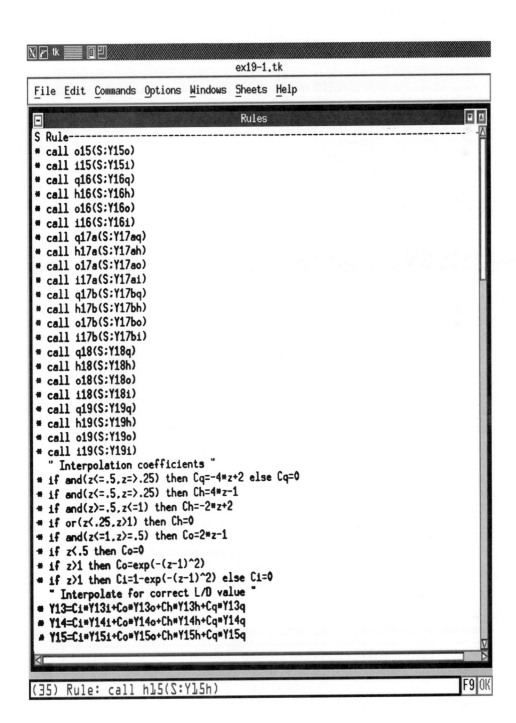

```
X ʔ tk ≣ ⊡ 凹
                              ex19-1.tk

 File  Edit  Commands  Options  Windows  Sheets  Help

⊟                              Rules                          ▣ ▲
 S Rule---------------------------------------------------------
 * call o15(S;Y15o)
 * call i15(S;Y15i)
 * call q16(S;Y16q)
 * call h16(S;Y16h)
 * call o16(S;Y16o)
 * call i16(S;Y16i)
 * call q17a(S;Y17aq)
 * call h17a(S;Y17ah)
 * call o17a(S;Y17ao)
 * call i17a(S;Y17ai)
 * call q17b(S;Y17bq)
 * call h17b(S;Y17bh)
 * call o17b(S;Y17bo)
 * call i17b(S;Y17bi)
 * call q18(S;Y18q)
 * call h18(S;Y18h)
 * call o18(S;Y18o)
 * call i18(S;Y18i)
 * call q19(S;Y19q)
 * call h19(S;Y19h)
 * call o19(S;Y19o)
 * call i19(S;Y19i)
    " Interpolation coefficients "
 * if and(z<=.5,z=>.25) then Cq=-4*z+2 else Cq=0
 * if and(z<=.5,z=>.25) then Ch=4*z-1
 * if and(z)=.5,z<=1) then Ch=-2*z+2
 * if or(z<.25,z>1) then Ch=0
 * if and(z<=1,z=.5) then Co=2*z-1
 * if z<.5 then Co=0
 * if z>1 then Co=exp(-(z-1)^2)
 * if z>1 then Ci=1-exp(-(z-1)^2) else Ci=0
    " Interpolate for correct L/D value "
 * Y13=Ci*Y13i+Co*Y13o+Ch*Y13h+Cq*Y13q
 * Y14=Ci*Y14i+Co*Y14o+Ch*Y14h+Cq*Y14q
 * Y15=Ci*Y15i+Co*Y15o+Ch*Y15h+Cq*Y15q
                                                             ▽
◁                                                          ▷

(35) Rule: call h15(S;Y15h)                          F9 OK
```

---

ex19-1.tk

| File | Edit | Commands | Options | Windows | Sheets | Help |

```
□                                Rules                                ▣ ▣
S Rule--------------------------------------------------------------------------
* Y16=Ci*Y16i+Co*Y16o+Ch*Y16h+Cq*Y16q
* Y17a=Ci*Y17ai+Co*Y17ao+Ch*Y17ah+Cq*Y17aq
* Y17b=Ci*Y17bi+Co*Y17bo+Ch*Y17bh+Cq*Y17bq
* Y18=Ci*Y18i+Co*Y18o+Ch*Y18h+Cq*Y18q
* Y19=Ci*Y19i+Co*Y19o+Ch*Y19h+Cq*Y19q

* ho=Y13*C
* f=Y14*C/r
* Pmax=P/(Y15)
* phi=(Y16)
* thetapo=(Y17a)
* thetapm=(Y17b)
* Q=Y18*r*C*n*L
* Qs=Y19*Q
* delT=(KU*P*f*r/C)/((1-.5*(Qs/Q))*(Q/(r*C*n*L)))
* Tf=Ti+delT
* Tgav=Ti+delT/2
   " NOTE: The following four lines determine the value of a performance
   "       factor defined to quantify the performance curves found in
   "       Figure 13.13.  This is done by calculating a maximum and minimum
   "       suggested S value, and then comparing the actual S value with
   "       these by calcualting Pf which will be -1 when S=Smin and +1 when
   "       S=Smax.  For values between Smin and Smax, Pf will be between
   "       -1 and +1, while values outside of -1 to +1 indicate that the
   "       calculated S is outside of the optimal zone.
* call Smax(ln(z):Smax)
* call Smin(ln(z):Smin)
* Pf=2*(log(Smin)-Log(S))/(Log(Smin)-log(Smax))-1
* aPf=abs(Pf)
   "  The following command is only used when list solving. It stores the
   "  value of Tgav so that by typing LG into the status field of Tgav
   "  in the variable sheet, the calculated Tgav from one solution is used
   "  as the first guess in the following solution. If a guess is required
   "  for some other variable, this line can be changed accordingly.
* place('Tgav.elt()+1)=Tgav
◁                                                                     ▷
```

(70) Status: * Unsatisfied   F9 OK

VARIABLES SHEET

```
X 𝚪 tk ▦ 🔲🔳
```

ex19-1.tk

File Edit Commands Options Windows Sheets Help

```
─ ┌──────────────────────── Variables ────────────────────────┐ 🔲🔺
St Input----  Name---  Output---  Unit-----  Comment---------------------------- 🔺
              n        30         rps        revolutions per second (rps)
    1800      RPM                 RPM        revolutions per minute (RPM)
    1500000   P                   N/m^2      Average film pressure (N/m^2 or psi)
    2         LOAD                 kN         Applied Load (N or lb)
              L        36.514837  mm         Length (m or in)
              D        36.514837  mm         diameter of the shaft (m or in)
    1         z                               L/D (none)
              r_C      690.84928              r/C (none)
              r        .01825742  m          radius of the shaft (m or in)
    30        GRADE                           SAE GRADE eg:30,40,etc. (none)
L             C        2.6427E-5  m          radial clearance (m or in)
L   65        Tgav                degC       average temperature (degC or degF)
              Ti       54.722168  degC       inlet temperature (degC or degF)

L             Q        2.1543E-6  m^3/s      total oil flow rate (m^3/s or in^3/s)
L             Tf       75.277832  degC       outlet temperature (degC or degF)
L             ho       1.4218E-5  m          minimum film thickness (m or in)
              f        .00707202             bearing coefficient of friction (none)
L             fPower   .04867741  kW         friction power loss (kW, hp, or btu/s)
              Torque   .25823366  Nm         friction torque (Nm or inlb)
L             Pmax     3263689.8  N/m^2      maximum film pressure (N/m^2 or psi)
              phi      59.76125   deg        minimum film thickness position (deg)
              thetapo  85.0765    deg        film termination position (deg)
              thetapm  17.435675  deg        maximum film pressure location (deg)
              mu       22         cp         viscosity (cp or mmreyn)
              Qs       1.1805E-6  m^3/s      oil side leakage (m^3/s or in^3/s)
              delT     20.555664  deldegC    temperature change(deldegC or deldegF)

    .0000083  KU                              conversion constant (do not change)

              Smin     .082                  Minimum suggested S value (none)
              S        .21                   bearing characteristic number (none)
                                             Note: S must be between 0 and 10
              Smax     .21                   Maximum suggested S value (none)

    ─                                                                            🔻
◁                                                                              ▷
```

(35) Input: F9 OK

$$S = 0.08 = (R/c)^2 \left(\frac{\mu n}{P}\right) = \left(\frac{25}{0.083}\right)^2 \left(\frac{\mu(20)}{5000/(0.050)^2}\right)$$

so that

$\mu = 91.01 \ cP$

From Fig. 19-12 for $S = 0.08$, we have $(R/c) \ f = 2.4$, or $f = 2.4 \ (0.083/25) = 0.0078$. Hence, we have

Power loss $= fW(2\pi Rn) = 0.0078(5000)(2\pi(0.025)(1200/60)$

$= 122$ watts or 0.18 HP

Example 19-2 was solved using the lead model called *ex19-2.tk*. The results are shown in the Variables Sheet on page 425.

19.6 Study of Temperature Rise in Journal Bearings

Up to now, we have assumed that we knew the average temperature of the fluid we were working with. In design, however, one knows, at best, the temperature of the fluid only as it enters the bearing. In this section, we present the mathematical theory developed to determine the exit temperature of a fluid if the entrance temperature is known.

As mentioned earlier, the inlet temperature and the exit temperature of a lubricant are different, due to a rise in temperature as the lubricant is used. Such a temperature increase occurs because of friction, which itself is caused by viscous behavior of the lubricant. Since the viscosity remains constant, the average of the inlet and exit temperatures has to be taken to perform the mathematical calculations. Let us find the exit temperature using certain assumptions. One assumption is that the rise in temperature is due solely to the heat generated by friction and that all such heat is being absorbed by the lubricant. Such an assumption is reasonable and will yield a design that is conservative.

To understand heat generation and heat absorption, we define the following terms: The *mechanical equivalent of heat*, M_e, is the work input required to generate 1 BTU. This turns out to be 9,336 in-lb of work. A BTU is simply energy (work). The mechanical equivalent of work has the units of in-lb/BTU.

The *specific heat* C_H is the energy required to raise the temperature of 1 pound of fluid through 1°F. Therefore, C_H has units of BTU/°F lb.

Each lubricant has a specific weight γ, defined as the weight of the fluid in pounds or newtons, per cubic inch or cubic meter, respectively. Therefore, the units of specific weight are lb/in³ or N/m³.

Now let $\Delta T = T_f - T_i$, where ΔT is the rise in temperature of a lubricant, T_f is the exit temperature of the lubricant, and T_i is the lubricant's inlet temperature. Then the work done on the fluid per second is $2\pi n T$, where n is revolutions

VARIABLES SHEET

```
X 🗎 tk      🔲 🗗
                              ex19-2.tk

 File  Edit  Commands  Options  Windows  Sheets  Help

🔲                            Variables                        🔲 🔼
                                                               🔼
 St Input---- Name--- Output--- Unit----- Comment-------------------------
              n        20        rps       revolutions per second (rps)
      1200    RPM                RPM       revolutions per minute (RPM)
              P        2000000   N/m^2     Average film pressure (N/m^2 or psi)
      5000    LOAD               N         Applied Load (N or lb)
      .05     L                  m         Length (m or in)
      .05     D                  m         diameter of the shaft (m or in)
              z        1                   L/D (none)
              r_C      300.16              r/C (none)
              r        .025      m         radius of the shaft (m or in)
              GRADE                        SAE GRADE eg:30,40,etc. (none)
 L            C        .08328891 mm        radial clearance (m or in)
              Tgav     80.719889 degC      average temperature (degC or degF)
              Ti       73.717996 degC      inlet temperature (degC or degF)

 L            Q        9.2857E-6 m^3/s     total oil flow rate (m^3/s or in^3/s)
 L            Tf       87.721782 degC      outlet temperature (degC or degF)
 L    .025    ho                 mm        minimum film thickness (m or in)
              f        .00779584           bearing coefficient of friction (none)
 L            fPower   .12246061 kW        friction power loss (kW, hp, or btu/s)
              Torque   .97448028 Nm        friction torque (Nm or inlb)
 L            Pmax     5368724   N/m^2     maximum film pressure (N/m^2 or psi)
              phi      45.416    deg       minimum film thickness position (deg)
              thetapo  65.064    deg       film termination position (deg)
              thetapm  18.8656   deg       maximum film pressure location (deg)
              mu       91.014004 cp        viscosity (cp or mreyn)
              Qs       .00000702 m^3/s     oil side leakage (m^3/s or in^3/s)
              delT     14.003785 deldegC   temperature change(deldegC or deldegF)

      .0000083 KU                          conversion constant (do not change)

              Smin     .082                Minimum suggested S value (none)
              S        .082                bearing characteristic number (none)
                                           Note: S must be between 0 and 10
              Smax     .21                 Maximum suggested S value (none)
                                                               🔽
◁                                                             ▷ 🔽

 (49) Comment: based on the Raimondi and Boyd                 │F9│OK│
```

per second and T is the torque in in-lb or N-m. Consequently, the heat generated per second is

$$H = \frac{2\pi n T}{M_e}$$

where H has units of BTU/sec.

If the lubricant has a flow rate of Q in^3/sec, one has to multiply Q by γ, the specific weight, which yields $Q\gamma$, the weight flow per second. If one then multiplies $Q\gamma$ by C_H (the specific heat of the lubricant), one gets

$$Q\gamma C_H \Rightarrow \frac{in^3}{sec} \frac{lb}{in^3} \frac{BTU}{F\ lb} = \frac{BTU}{sec\,°F}$$

It is now easy to see that if H is the heat generated per second (BTU/sec), then

$$Q\ \gamma\ C_H \Delta T_f\ = H$$

Therefore,

$$\Delta T_F = H/(Q\ \gamma\ C_H)$$

$$= 2\pi n T/(M_e Q\ \gamma\ C_H)$$

In terms of units,

$$\frac{H}{Q\gamma C_H} \Rightarrow \frac{BTU/sec}{\dfrac{BTU}{sec\,°F}} = °F$$

Thus, the units of ΔT_f, which are °F, match with those of the right side. Here, ΔT_F is the temperature change due to friction, f, and it is given in degrees Fahrenheit.

If ΔT_f is known, then one can determine the outlet temperature $T_F = T_i + \Delta T_f$, as long as T_i is known.

How does one calculate the torque required to overcome friction due to the viscous lubricant? If one assumes that the overall coefficient of friction is f between the shaft and the lubricant then the torque based on freebody diagram of shaft is

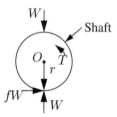

$$T = fWr$$

(*Note*: fWr is a moment of the external forces about 0.)

where W is the external weight on a bearing and r is the radius of the shaft or bearing. The Raimondi–Boyd graph (Fig. 19–12) yields Sommerfeld's number versus $(r/c)\,f$, called the friction variable value, which we may denote by $F_{\text{Frict.}}$ Also, the graph shown in Fig. 19–17 yields $Q/(rcnL)$, the flow variable as a function of Sommerfeld's number. Let us call this variable F_{Flow}. Then, rewriting, we have

$$\Delta T_F = \frac{H}{\gamma Q C_H}$$

But

$$H = \frac{2\pi n}{M_e}(fWr)$$

Therefore,

$$\Delta T_F = \frac{2\pi n fWr}{M_e \gamma Q C_H}$$

Rearranging, we obtain

$$\Delta T_F = \frac{4\pi P}{M_e \gamma C_H} \frac{F_{\text{Frict.}}}{F_{\text{Flow}}}$$

where

$$P = \frac{W}{2rL}$$

Now, if we let

$M_e = 9336$ in-lb/BTU

$\gamma = 0.86(62.4)$ (for oil)

and

$C_H = 0.42$ BTU per lb per °F (a typical value for oil)

then

$$\Delta T_F = 0.103 P \left(\frac{F_{\text{Frict.}}}{F_{\text{Flow}}} \right) \tag{19–7}$$

where P is in psi. Similarly, it can be shown that, for SI units,

$$\Delta T_C = 8.30 \times 10^6 P \left(\frac{F_{\text{Frict.}}}{F_{\text{Flow}}} \right) \tag{19-8}$$

where P is in pascals. Since one does not know ΔT unless $F_{\text{Frict.}}$ and F_{Flow} are known, and since $F_{\text{Frict.}}$ and F_{Flow} depend on temperature, then calculating for ΔT requires an iterative type of calculation. That is, we first assume an average temperature and find ΔT, and then we compare ΔT with the assumed value. If it is nearly equal to the assumed value, then the assumption is correct. If not, we assume a fresh value and iterate.

The iterative process is carried out using the lead model *ex19-4.tk*. This model is extensive and should not be rewritten. It could be modified or added to; however, if you copy it under a different name, you can make whatever changes you want. Use the model the way it is, and it will carry on the analysis or design, and the optimization of journal bearings. This model saves all graphical solutions as digitized data, and therefore, one need not go to graphs to determine any values for givens.

To account for side flow, we must modify Equations (19–7) and (19–8). Any use of a journal bearing will yield a side flow. The modified equations are

$$\Delta T_F = \frac{H}{\gamma C_H Q \left[1 - \frac{1}{2} \left(\frac{Q_s}{Q} \right) \right]} \tag{19-9}$$

where Q_s is the side flow, so that

$$\Delta T_F = \frac{0.103 P}{\left[1 - \frac{1}{2} \left(\frac{Q_s}{Q} \right) \right]} \left(\frac{F_{\text{Frict.}}}{F_{\text{Flow}}} \right) \quad \text{(P is in psi)}$$

and

$$\Delta T_C = \frac{8.30 P}{\left[1 - \frac{1}{2} \left(\frac{Q_s}{Q} \right) \right]} \left(\frac{F_{\text{Frict.}}}{F_{\text{Flow}}} \right) \quad \text{(P is in MPa)} \tag{19-10}$$

This concludes all mathematical equations required for the analysis, design, or optimization of a journal bearing.

19.7 Summary for Design of Journal Bearings

This section summarizes the equations and steps required for design, presented for the purpose of reviewing the material. Recall that we know only the inlet temperature T_i and never really know the average temperature T_{av} of a lubricant. To determine the change in temperature, we derive the equations

$$\Delta T_F = (K_{uf})\frac{P}{z}\left(\frac{F_{Frict.}}{F_{Flow}}\right)$$

$$\Delta T_C = (K_{uc})\frac{P}{z}\left(\frac{F_{Frict.}}{F_{Flow}}\right) \tag{19–11}$$

where

$K_{uf} = 0.103$

$z = L/D$

$K_{uc} = 8.33 \times 10^{-6}$ (if P is in Pa) or 8.33 (if P is in MPa)

$P = \dfrac{W}{DL}$ (in psi or MPa)

$F_{Frict.} = (r/c)f$ (see Fig. 19–12)

$F_{Flow} = Q/(rcnL)$ (see Fig. 19–17)

$z = \left(1 - 0.05\dfrac{Q_s}{Q}\right)$ (see Fig. 19–18)

If no lead model is used, the steps in the design of a journal bearing are as follows:

1. Given T_i, assume T_{av} and find $\Delta T_{assumed} = 2(T_{av} - T_i)$. It is assumed that we are also given R, c, $P = W/DL$ and n in rps.

2. Now, using T_{av}, find μ from Fig. 19–5 or Fig. 19–6. (Watch for °C or °F.)

3. Knowing r, c, μ, n, and P, determine Sommerfeld's number

$$S = \frac{\mu n}{P}(r/c)^2$$

where c is the clearance and n = rpm/60 (since n is required in rps).

4. Determine $F_{Frict.}$ and F_{Flow}, knowing z and using Figs. 19–12, 19–17, and 19–18.

5. Determine $\Delta T_{calculated}$ using Equation (19–11) or Equation (19–12), depending upon the units used.

6. Compare $\Delta T_{\text{assumed}}$ with $\Delta T_{\text{calculated}}$ in Step 5. If they are almost equal to each other, then proceed to the next step. If not, assume a new T_{av}, repeat Steps 2 through 5, and then compare again. Ultimately, $\Delta T_{\text{assumed}} = \Delta T_{\text{calculated}}$. Then find the exact μ value and Sommerfeld's number. Now you can find the other unknown variables using the next two steps.

7. Determine h_0, P_{max}, $\theta_{p\text{max}}$, θ_{h0}, f, T_f, Q_s, Q, etc., using the charts. (See Figs. 19–11 through 19–18.)

However, if you want to design journal bearings in the future, then use lead model *ex19-3.tk*.

Once you guess a value of T_{av} and the solve command is given, the model will yield all required outputs. This will happen, however, only if all the other required inputs and units are presented correctly. Try it. It is fun and does a lot for you in a very short time. You no longer need to carry out iterative calculations; the lead model does it for you.

Next, we present an example that uses hand calculations. Then the same problem is solved using the TK lead model *ex19-3.tk*.

Example 19-3

A journal bearing is used on a high-speed shaft. SAE20 oil is used, and the sleeve bearing has the dimensions $L = 3"$, $D = 3"$. If the inlet temperature $T_i = 100°F$, and if the shaft is rotating at $n = 1,200$ rpm and carries a load $W = 1,500$ lb, then, for a concentric clearance, $c = 0.0015"$.

a. Find the magnitude and location of the minimum oil film thickness.

b. Find the coefficient of friction.

c. Compute the side flow and the total oil flow.

d. Find the maximum oil film pressure and its location.

e. Find the Q and Q_s – side flow.

Solution

We have

$$z = L/D = \frac{3}{3} = 1, \text{ average pressure} = P = \frac{W}{DL} = \frac{1,500}{(3)(3)} = 167 \text{ psi}$$

$$n = \frac{1,200}{60} = 20 \text{ rps}$$

$$r/c = 1.5/0.0015 = 1,000$$

First, we guess that $T_{av} = 130°F$. Then, from the graph (Fig. 19–5), we find that

$\mu = 3.9$ μreyn

$$S = \left(\frac{r}{c}\right)^2 \frac{\mu n}{P} = (1,000)^2 \cdot \frac{3.910^{-6} \times 20}{167} = 0.466$$

$F_{Frict.} = (r/c)f = 9.5$ (see Fig. 19–12)

Now we use Fig. 19–17 to get

$F_{Flow} = Q/rcnL = 3.70$

and, using Fig. 19–18, we obtain

$$\frac{Q_s}{Q} = 0.35$$

Therefore,

$$\Delta T_F = \left(\frac{0.103P}{1 - 0.5Q_s/Q}\right)\frac{(r/c)f}{Q/rcnf}$$

$$= \left(\frac{0.(103)(67)}{1 - 0.5(0.35)}\right)\frac{(9.5)}{3.70}$$

$$= 53.5°F$$

Thus,

$$T_{av} = 100 + \frac{53.5}{2} = 126°F$$

which is not the same as the assumed 130°F; hence, we use

$$\frac{130 + 126}{2} = 128°F$$

as the new average temperature. Then $\mu = 4.00$ microreyn (from Fig. 19–5 for 128 °F), and we have

$$S = \left(\frac{r}{c}\right)^2 \frac{2\mu n}{P} = \frac{(1,000)^2 4.0 \times 10^{-6} \times 20}{167} = 0.478$$

$$\frac{h_0}{c} = 0.72 \qquad (r/c)f = 10 \quad \text{(see Figs. 19–11, 19–12)}$$

$$\frac{Q}{rcnl} = 3.6 \qquad \frac{Q_s}{Q} = 0.34 \quad \text{(see Figs. 19–17, 19–18)}$$

$$\frac{P}{P_{max}} = 0.52$$

$$\Delta T = \frac{0.103P}{1 - 0.5[Q_s/Q]}\frac{(r/c)f}{Q/rcnL} = \frac{0.103(167)(10)}{1 - 0.5(0.34)3.6}$$

$$= \frac{0.103(167)(10)}{0.83(3.6)} = 57.56°$$

Therefore,

$$T_{av} = 100 + \frac{57.56}{2} = 128.75$$

$$= 128.75 \approx (128) \text{ ok}$$

which is a usable value because the exact $T_{av} = 125.44°F$. (See the Variables Sheet on page 434.)

The following dimensionless terms are obtained from Figs. 19–11 through 19–18:

$$\frac{h_0}{c} = 0.72 \text{ (from 19–11)}$$

So find $h_0 = 0.72(0.0015) = 1.08 \times 10^{-3}$

Similarly,

$$(r/c)\ f = 10; \ f = 10/1000 = 1 \times 10^{-2}$$

$$Q/rcn\ L = 3.6; \ Q = (3.6)1.5(0.015)20(3) = 0.487 \text{ in}^3/\text{sec}$$

$$Q_s/Q = 0.35; \ Q_s = (0.35Q)$$

But

$$Q = .487$$

So

$$Q_s = 0.170 \text{ in}^3/\text{sec}$$

Also,

$$\frac{P}{P_{max}} = 0.52$$

Therefore,

$$P_{max} = \frac{167}{0.52} = 321.15 \text{ psi}$$

The lead model *ex19-3.tk* is used to solve the preceding problem. You will find that this model, along with TK Solver, carries on an iterative process

internally and eliminates the need to calculate one step at a time, as in
Example 19-3. In sum, the lead model

    1.   Avoids the need to read graphs

    2.   Eliminates an iterative process by hand (as it does it inside the
        computer)

    3.   Allows one to analyze, design, and optimize the design of jour-
        nal bearings using the same model

    4.   Above all, allows one to ask "What if?" and "How can I?" ques-
        tions.

For the solution to the foregoing problem, see the Variables Sheet on
page 434.

19.8 Details of Lead Model

Example 19-4, presented shortly, is solved using a lead model called *ex19-4.tk*, or
jbr.tk. The design parameters and nomenclature used in the model are described
and shown in Fig. 19–10. We have seen that the design task is lengthy and
requires the use of an iterative process, which in turn requires reading many
graphs and charts and therefore is susceptible to errors.

    We also know that the Reynolds equation for two-dimensional flow was
solved by A. A. Raimondi and John Boyd using numerical techniques on a digital
computer. They presented their solutions in the form of graphs and charts that
had to be read one at a time while carefully recording information based on the
Sommerfeld's number. Reading these graphs is time consuming and is susceptible to error, especially when interpolation or extrapolation is required. Design
analysis is based on an average constant viscosity of the lubricant; however, at
best, one knows only an inlet temperature of a fluid entering a journal bearing.
Thus, an iterative process is required to arrive at the average temperature,
which then can be used to determine the viscosity of the fluid in the journal bearing. This, again, is time consuming and susceptible to error.

    To use the lead model effectively, we also need to have a clear background in
designing or analyzing a journal bearing. This is achieved by understanding that
one does not have control over every variable in design calculations. It is, however, possible for the designer to have control over variables such as the following:

    1.   the viscosity μ

    2.   the load per unit of projected area, P

    3.   the speed n

    4.   the bearing dimensions r, c, and L

VARIABLES SHEET

```
X tk          OP
                              ex19-3.tk

File  Edit  Commands  Options  Windows  Sheets  Help

─  ▐                       Variables                       ▐ ▐ ▐

St Input---- Name--- Output--- Unit----- Comment----------------------------
             n        20         rps       revolutions per second (rps)
    1200      RPM                 RPM       revolutions per minute (RPM)
             P        166.66667  psi       Average film pressure (N/m^2 or psi)
    1500      LOAD               lb        Applied Load (N or lb)
    3         L                  in        Length (m or in)
    3         D                  in        diameter of the shaft (m or in)
             z        1                    L/D (none)
             r_C      1000                 r/C (none)
             r        1.5        in        radius of the shaft (m or in)
    20        GRADE                        SAE GRADE eg:30,40,etc. (none)
L   .0015     C                  in        radial clearance (m or in)
             Tgav     125.43529  degF      average temperature (degC or degF)
    100       Ti                 degF      inlet temperature (degC or degF)

L             Q        .5040644   in^3/s    total oil flow rate (m^3/s or in^3/s)
L             Tf       150.87058  degF      outlet temperature (degC or degF)
L             ho       .0010752   in        minimum film thickness (m or in)
             f        .00911208            bearing coefficient of friction (none)
L             fPower   .29108637  kW        friction power loss (kW, hp, or btu/s)
             Torque   2.3163198  Nm        friction torque (Nm or inlb)
L             Pmax     323.12294  psi       maximum film pressure (N/m^2 or psi)
             phi      70.601818  deg       minimum film thickness position (deg)
             thetapo  99.614299  deg       film termination position (deg)
             thetapm  13.027996  deg       maximum film pressure location (deg)
             mu       3.7424084  mmreyn    viscosity (cp or mmreyn)
             Qs       .17787426  in^3/s    oil side leakage (m^3/s or in^3/s)
             delT     28.261434  deldegC   temperature change(deldegC or deldegF)
    .0000083  KU                           conversion constant (do not change)

             Smin     .082                 Minimum suggested S value (none)
             S        .44922827            bearing characteristic number (none)
                                           Note: S must be between 0 and 10
             Smax     .21                  Maximum suggested S value (none)

L             Pf       2.6172546            Performance Factor (none)

─
(34) Status:                                                        F9 OK
```

Those variables which are independent and which the designer cannot control, except indirectly by changing one of the preceding variables, are:

1.   the coefficient of friction, f
2.   the temperature rise ΔT
3.   the flow of oil, Q
4.   the minimum film thickness h_0

The latter variables are called performance variables, and they suggest how well the bearing is performing. The designer needs to define satisfactory limits on these so-called performance factors. There are no unique values that can define satisfactory limits; however, such values can be left to the judgment of the designer, who must try to optimize performance factors as much as possible, requiring answers based on repetitive solutions. The generation of such solutions is time consuming and susceptible to error. Also, to study the performance of a journal bearing, a new factor, called a performance factor, must be defined. We do this in the next section.

19.9 Performance Factor

The performance factor allows for the quantification of the performance curves found on the Raimondi and Boyd graphs indicating the positions of maximum load and minimum friction for a given ratio of length to diameter. (See Fig. 19–11.) This factor, abbreviated Pf, is defined to have a value of –1 for points on the minimum friction line and 1 for points on the maximum load line. Therefore, any value between –1 and 1 will be in the optimal zone, while those solutions for which PF lies outside of this range should be modified.

As with any design, when the designer tries to ask "What if?" and "How can I?" questions and tries to optimize the design, the process becomes tedious, time consuming, and, again, susceptible to errors. All of these problems are minimized by developing a journal bearing analysis–design–optimization lead model.

19.10 Sample Solution Using the Lead Model

To use a lead model for solving basic problems, one begins by entering all of the relevant constraint equations into the Rules Sheet, as in Example 19-2. When this is done, the variables that are used will appear in the Variables Sheet. Next, all the known input values must be entered in the input field of the Variables Sheet. At this point, the number of equations in the Rules Sheet should equal the number of variables that do not have specific values. User-defined functions to be entered into the Rules Sheet can also be defined by entering additional equations into the Function Sheet.

Pressing F9 at this juncture will cause the computer to attempt to solve the system by direct substitution. For some well-structured systems of equations, this will produce a solution. However, in many cases it will not. Then the Iterative Solver must be invoked. To do this, a guess must be provided for the value of one or more of the unknown variables. This is done by typing a G into the status field for a variable, which will cause a number to appear in the input field. This value can be used, or it can be replaced with a new guessed value. It is best to guess a value for a variable that, if it were known, would allow many other values to be computed directly.

The lead model developed for the analysis, design and optimization of journal bearings is extensive. The heart of the model is the equations in the Rules Sheet and an extensive set of list functions with cubic mapping. Still, all that is needed to use the model is the Variables Sheet and, occasionally, lists and plots. From the Variables Sheet, any known values can be specified according to the problem at hand. Each of the variables is clearly labeled in the Comment Sheet to identify its physical meaning. SI calculation units have been specified for each, but either the SI or English units, which are listed for each variable in the comment field, can be used for display purposes.

Example 19-4

Consider a 360° hydrodynamic journal bearing 4 cm in diameter and 3 cm long, intended to carry a load of 2,000 N at a rotational speed of 1,800 rpm. If SAE30 oil with an inlet temperature of 75°C is used, find the values of Pf, ΔT_F, h_0, f, the power, and Q, for radial clearance (c) values from 0.000005 to 0.000075 m.

To solve this problem, the first step is to enter the input values into the Variables Sheet. While we are doing this, we must use the correct units. Since values are given in SI units, all of the variables should first be set to appropriate SI units. Then the numbers can be entered as shown in the Variables Sheet. As can be seen from this sheet, the length and diameter measurements are entered in meters, as centimeters are not defined in this model.

Next, an L and an O are typed into the status field corresponding to each of the output values to be listed: Pf, T_F, h_0, f, the power, and Q. Also, an L and an I must be typed into the status field of the input list variable c (clearance). (You will not see the O or I.) Then, a list of numbers must be put into the list for c. To do this, the list must be opened and the "Fill List" command selected from the commands menu. At this point, the range of values to be solved for are entered (0.000005, 0.000075), and the step size (0.000005) is selected.

In addition to these lists, Tg_{av} will be designated as a guess list input. This is done by typing LIG into the Tg_{av} status field. That field will then use the values in the Tg_{av} list as guesses to initiate the iterative solver. The list will be continuously updated by the last line in the Rules Sheet, so that each time the iterative solver is started, the first guess will be the solution from the previous element.

VARIABLES SHEET

```
X ┌ tk    ▢ ◿
                          ex19-4.tk

 File  Edit  Commands  Options  Windows  Sheets  Help

┌─────────────────────────── Variables ──────────────── ▢ ▣ ▲
│St Input---- Name--- Output--- Unit----- Comment-------------------
│              n       30        rps       revolutions per second (rps)
│    1800     RPM                RPM       revolutions per minute (RPM)
│              P       1666666.7 N/m^2     Average film pressure (N/m^2 or psi)
│    2000     LOAD               N         Applied Load (N or lb)
│    .03      L                  m         Length (m or in)
│    .04      D                  m         diameter of the shaft (m or in)
│              z       .75                 L/D (none)
│              r_C                         r/C (none)
│              r       .02       m         radius of the shaft (m or in)
│    30       GRADE                        SAE GRADE eg:30,40,etc. (none)
│L   0        C ‾               m         radial clearance (m or in)
│G 98.969531  Tgav              degC       average temperature (degC or degF)
│    75       Ti                degC       inlet temperature (degC or degF)
│
│L            Q                 m^3/s      total oil flow rate (m^3/s or in^3/s)
│L            Tf                degC       outlet temperature (degC or degF)
│L            ho                m          minimum film thickness (m or in)
│             f                            bearing coefficient of friction (none)
│L            fPower            kW         friction power loss (kW, hp, or btu/s)
│             Torque            Nm         friction torque (Nm or inlb)
│L            Pmax    3608776.6 N/m^2      maximum film pressure (N/m^2 or psi)
│             phi               deg        minimum film thickness position (deg)
│             thetapo           deg        film termination position (deg)
│             thetapm           deg        maximum film pressure location (deg)
│             mu                cp         viscosity (cp or mmreyn)
│             Qs                m^3/s      oil side leakage (m^3/s or in^3/s)
│             delT              deldegC    temperature change(deldegC or deldegF)
│
│    .0000083 KU                           conversion constant (do not change)
│
│             Smin                         Minimum suggested S value (none)
│             S                            bearing characteristic number (none)
│                                          Note: S must be between 0 and 10
│             Smax                         Maximum suggested S value (none)
│
◿                                                                          ◺
(10) Name: GRADE                                                      F9 OK
```

At this point, pressing SHIFT-F9 or selecting "List Solve" from the commands menu should produce a list of solutions. The results can best be seen by creating a table. To do this, the Tables Sheet is selected, and the name of the table is entered into the sheet (see Table 19–2). Then the names of the lists that are to appear in the table are entered into the Table subsheet. Finally, the table appears as seen in Table 19–2.

In addition to creating the table, it may be useful to plot some of the parameters. For example, to better understand the changes occurring in h_0 as c changes, they can be plotted against each other. This is done by simply entering the names of the lists (c, h_0) in the plots subsheet and then selecting "Display Plot" from the menu. (See Plot 1 of c vs. h_0.) Similarly, Plot 2 shows the graph of c vs. Pf. It is obvious from the two plots that a value of c in the range of 0.00002 to 0.00003 is optimum for h_0 to be maximum (See Plot 1), and similarly, a value of c in the range of 0.000015 to 0.00004 is optimum for Pf to be in the range of +1 to –1 for better performance. (See Plot 2.)

19.11 Conclusions

By using a lead model, the analysis, design, and/or optimization of a journal bearing can be done in a very short period of time. The single set of equations in the lead model can be used to analyze, design, and optimize journal bearings, even if the input and the output variables are reversed in their roles. With the use of the lead model, the tedious iterative approach to solving journal bearing problems is avoided. We remain with the following conclusions:

1. The use of a "performance factor" allows one to design a journal bearing for optimum conditions.

2. The lead model permits designers to get answers to "What if?" and "How can I?" questions without writing elaborate programs.

3. The equations used in analysis, design, and optimization must be well understood by the engineer; TK does not allow the lead model to be used as a "black box."

4. A good understanding of the lead model enables the user to modify, add, or build a new model in a very short time.

5. The ideas developed in this work can be extended to analyze, design, or optimize other machine elements. All that one is required to do is create a lead model.

6. A lead model allows engineers to do more and better design work in less time.

Table 19–2

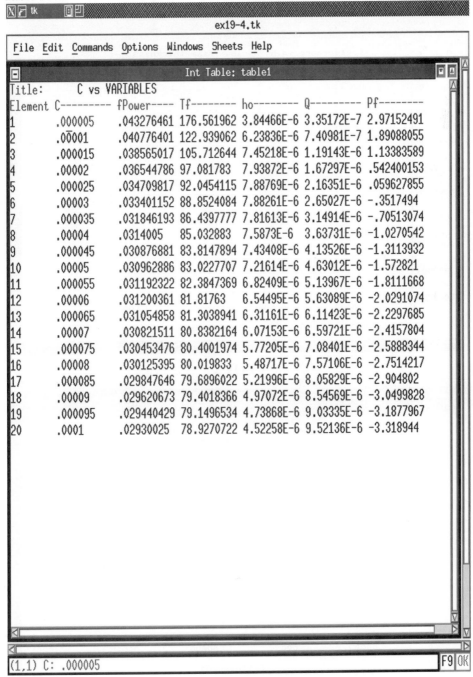

ex19-4.tk

File Edit Commands Options Windows Sheets Help

Int Table: table1

Title: C vs VARIABLES

Element	C---------	fPower----	Tf--------	ho--------	Q---------	Pf--------
1	.000005	.043276461	176.561962	3.84466E-6	3.35172E-7	2.97152491
2	.00001	.040776401	122.939062	6.23836E-6	7.40981E-7	1.89088055
3	.000015	.038565017	105.712644	7.45218E-6	1.19143E-6	1.13383589
4	.00002	.036544786	97.081783	7.93872E-6	1.67297E-6	.542400153
5	.000025	.034709817	92.0454115	7.88769E-6	2.16351E-6	.059627855
6	.00003	.033401152	88.8524084	7.88261E-6	2.65027E-6	-.3517494
7	.000035	.031846193	86.4397777	7.81613E-6	3.14914E-6	-.70513074
8	.00004	.0314005	85.032883	7.5873E-6	3.63731E-6	-1.0270542
9	.000045	.030876881	83.8147894	7.43408E-6	4.13526E-6	-1.3113932
10	.00005	.030962886	83.0227707	7.21614E-6	4.63012E-6	-1.572821
11	.000055	.031192322	82.3847369	6.82409E-6	5.13967E-6	-1.8111668
12	.00006	.031200361	81.81763	6.54495E-6	5.63089E-6	-2.0291074
13	.000065	.031054858	81.3038941	6.31161E-6	6.11423E-6	-2.2297685
14	.00007	.030821511	80.8382164	6.07153E-6	6.59721E-6	-2.4157804
15	.000075	.030453476	80.4001974	5.77205E-6	7.08401E-6	-2.5888344
16	.00008	.030125395	80.019833	5.48717E-6	7.57106E-6	-2.7514217
17	.000085	.029847646	79.6896022	5.21996E-6	8.05829E-6	-2.904802
18	.00009	.029620673	79.4018366	4.97072E-6	8.54569E-6	-3.0499828
19	.000095	.029440429	79.1496534	4.73868E-6	9.03335E-6	-3.1877967
20	.0001	.02930025	78.9270722	4.52258E-6	9.52136E-6	-3.318944

(1,1) C: .000005   F9 OK

PLOT 1

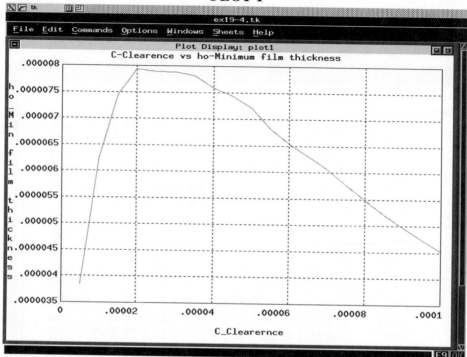

PLOT 2

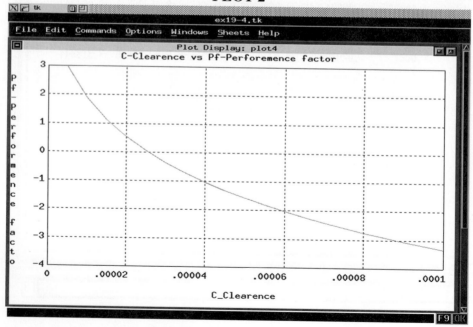

PROBLEMS

19–1) Design a journal bearing that will carry 300 lb of external load for minimum friction if $L/D = 1$, the shaft diameter is 1˝, and the design rotating speed is 1,200 rpm. The average temperature of the fluid is 75°F. Remember, the design requires a type of lubricant, clearance, horsepower required to drive the mechanism, and minimum film thickness.

19–2) Design the bearing of Problem 1 for maximum weight-carrying capacity.

19–3) Design a journal bearing that will carry a 5-kN load for a shaft 10 cm in diameter, rotating at 1,800 rpm. SAE30 oil is used with an inlet temperature of 45°C. $D/L = 0.50$.

    a.   Design for minimum friction.

    b.   Design for maximum load-carrying capacity.

19–4) Use lead models for the design of a journal bearing to solve Problems 19–1 to 19–3, and check your hand-calculated results against the solution obtained from the lead model.

19–5) A single piston develops 12 hp at 900 rpm. Design bearings at the supports A and B and at the center C if the oil used is SAE20 and the inlet temperature is 80 °F. Use the following diagram:

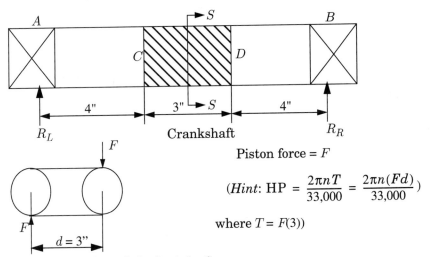

(Hint: $\text{HP} = \dfrac{2\pi n T}{33{,}000} = \dfrac{2\pi n (Fd)}{33{,}000}$)

where $T = F(3)$)

Cross section of crankshaft at $S - S$

19–6)   Do Problem 19–3 using an inlet temperature of 50°C.

19–7)   Do Problem 19–6, but use a lead model.

19–8)   Do Problem 19–6 if Pf = 0. What does Pf = 0 mean?

19–9)   In the following diagram, the maximum radial loads on the shaft at the
        bearings are 20 lb at R_1 and 80 lb at R_2. The shaft diameters at R_1 and R_2
        are 0.6 inch. The shaft speed is 1,725 rpm. Use a clearance ratio of 0.0017
        and L/D 1.0. Design a journal bearing for minimum friction. Also,
        determine the bearing eccentricity ratio, the maximum pressure and its
        location, the minimum film thickness, the coefficient of friction, the torque,
        and the power loss in the bearing for the given values. Select a suitable
        lubricant for the bearing to operate at an average temperature of 163°F.

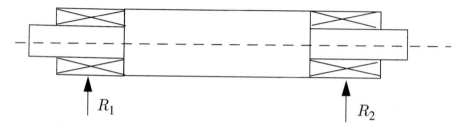

I N D E X

END USER LICENSE AGREEMENT

You should carefully read the following terms and conditions before breaking the seal on the disk envelope. Among other things, this Agreement licenses the enclosed software to you and contains warranty and liability disclaimers. By breaking the seal on the disk envelope, you are accepting and agreeing to the terms and conditions of this Agreement. If you do not agree to the terms of this Agreement, do not break the seal. You should promptly return the package unopened.

LICENSE

Prentice-Hall, Inc. (the "Company") provides this Software to you and licenses its use as follows:

a.   use the Software on a single computer of the type identified on the package;

b.   make one copy of the Software in machine-readable form solely for back-up purposes.

LIMITED WARRANTY

The Company warrants the physical diskette(s) on which the Software is furnished to be free from defects in materials and workmanship under normal use for a period of sixty (60) days from the date of purchase as evidenced by a copy of your receipt.

DISCLAIMER

THE SOFTWARE IS PROVIDED "AS IS" AND COMPANY SPECIFICALLY DISCLAIMS ALL WARRANTIES OF ANY KIND, EITHER EXPRESS OR IMPLIED, INCLUDING, BUT NOT LIMITED TO, THE IMPLIED WARRANTIES OF MER-CHANTABILITY AND FITNESS FOR A PARTICULAR PURPOSE. IN NO EVENT WILL COMPANY BE LIABLE TO YOU FOR ANY DAMAGES, INCLUDING ANY LOSS OF PROFIT OR OTHER INCIDENTAL, SPECIAL OR CONSEQUENTIAL DAMAGES EVEN IF COMPANY HAS BEEN ADVISED OF THE POSSIBILITY OF SUCH DAMAGES.

SOME STATES DO NOT ALLOW THE EXCLUSION OF IMPLIED WARRANTIES OR LIMITATION OR EXCLUSION OF LIABILITY FOR INCIDENTAL OR CONSEQUENTIAL DAMAGES, SO THE ABOVE EXCLUSIONS AND/OR LIMITATIONS MAY NOT APPLY TO YOU.

LIMITATIONS OF REMEDIES

The Company's entire liability and your exclusive remedy shall be:

1.   the replacement of such diskette if you return a defective diskette during the limited warranty period, or

2.   if the Company is unable to deliver a replacement diskette that is free of defects in materials or workmanship, you may terminate this Agreement by returning the Software.

GENERAL

You may not sublicense, assign, or transfer the license of the Software or make or distribute copies of the Software. Any attempt otherwise to sublicense, assign, or transfer any of the rights, duties, or obligations hereunder is void.

Should you have any questions concerning this Agreement, you may contact Prentice-Hall, Inc. by writing to:

Prentice Hall
Engineering
One Lake Street
Upper Saddle River, NJ 07458
Attention: Mechanical Engineering Editor

YOU ACKNOWLEDGE THAT YOU HAVE READ THIS AGREEMENT, UNDERSTAND IT, AND AGREE TO BE BOUND BY ITS TERMS AND CONDITIONS. YOU FURTHER AGREE THAT IT IS THE COMPLETE AND EXCLUSIVE STATEMENT OF THE AGREEMENT BETWEEN US THAT SUPERSEDES ANY PROPOSAL OR PRIOR AGREEMENT, ORAL OR WRITTEN, AND ANY OTHER COMMUNICATIONS BETWEEN US RELATING TO THE SUBJECT MATTER OF THIS AGREEMENT.